Springer Series in Reliability Engineering

Series Editor

Hoang Pham, Department of Industrial and Systems Engineering, Rutgers University, Piscataway, NJ, USA

More information about this series at http://www.springer.com/series/6917

Roberto Galeazzi · Hilmar Kjartansson Danielsen ·
Bjarne Kjær Ersbøll · Dorte Juul Jensen ·
Ilmar Santos
Editors

Intelligent Quality Assessment of Railway Switches and Crossings

Springer

Editors
Roberto Galeazzi
Department of Electrical Engineering
Technical University of Denmark
Kongens Lyngby, Denmark

Bjarne Kjær Ersbøll
Department of Applied Mathematics
and Computer Science
Technical University of Denmark
Kongens Lyngby, Denmark

Ilmar Santos
Department of Mechanical Engineering
Technical University of Denmark
Kongens Lyngby, Denmark

Hilmar Kjartansson Danielsen
Department of Wind Energy
Technical University of Denmark
Roskilde, Denmark

Dorte Juul Jensen
Department of Mechanical Engineering
Technical University of Denmark
Kongens Lyngby, Denmark

ISSN 1614-7839 ISSN 2196-999X (electronic)
Springer Series in Reliability Engineering
ISBN 978-3-030-62471-2 ISBN 978-3-030-62472-9 (eBook)
https://doi.org/10.1007/978-3-030-62472-9

This Springer imprint is published by the registered company Springer Nature Switzerland AG
The registered company address is: Gewerbestrasse 11, 6330 Cham, Switzerland

Preface

Railway infrastructure managers daily face the challenge of securing and yet improving the network capacity while reducing the operational cost under the non-negotiable constraint of passenger safety. Trading-off between these contrasting needs has led to the adoption of preventive maintenance policies and contributed significantly to the overall operational expenditure of the infrastructure.

Switches and crossings (S&Cs), also known as turnouts, are an essential part of railway networks as they enable trains to be directed from one track onto another at railway junctions, allowing for necessary flexibility during train operations. At the same time they are also the most vulnerable part of the railway network, suffering from peak stresses because of moving parts and their greater geometric complexity compared to normal tracks. As a critical part of the railway, poorly maintained S&Cs have a detrimental impact on railway operations.

The overall performance of S&Cs is determined by the concomitant action of operational and environmental factors, which determine simultaneous degradation at different levels of the railway track. Therefore a complete characterization of the health state of S&Cs is achievable by analyzing the degradation processes from multiple perspectives in order to fully understand their origins and developing mechanisms. A holistic approach is hence needed to establish complete analysis tools and models describing the failure mechanism of the different S&C components.

Digitalization of the railway infrastructure introduces new prospects for network managers and operators, who now have the possibility to gain track performance information previously unavailable. Ubiquitous sensing, distributed data collection, almost unlimited data storage and ever growing computational capabilities enable a more thorough and systematic analysis of the infrastructure health state. Field data, acquired through both novel track side measuring systems and more conventional tools, combined with high-fidelity simulation data will enable the development of comprehensive models describing the degradation and failure mechanisms of S&C's components. These models will be the cornerstone for the creation of S&C condition monitoring systems, whose ultimate availability will pave the way to the implementation of predictive maintenance in the railway industry.

This book contributes to the transition towards predictive maintenance for railway infrastructure by offering a holistic approach to the scientific investigation of the factors and mechanisms determining the performance degradation of S&Cs, and the consequent development of condition monitoring systems. The book gathers scientific contributions of latest advances in three main technological areas, namely, *Design and Modelling*, *Metallurgical Characterization* and *Condition Monitoring and Asset Management*. With the ambition of becoming a reference for the railway industry, each chapter begins with an executive summary explaining in layman terms the key findings and how railway practitioners may leverage this new knowledge.

First Part *Design and Modelling* addresses the development of models of different complexity for the evaluation of the dynamic behaviour of S&Cs during train passage in the presence of different defects affecting one or more track components. Chapter "Modelling of the Dynamic Train-Track-Substructure Interaction for Track Monitoring" addresses the analysis of vibrations of track components in the presence of different track irregularities exploiting high-fidelity modelling of the vehicle-track-substructure system by means of finite elements and multibody simulation methods. Such vibration analysis allows the identification of track sensitive locations suitable for installation of vibration sensors for track monitoring, as well as the estimation of the frequency range where monitoring should be performed. Chapter "Statistical Approach to Model Track Dynamics Towards the Monitoring of Railway Turnouts" proposes a statistical approach for the generation of low-complexity behavioural models describing the dominant track dynamics. The designed method generates robust estimates of the first and second track resonances, which can be utilized to monitor the development of degradation phenomena on key track components. Chapter "Improving Switches and Crossings Performance and Reliability" presents the use of modelling and site validation to solve track maintenance issues leading to improvement in performance and reliability of switches and crossings. Two case studies from the UK railway network are described and the benefit of utilizing modelling and analysis tools, traditionally adopted in the design phase, to solve maintenance problems is demonstrated.

Second Part *Metallurgical Characterization* addresses the experimental characterization and analysis of rail deformation and degradation processes induced by repeated loading due to train passage. An investigation concerning the failures of screws of the fastening system is also presented. Chapter "Damage and Microstructure Evolution in Cast Hadfield Steels Used in Railway Crossings" investigates the evolution of damages in railway steels utilized for turnout crossings by assessing the microstructural changes in the material through methods as electron channelling contrast imaging and transmission electron microscopy. Twinning substructures at nanoscale level are identified, and their impact on damage development is discussed. Chapter "Metallurgical Investigation of Crossing Noses" proposes novel characterization techniques for railway components based on optical microscopy, X-ray tomography and synchrotron X-ray diffraction. Such techniques are adopted to provide insightful information about the properties and degradation mechanisms of railway crossing noses.

Chapter "Analysis of Sleeper Screw Failures" presents a state-of-the-art combination of advanced complementary laboratory techniques suited for a comprehensive multiscale analysis of failure of rail screws on scales from meters to nanometres.

Third Part *Condition Monitoring and Asset Management* addresses the design, implementation and testing of diagnostic and prognostic algorithms able to detect faults and predict failures of the different components constituting the S&C Chapter "Data Sources and Research Models for Turnouts" performs a technological scoping of the systems currently available to perform inspection of S&Cs to assess the overall readiness level in carrying out an automatic and objective assessment of S&Cs quality. The Standard Elements analysis is also applied to gain insight into which turnout components are mainly affecting the lifetime of the S&C itself in the light of maintenance and renewal cost. Chapter "CoMPAcT-Data Based Condition Monitoring and Prediction Analytics for Turnouts" presents an approach to condition monitoring for S&Cs based on the statistical and spectral analysis of turnout geometry data collected in loaded conditions. First, a post-positioning method for the alignment of the different geometry data across consecutive measurement campaigns is developed and validated on 15 turnouts of the Austrian railway infrastructure. Then, the standard deviation of the track longitudinal level and the average power spectral density of the same signal are proposed as tools to assess the condition of the ballast layer. Chapter "Monitoring of Switches and Crossings/Tracks Using Smart Sensors" discusses the design and field testing of a novel wireless vibration sensor specifically developed for continuous measurement of railway track accelerations. The design phase accounted for several intrinsic and extrinsic parameters, which potentially affect the long-term reliability and performance of the sensor. Field testing on Norwegian and German railways lines demonstrated the feasibility of the conceptual design for sensing vibration during train passage. Chapter "Labelling the State of Railway Turnouts Based on Repair Records" proposes the use of unsupervised machine learning methods, as principal component analysis and clustering, to automatically label the condition of railway turnouts exploiting the repair records. The outcome of the labelling is investigated through the comparison with the turnout's loaded geometry measurements. Chapters "Automatic Detection of Rail Defects from Images" and "Deep Learning for Automatic Railway Maintenance" investigate the use of different computer vision methods to detect defects on rails. The former proposes three different approaches based on statistical measures of the pixel intensity in the image regions containing the rails. The latter adopts a deep learning method for supervised training of a defect detector based on images collected by the track measurement car. The verification of the methods carried out on a limited data set shows promising results in detecting surface defects both in straight track and turnouts. Chapter "The Effects of Renewal and Tamping on Ballast and Track Geometry in Turnouts" addresses the systematic understanding of the impact of maintenance and renewal actions on turnout's performance. A model for track geometry degradation accounting for ballast tamping and renewal is developed using fractal analysis and regression modelling on turnout geometry data like the longitudinal track level. The method is tested on a large number of switches and crossings of the Danish railway infrastructure. Chapters "The Organic Growth of an Asset Management System at ProRail" and "Ways to Improve the Performance (And Cost) of Switches" tackle the key aspect of establishing and exploiting a holistic asset management system, which accounts for maintenance strategies, risk modelling, design, monitoring and control of assets. Both chapters bring forward the lessons learned at ProRail, the Dutch railway infrastructure manager, which underwent the process of creating such asset management system during a period of twenty years from 1995 to 2015.

Kongens Lyngby, Denmark — Roberto Galeazzi
Roskilde, Denmark — Hilmar Kjartansson Danielsen
Kongens Lyngby, Denmark — Bjarne Kjær Ersbøll
Kongens Lyngby, Denmark — Dorte Juul Jensen
Kongens Lyngby, Denmark — Ilmar Santos
July 2020

Acknowledgements

The editorial work and the research studies described in selected chapters of this contributed book have been performed as part of the INTELLISWITCH project. The financial support by Innovation Fund Denmark, with grant number 4109-00003B, is gratefully acknowledged.

The scientific and technological expertise provided during the reviewing process by colleagues from the Technical University of Denmark, the Norwegian University of Science and Technology, the University of Birmingham, Chalmers University of Technology, Ghent University, Graz University of Technology and Banedanmark is thankfully appreciated by the editors.

Contents

Contributors

Seyed Mohammad Asadzadeh Department of Electrical Engineering, Technical University of Denmark, Lyngby, Denmark

Pegah Barkhordari Department of Electrical Engineering, Technical University of Denmark, Lyngby, Denmark

Xiaopei Cai School of Civil Engineering, Beijing Jiaotong University, Beijing, China

Line H. Clemmensen Department of Applied Mathematics and Computer Science, Technical University of Denmark, Lyngby, Denmark

André Filipe da Silva Rodrigues Banedanmark, Copenhagen, Denmark

Vedrana Andersen Dahl Department of Applied Mathematics and Computer Science, Technical University of Denmark, Lyngby, Denmark

Hilmar Kjartansson Danielsen Department of Wind Energy, Technical University of Denmark, Roskilde, Denmark

Alejandro de Miguel Tejada Banedanmark, Copenhagen, Denmark

Somrita Dhar Department of Wind Energy, Technical University of Denmark, Roskilde, Denmark

Pavol Duroska Department of Applied Mathematics and Computer Science, Technical University of Denmark, Lyngby, Denmark

Niall Fagan Network Rail, London, UK

Michael Fellinger Institute of Railway Engineering and Transport Economy, Graz University of Technology, Graz, Austria

Roberto Galeazzi Department of Electrical Engineering, Technical University of Denmark, Lyngby, Denmark

Helena Hansen Department of Applied Mathematics and Computer Science, Technical University of Denmark, Lyngby, Denmark

Emil Hovad Department of Applied Mathematics and Computer Science, Technical University of Denmark, Lyngby, Denmark

Sin Sin Hsu Network Rail, London, UK

Dorte Juul Jensen Department of Mechanical Engineering, Technical University of Denmark, Lyngby, Denmark

Elias Kassa Department of Civil and Environmental Engineering, Norwegian University of Science and Technology, Trondheim, Norway

Maxim Khomiakov Department of Applied Mathematics and Computer Science, Technical University of Denmark, Lyngby, Denmark

Ankit Kumar Department of Materials Science and Engineering, Delft University of Technology, Delft, The Netherlands

Martin Tandrup Lange Banedanmark, Copenhagen, Denmark

Stefan Marschnig Graz University of Technology, Institute of Railway Engineering and Transport Economy, Graz, Austria

Roumen Petrov Department of Electrical Energy, Metals, Mechanical Constructions and Systems, Ghent University, Ghent, Belgium;
Department of Materials Science and Engineering, Delft University of Technology, Delft, The Netherlands

Carsten Jørn Rasmussen Banedanmark, Copenhagen, Denmark

Jan Saramota Department of Mechanical and Industrial Engineering, Norwegian University of Science and Technology, Trondheim, Norway

Jilt Sietsma Department of Materials Science and Engineering, Delft University of Technology, Delft, The Netherlands;
Department of Electrical Energy, Metals, Mechanical Constructions and Systems, Ghent University, Belgium

Amund Skavhaug Department of Mechanical and Industrial Engineering, Norwegian University of Science and Technology, Trondheim, Norway

Jan Swier Swier Advice Asset Management, VE Leusden, The Netherlands

Camilla Thyregod Department of Applied Mathematics and Computer Science, Technical University of Denmark, Lyngby, Denmark

Georgios Vassos Department of Applied Mathematics and Computer Science, Technical University of Denmark, Lyngby, Denmark

Peter Veit Institute of Railway Engineering and Transport Economy, Graz University of Technology, Graz, Austria

Petra A. Wilfling Institute of Railway Engineering and Transport Economy, Graz University of Technology, Graz, Austria

Thomas Wix Department of Applied Mathematics and Computer Science, Technical University of Denmark, Lyngby, Denmark

Xiaodan Zhang Department of Mechanical Engineering, Technical University of Denmark, Lyngby, Denmark

Yubin Zhang Department of Mechanical Engineering, Technical University of Denmark, Lyngby, Denmark

Acronyms

AM	Asset Management
APS	Advanced Photon Source
CAD	Computer Aided Design
CF	Capitalizing Factor
CNN	Convolutional Neural Network
CT	Computerized Tomography
DTW	Dynamic Time Warping
EBSD	Electron Backscatter Diffraction
ECCI	Electron Channelling Contrast Imaging
EDM	Euclidean Distance Method
EDS	Energy Dispersive X-ray Spectroscopy
EFQM	European Foundation for Quality Management
EMD	Empirical Mode Decomposition
ERA	Eigenstructure Realization Algorithm
FE	Finite Element
FFT	Fast Fourier Transform
FIB	Focused Ion Beam
FMECA	Failure Mode Effect and Criticality Analysis
GEV	Generalized Extreme Value
HAZ	Heat Affected Zone
IMF	Intrinsic Mode Function
IMU	Inertial Measurement Unit
IRJ	Insulated Railway Joint
LCC	Life Cycle Cost
M&R	Maintenance and Renewal
MAC	Medium Access Control
MBS	Multi Body Simulation
MGT	Million Gross Tons
MPI	Maintenance Performance Indicator
N4SID	Numerical algorithms for Subspace State Space System Identification
NPV	Net Present Value
OM	Optical Microscopy

PCA	Principal Component Analysis
PCB	Printed Circuit Board
RAMS	Reliability Availability Maintainability Safety
RCF	Rolling Contact Fatigue
RCM	Reliability Centred Maintenance
RVDS	Railway Visual Detection System
S&C	Switch and Crossing
SAED	Selected Area Electron Diffraction
SAM	System Asset Management
SEM	Scanning Electron Microscopy
SFE	Stacking Fault Energy
TDMA	Time Division Multiple Access
TEM	Transmission Electron Microscopy
TMC	Track Measurement Car
TWIP	Twinning Induced Plasticity
VSM	Vibrating Sample Magnetometer
YOLO	You Only Look Once

Design and Modelling

Modelling of the Dynamic Train-Track-Substructure Interaction for Track Monitoring

Elias Kassa and Xiaopei Cai

Abstract A vehicle-track-substructure dynamic system model is developed using FE tool Abaqus coupled to an MBS code Gensys. A vehicle dynamics is simulated in Gensys and the contact force and contact patch from the MBS simulation are used as input file in a detailed FE simulation. Vibration analysis of the rail component has been studied using the FE model of a track representation. To identify the range of the rail acceleration, different track irregularity conditions and train speeds are considered. The rail acceleration sampled at different sampling location along and across the rail indicated a considerable drop of the acceleration from the rail head sampling location to the rail foot. The acceleration of the rail head is most sensitive to variation of loading magnitude and loading position, as well as to the train speed. For an operation speed increasing from 100 to 160 km/h, the accelerations of rail head increase from 737 to 1000 m/s^2, while the increment at the rail foot and web is moderate. This investigation is part of a study towards a continuous monitoring of railway tracks/switches and crossing conditions.

Keywords Modelling of train-track interaction · MBS simulation · Rail acceleration sampling · Condition monitoring

Executive Summary
This chapter focused on the modelling of the vehicle-track-substructure system to provide input for the track monitoring system development. Through the modelling of the vehicle-track-substructure system, vibration sensitive locations across the rail structure as well as along the track has been identified for a plain line track as to where to place the acceleration sensors. A similar sensitive area across the stock/switch rail is expected in switches and crossings. In addition, the range and frequency of

E. Kassa (✉)
Department of Civil and Environmental Engineering, Norwegian University of Science and Technology, 7491 Trondheim, Norway
e-mail: elias.kassa@ntnu.no

X. Cai
School of Civil Engineering, Beijing Jiaotong University, Beijing, China

R. Galeazzi et al. (eds.), *Intelligent Quality Assessment of Railway Switches and Crossings*, Springer Series in Reliability Engineering,
https://doi.org/10.1007/978-3-030-62472-9_1

the vibration accelerations under a train dynamic loading is used as input to the monitoring system. The effects of different variables and track condition on the vibration level has been included in this study. The railhead is more sensitive and an acceleration range in the order of 100–200 g may be required for a monitoring system. This location is close to the wheel-rail load application and it is very sensitive for train speed change, inclusion of small rail irregularity and location of the contact patch. In contrary, the rail foot and rail web have shown more stable acceleration and the change in the acceleration value is minimal while offsetting the contact patch as well as including moderate track irregularity. This will enable to differentiate the vibration from the normal operation with vibrations due to significant defects on the track or the train. Different factors were included in the assessment to analyse their effects on the rail vibration acceleration range and frequency as a guide for the subsequent tasks. The acceleration range to monitor the vibration at the rail web and rail foot is much lower than the railhead and a range of acceleration in the order of 20–40 g would be sufficient. Further, for practical field application, the location of the sensor position should not interfere with the rail regulations for structure gauge and clearance zone, hence, the rail web or rail foot are identified to be the location most suitable to mount the sensor for track monitoring.

1 Introduction

The track structure consists of a rail discretely supported by sleepers or concrete blocks, through fastening systems, which is supported by a ballast or concrete slab structure. The rail is a spatial structure that consists of rail head, rail web and rail foot and it has functions of guiding wheels through the interaction of the wheel profile and the rail profile, supporting wheel loads and distributing the loads over several sleepers. In vehicle-track dynamic interaction system, the rail is a key connecting part between the vehicle and the track substructure systems [1, 2]. The vibration on the rail and wheel has impact on the wheel-rail contact and it can be reflected by the track condition and track quality. Hence, the rail acceleration, and in some cases the wheelset acceleration, is usually used as a main index for investigation of track condition and vehicle condition in theoretical analysis and in the field test [3, 4].

In existing dynamic analysis of the vehicle-track system, the rail is often simulated as Euler beam or Timoshenko beam supported continuously or discretely [1, 4, 5]. With the development of vehicle-track dynamic theory [6], the vehicle and the track models are considered in more detail, but the rail is still simulated simply and only single point vibration acceleration is calculated. However, various positions of the spatial rail will attain different accelerations. If this factor is not taken into account during the numerical analysis of track dynamics and dynamic tests, it will increase the possibility of assessment error.

The rail acceleration is the main content of the railway dynamic test, and it is used to evaluate the vibration characteristics of track structure or wheel-rail impact effect [7–9]. Remington [10] installed six acceleration sensors across the rail between two

Fig. 1 Ballasted track

sleepers, and measured rail acceleration to predict wheel-rail noise. Zhao and Wang [11] measured rail vibrations in transition zone between slab and ballasted track through setting acceleration sensors on top of rail head and foot above the sleepers.

Recently Shen et al. [12] presented a method for railway crossing monitoring through measurement of wheel-crossing impact and ambient vibration. Field dynamic tests, which are carried out to measure acceleration of a rail at a single point or in an arbitrary section without considering the sensitive area, will have influence on the accuracy of the test result. Thus, it is important to identify sensitive locations across the rail as well as along the track under train dynamic loading.

The rail vibration characteristics differ with different types of track structure, different fastening stiffness and different operation speed and train types used. The vibration level for ballasted track will be different from the corresponding vibration on slab track. A ballasted track used for testing in the later stage of the study is shown in Fig. 1. In order to understand the state-of-the-art track dynamics and train-track interaction, relevant literatures have been referred and some important and useful articles are summarized in this paper.

The rail acceleration is directly related to the dynamic interaction between the wheel and rail. This study focuses on modelling of a train-track-substructure dynamic system using Finite Element (FE) tool ABAQUS and a multi-body-dynamic (MBS) code Gensys. The contact force and contact patch parameters from the MBS simulation are used as input in the FE simulation. The refined simulation to the track dynamics is realized using the FE model with solid elements for rail representation. Based on the FE model, rail vibration analysis has been performed and different types of track structure, track irregularity conditions and train speeds are considered. Results of the rail acceleration under certain track irregularity and different train speeds are calculated to determine the sensitive area which has the maximum accelerations.

2 Vehicle-Track-Substructure Model

This study focuses on modelling of a vehicle-track-substructure dynamic system by coupling FE tool ABAQUS and MBS software Gensys [13]. The methodology involves three steps, see Fig. 2. The first step involves vehicle dynamics simulations to calculate wheel–rail contact forces, creepages, and contact positions. The influence of track geometric irregularity in wheel-rail contact is considered. Step two involves loading application as sliding load platform on the surface of the rail head. The last step is dynamic simulation of track-substructure to calculate the track vibration at different sampling location. There are similar co-simulation approaches through combining MBS models for vehicle dynamics with FE models describing the track dynamics in more detail. Kassa and Johansson [14] demonstrated simulations of train-turnout interaction using Gensys and plastic deformation of rail profiles using FE model. Nicklisch et al. combined MBS model with FE model for simulation of material degradation in switches and crossings [15]. Vehicle dynamics simulation using MBS software Gensys coupled with FE software ABAQUS for predicting damage of S&C components are presented by Johansson et al. [16]. Vehicle dynamics simulation using the commercial software Gensys is used to calculate wheel-rail contact forces and contact position. From the dynamic analysis of vehicle-track interaction, the time history of the contact force and contact point location are used as input in the FE simulation. Simulation of track dynamics with finite element software ABAQUS is used to obtain realistic wheel-rail contact patch and contact pressure distribution. The accelerations of rail head, rail web and rail foot are obtained from the FE simulation.

2.1 Vehicle-Track Dynamic Model

An MBS code Gensys is used for modelling the vehicle system with all the components represented in detail while the track system is represented by simple mass-spring-damper system coupled to each wheel. Gensys is a three-dimensional multi-body-dynamic program for modelling rail vehicles running on track. Gensys is widely applied in analyzing dynamic behaviour of vehicles and it enables to calculate

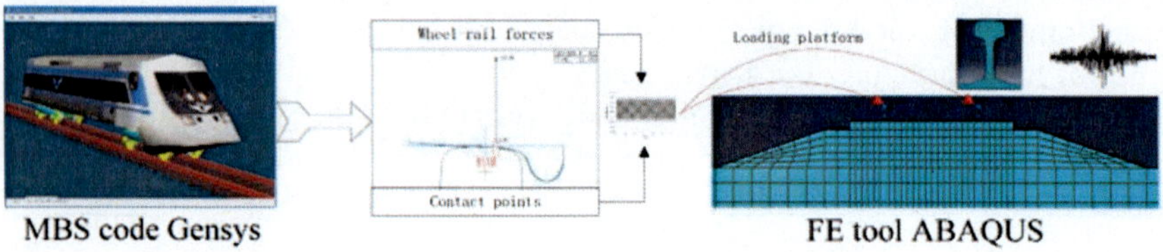

Fig. 2 Methodology for simulation of vehicle-track-substructure dynamics through coupling MBS code Gensys with FE tool ABAQUS

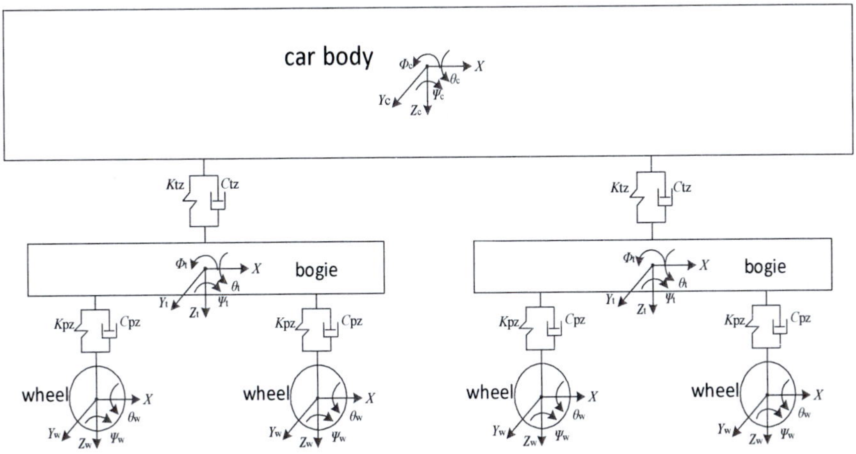

Fig. 3 The vehicle model

vehicle vibration, wheel-rail force and wheel-rail contact patch, etc. The vehicle-track coupling dynamic model which is established in this paper consists of vehicle model, track model and wheel-rail contact model.

The vehicle is modelled as a multi-body system composed of a car body, two bogie frames and four wheelsets, as shown in Fig. 3. The car-body, bogie frames and wheelsets as rigid bodies is connected by suspension system which is simulated by spring-damper element. The entire vehicle model has 38 degrees of freedom. However, for this study a simple vehicle model, only considering the unsprung mass and the bogie model with its primary suspension would be sufficient. The flexibility in the wheelset and axle is not considered in this study.

A FE tool ABAQUS is used to model the track. The track is taken as continuous elastic support model. The rail structure is simulated by Euler beam and the support type of foundation such as ballast bed, sleeper, etc. is simulated by single continuous support model. The rail and slab are connected by several fasteners which can be modelled by springs with linear property and dampers with viscous property. The wheel-rail contact relationship which is tied between vehicle and track includes normal interaction and tangent interaction. The non-linear Hertz elastic contact theory is used to calculate the normal contact forces. Normal force *p(t)* would generate by the compression of the wheel and rail, and it would disappear when they separate. The force is defined as follows:

$$p(t) = \left[\frac{1}{G}\Delta Z(t)\right]^{\frac{3}{2}} \tag{1}$$

where G is the wheel-rail contact coefficient and $\Delta z(t)$ is the normal elastic compressing deformation at wheel-rail contact patch.

The analysis of tangent interaction takes the creep ratio of wheel-rail into account. In the calculation of creep forces, Kalker creep theory is used for the calculation firstly, and then Shen's theory adopted in the amended results. The creep force is nonlinear which contains longitudinal and lateral forces.

2.2 *Track-Substructure Model*

In ballasted track, the track-substructure FE model includes rail, fastening system, sleeper, ballast bed and subgrade. The fastening system is simulated by spring-damper element and others are all simulated by deformable solid structure. The rail section is 60E1 (UIC60) and the section area is 76.86 cm^2, the inertia moment to horizontal axis is 3055 cm^4 and the inertia moment to vertical axis is 512.9 cm^4. The dimensions of the rail 60E1 is given in the European standard EN 13674-1. Considering the average support of the rail pad, three spring-damper elements distribute in longitudinal direction and three spring-damper elements distribute in lateral direction in each fastener, as shown in Fig. 4. The vertical stiffness of the fastener is 7.5×10^7 N/m and the lateral stiffness of the fastener is 3.9×10^7 N/m. The size of the concrete sleeper is 2.6 m × 0.2 m × 0.3 m. The thickness of the ballasted bed is 0.3 m which is simulated by C3D8R solid element in Abaqus. The rail and fastening

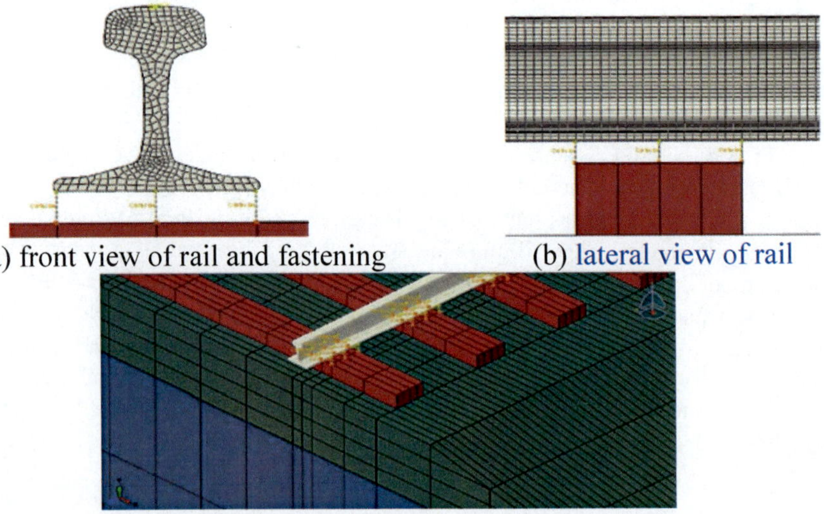

(a) front view of rail and fastening (b) lateral view of rail

(c) 3D view of rail-sleeper-substructure model

Fig. 4 3D model of the track-substructure and model of a Fastening System as 9 spring-damper systems system in FE model and whole model of track-substructure FE model is shown in Fig. 4.

To some extent, the track vibration is affected by foundation structure, so the vibration of the subgrade is considered in the model. The height of the subgrade is 3 m and the slop is 1:1.75. In order to eliminate influence of the boundary condition from the track end condition, the length of the integral model is 100 m and the measuring point of rail acceleration is in the center of the model. Further, in order to prevent vibration reflection, the spring-damper element as absorbing boundary is adopted in each side of the model.

2.3 *Simulation of Wheel-Rail Force Transfer*

The dynamic interaction of vehicle and track is transmitted by wheel-rail contact. The rail and wheelset are elastomer which have large modulus. According to Hertz contact theory, the shape of the wheel-rail contact patch is elliptical which is on the top surface of the rail head, as is shown in Fig. 5. In process of load application, the uneven distributing stress in elliptical contact patch is neglected taking contact patch as sliding load platform on the surface of rail head.

While the vehicle running in straight line, according to calculated results of Gensys, the slight lateral displacement of contact patch on the surface of rail head cannot be taken into account. The effect of track irregularity in the rail acceleration and in the wheel rail forces is important to be considered. The dynamic interaction relationship between wheel and rail in the model is assumed a non-linear Hertz contact. In the model, the wheelset velocity is set as boundary conditions applying to load platform.

The values of the wheel-rail force which is the excitation source to FE model are calculated results from multi-body dynamic model in Gensys. The wheel-rail force is a discrete function of time and it is imported into the FE model in order to simulate real wheel-rail time history. Based on Abaqus preprocess environment, wheel-rail force is applied in the elliptical load platform in the FE model, moving with the load platform. In addition, the lateral force about 2–3 kN is much smaller compared to

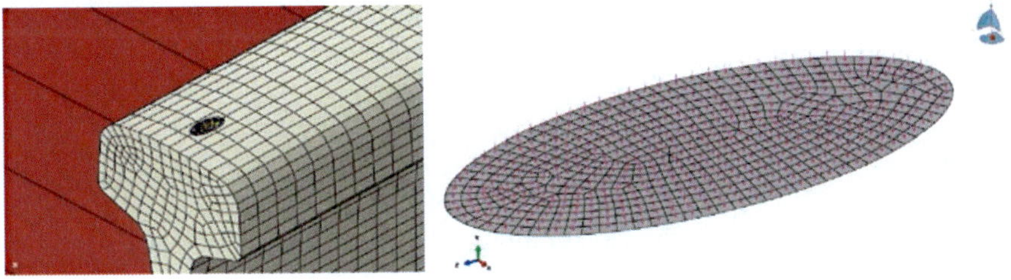

Fig. 5 Elliptic shape of the wheel-rail contact (left) and the load platform (right)

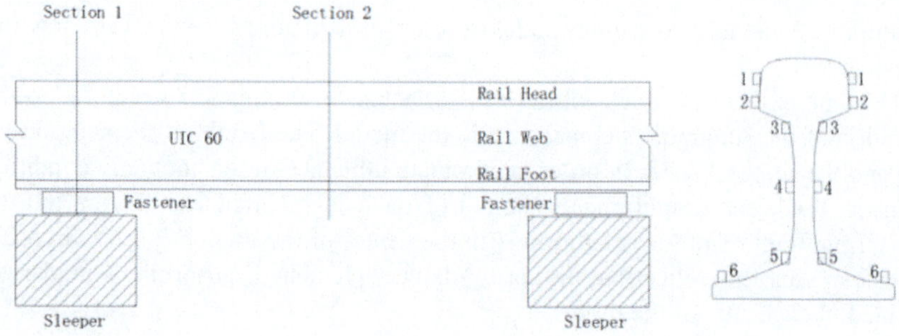

Fig. 6 Sampling section along the rail and sampling points across the rail

the vertical force and it is feasible to neglect the effect of lateral force without hugely compromising the final result.

2.4 *Rail Points for Analysis*

Considering the space distribution characteristics of rail accelerations, two sections along the track are chosen, one located above sleepers and the other between the two sleepers, as seen in Fig. 6. Each rail section has 12 acceleration output points, located in the rail head, rail web and rail foot respectively, as seen in Fig. 6. The inside and outside measuring points of the rail is symmetrical, and each side has 6 acceleration measuring points. Two measuring points of rail head are located below the top surface of 16 mm and 25 mm, and two measuring points of rail web are located at the upper part of the rail and the neutral axis, besides, measuring points of the rail foot are in the center of the bottom surface and the edge of upper surface.

3 Rail Acceleration of Ballasted Track

Ballasted track is the main form of the railway track, accounting for about 90% railway in the world. In the ballasted track, the rails and concrete sleepers are connected by fasteners forming the track panel structure. Along the track, supports for the rails by fasteners are point supports, and the restraints of the rail and transmitting characteristics are not the same in different support positions. In addition, the sections of rail head, rail web and rail foot are different along the rail and the distance is also different from wheel-rail excitation. Therefore, there would be some differences for vibration of the rail at different positions.

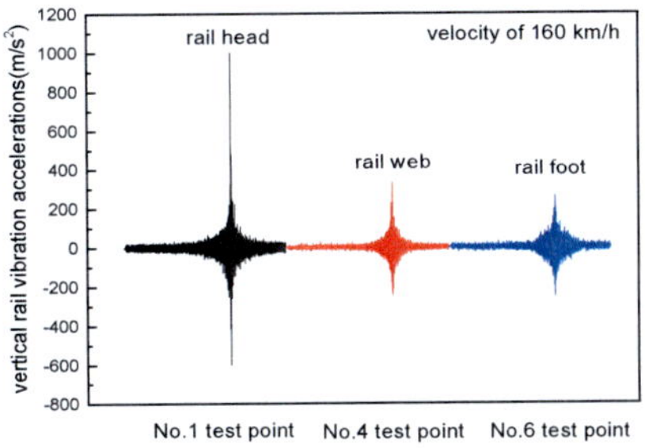

Fig. 7 Rail vertical acceleration

3.1 Distribution of Acceleration

The vertical rail acceleration for the vehicle with axle load 22 tones and travelling speed of 160 km/h running on the ballasted track is shown in Fig. 7. It is clearly seen that rail accelerations are increasing while the train dynamic load is getting close to the test section. When the load is in the test section, the rail accelerations reach a maximum. When the load departs the test section, rail vibration accelerations attenuate constantly, and gradually return to zero. Because of the high wheel-rail contact force on the rail head, the acceleration of the rail head is much bigger than rail web and rail foot and the values of the acceleration of the rail web and rail foot are much closer.

The measuring point of rail vertical acceleration peak value above sleepers is shown in Fig. 8. It is obvious that rail accelerations of inside and outside test point are almost the same. Due to wheel-rail contact force directly acting on the rail head, the acceleration of rail head is maximum and the acceleration of rail web is relatively small. For the test section above sleepers, because of the constraint of fastener plates and local buckling effects of elastic bars, the peak acceleration values of rail acceleration are much closer for the two points in the rail foot.

Test point of rail vertical acceleration peak value between sleepers is also shown in Fig. 8. It is clear that due to the location of the wheel-rail contact close to rail head

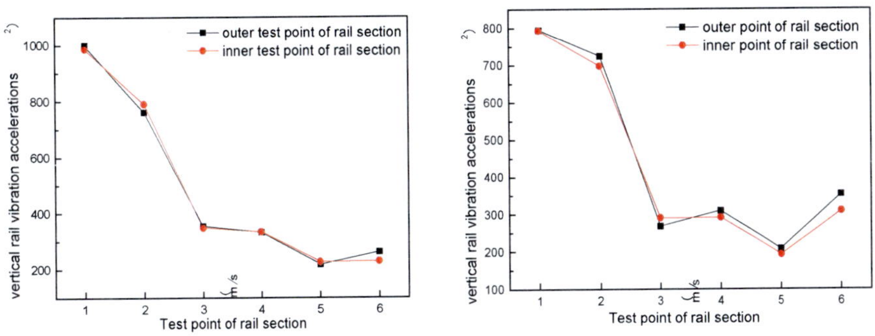

Fig. 8 Vertical rail acceleration above sleepers (left), between two sleepers (right)

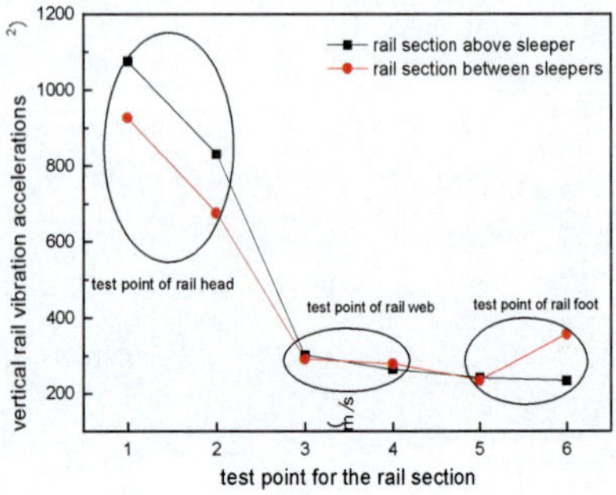

Fig. 9 The difference of the vertical rail acceleration in different two sections

sampling point and the damping characteristics of the rail material, rail accelerations from rail head to rail web gradually decay, because there is no constraint for sections between two sleepers. The accelerations of No.6 on the edge of rail foot are much bigger than that of No.5 in foot center point.

The rail acceleration above sleepers and rail acceleration between two sleepers are shown in Fig. 9. And time history curve of rail accelerations is shown in Fig. 10. For the rail head, the acceleration of rail section between two sleepers are less than that of the section above sleepers. For the rail web, the vibration accelerations of all points are relatively close. For the rail foot, the vibration accelerations between two sleepers are slightly larger than rail acceleration above sleepers.

3.2 *Effect of Vehicle Speed*

With the increase of train speed, the impact from wheel to rail becomes more apparent, and the magnitude of the wheel-rail contact forces became larger and larger. Based on multi-body dynamic software Gensys, the wheel-rail forces at the speed of 100, 120,

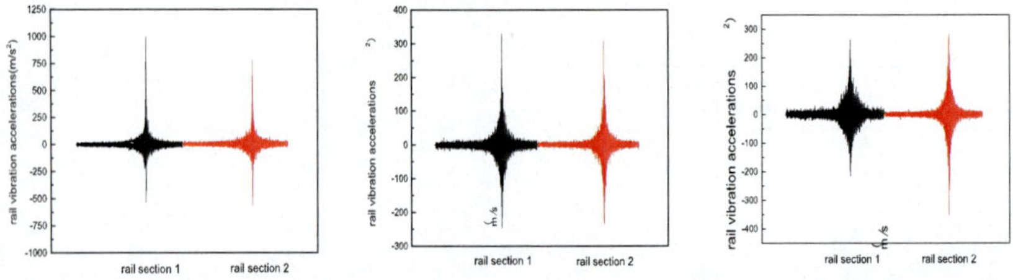

(a) vertical rail acceleration of Test point No.2

(b) vertical rail acceleration of Test point No.4

(c) vertical rail acceleration of Test point No.6

Fig. 10 Time history of rail acceleration at different test points

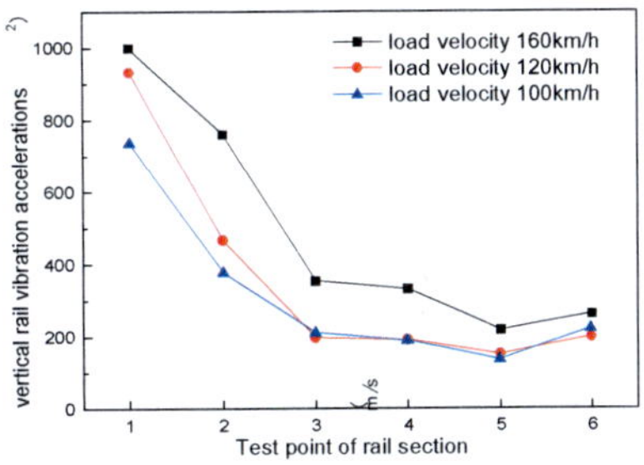

Fig. 11 Rail acceleration above sleepers

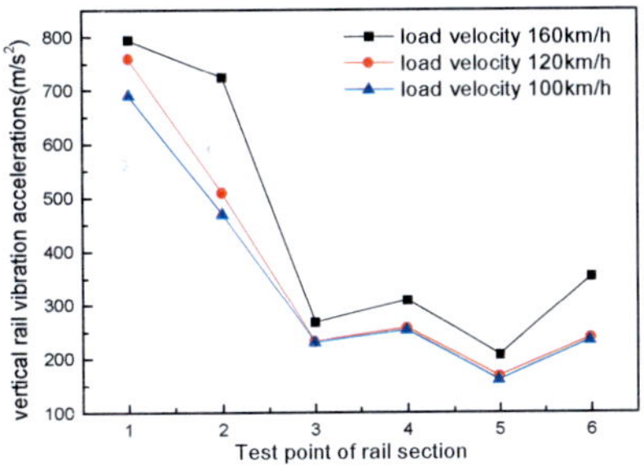

Fig. 12 Rail accelaration between two sleepers

160 km/h are obtained. Due to the magnitude of vertical rail acceleration is much larger than lateral rail vibration acceleration, only the vertical acceleration of rail is analyzed. The vertical acceleration results of rail in different sections are shown in Figs. 11 and 12, and time history curve of rail acceleration is shown in Fig. 13. It can be seen clearly that, with the improvement of train speed, the wheel-rail force fluctuates dramatically and peaks at a higher value. The results of wheel-rail force in Gensys as input data were applied to elliptical platform, which is on the rail head. The acceleration of rail head increases obviously and with the increase of load moving speed, acceleration of rail test section increases.

Considering the consistence between inside and outside rail accelerations, and considering the practicality of the sensor positioning in real field, only the outside vibration accelerations are analyzed to reduce the analysis work. The rail acceleration

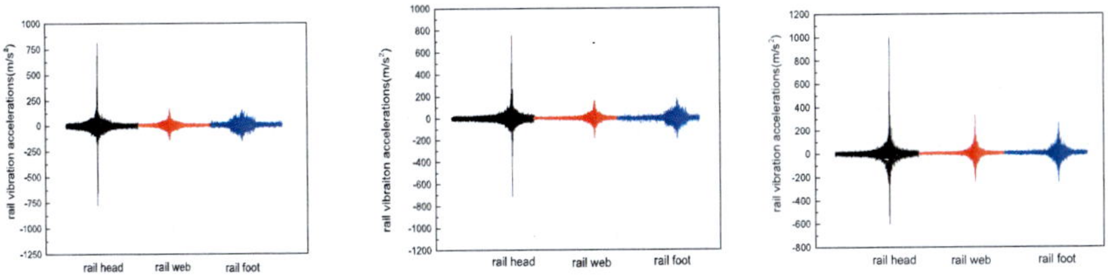

(a) Velocity of 100km/h (b) Velocity of 120km/h (c) Velocity of 160km/h

Fig. 13 Rail acceleration for various speeds

which is on the top of sleepers and rail acceleration that is between two sleepers are presented in Fig. 9. It is observed that the former rail acceleration is larger than the latter one at different speeds. Furthermore, increasing of vehicle's speed will result in a larger impact on the rail section, thus acceleration at different spots of the rail will be different.

3.3 *Effect of Track Irregularity*

Existing researches have shown that track irregularity and abnormal wheel-rail contact conditions are the main excitation source of wheel-rail impact vibration [12, 17], and track vertical irregularity has bigger influence on the wheel-rail vertical force. This effects are more significant with higher speeds. In multi-body dynamics model, track vertical irregularity spectrum as excitation source is imported into Gensys and calculation results show that wheel-rail vertical force increases significantly, whose peak increases from 122 to 140 kN. Based on the calculation of FE model which is imported wheel-rail force under track irregularities, the rail accelerations are analyzed.

Figure 14 shows that rail accelerations above sleepers are more sensitive to track irregularities than that section between sleepers. In section above sleepers, acceleration of rail head under track irregularities is larger than that without track irregularities, and acceleration of rail web and rail foot is almost the same. In section between two sleepers, the calculated data do not show evident laws about rail head. The discipline of acceleration of rail head, rail web and rail bottom is basically the same, while the rail head acceleration is more obvious.

Overall, because wheel rail impact vibration is more apparent, rail accelerations in existing lines under track irregularity in each position increase obviously, with the peak also increasing substantially. In railway operation, the track irregularity is more complex, moreover with defects on the wheel tread. Therefore, rail accelerations will be bigger. Acceleration of ballasted track under complex operating conditions can

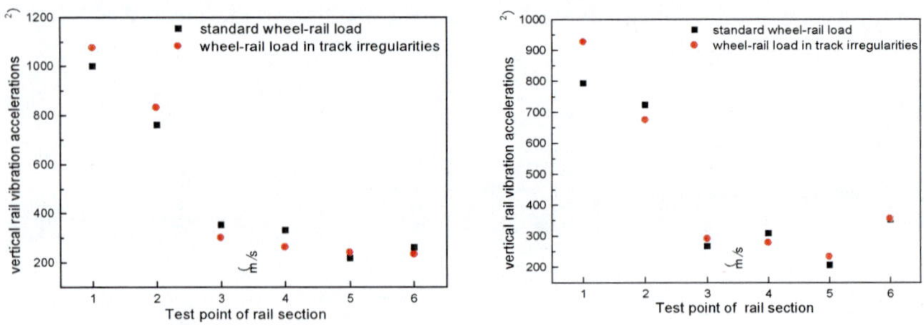

(a) Test section above sleepers (b) Test section between two sleepers

Fig. 14 Rail acceleration in the case of different track irregularities

Fig. 15 Difference of rail accelerations between outer points and inner points

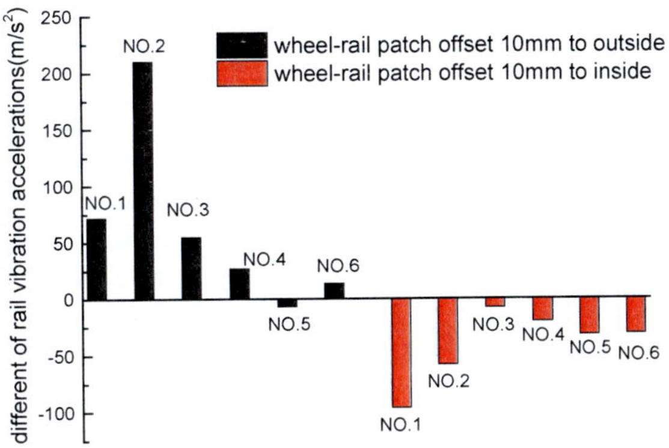

reach 1000 m/s^2 and even greater [12]. For monitoring rail accelerations of ballasted track, it is recommended that the range of sensors should be 1000–2000 m/s^2.

3.4 *Effect of Offset of Wheel-Rail Patch*

Phenomenon of vehicle hunting will appear in railway operation and it could cause lateral motion of wheel-rail contact patch. Because of the lateral offset of wheel-rail contact patch, the distribution of rail acceleration changes a lot. In order to study the effect of offset of wheel-rail contact patch, the contact patch in FE model is offset 10 mm to the left and right respectively. Two conditions are calculated and different rail accelerations test point of inside and outside are shown in Fig. 15.

Because of high wheel-rail contact force existing in rail section above sleepers, it could influence distribution of rail acceleration. Rail accelerations do not show obvious rules in the section above sleepers. In section between sleepers, the wheel-rail contact force weakens and the law of distribution of rail acceleration is shown in Fig. 15. It shows that wheel-rail offset point will influence the rail accelerations and rail accelerations of test points which near load platform are larger than that on other side.

4 Conclusions

A vehicle-track-substructure dynamic system model is introduced. The model contains a FE model of the track and an MBS model of a vehicle and track using a software Gensys. Numerical simulations run to study the vibration acceleration of the rail structure under train passing. The rail acceleration at the rail head and close to the loading location is much larger than that of at the rail web and rail foot. The rail acceleration data at the rail head shows a sharp increase in the magnitude with the increase of train speed and track irregularity. The flexibility in the wheelset and axle may have some effect on the acceleration levels, and future studies should account for these flexible components.

Due to the elastic damping characteristics of the fastening system, the acceleration of rail head above the sleeper is significantly larger than that of rail head between two sleepers. The acceleration of rail web in both sections are much closer. The acceleration of rail foot above sleepers is much smaller than that between sleepers. The rail vibration accelerations increase with the expediting of the train speed. For the operation speed increasing from 100 to 160 km/h, the accelerations of rail head increase from 737 to 1000 m/s^2.

In conclusion, the acceleration sensors should be installed in rail web or foot, as the rail head position is very sensitive for small perturbation and change in the wheel-rail contact location. The range of the accelerometer sensor depends on the train speed and the excitation loadings and track irregularities. Larger range acceleration sensors should be used for monitoring of track with higher train speeds.

Acknowledgements This work was part of the EC funded project DESTinationRAIL (No.SEP-210181866). Mr Ingemar Persson of DEsolver AB, and Mr Liangwu Guo of BJTU School of Civil Engineering are gratefully acknowledged for their contributions in the modelling work.

References

1. Zhai W, He Z, Song X (2010) Prediction of high-speed train induced ground vibration based on train-track-ground system model. Earthq Eng Eng Vibr 9(4):545–554
2. Zhai WM, Sun X (1994) A detailed model for investigating vertical interaction between railway vehicle and track. Veh Syst Dyn 23:603–615
3. Bracciali A, Cascini G (1997) Detection of corrugation and wheelflats of railway wheels using energy and cepstrum analysis of rail acceleration. Proc Instit Mech Eng Part F J Rail Rapid Transit 211(2):109–116
4. Thompson D, Kouroussis G, Ntotsios E (2019) Modelling, simulation and evaluation of ground vibration caused by rail vehicles. Veh Syst Dyn 57(7):936–983
5. Oscarsson J (2002) Dynamic train-track-ballast interaction with unevenly distributed track properties. Veh Syst Dyn 37(Supp 1):385–396
6. Wickens AH (2005) Fundamentals of rail vehicle dynamics. Swets & Zeitlinger, Lisse, Netherlands
7. Shi J, Burrow MPN, Chan AH et al (2013) Measurements and simulation of the dynamic responses of a bridge-embankment transition zone below a heavy haul railway line. Proc Instit Mechan Eng [F] J Rail Rapid Transit 227(3):254–268
8. Insa R, Salvador P, Inarejos J et al. (2011) Analysis of the influence of under sleeper pads on the railway vehicle/track dynamic interaction in transition zones. Proc Instit Mechan Eng [F] J Rail Rapid Transit, 0954409711430174
9. Barke D, Chiu WK (2005) Structural health monitoring in the railway industry: a review. Struct Health Monit 4(1), 81–93
10. Remington PJ (1987) Wheel/rail rolling noise, II: validation of the theory. J Acoust Soc Am 81(6):1824–1832
11. Zhao C, Wang P (2015) Application of polyurethane polymer and assistant rails to settling the abnormal vehicle-track dynamic effects in transition zone between ballastless and ballasted track. Shock Vibr

12. Shen C, Li Z, Dollovoet R (2019) A novel method for railway crossing monitoring based on ambient vibration caused by train-track interaction. In: Proceedings IAVSD2019, 12–16 Aug 2019. Gothenburg, Sweden pp 1–9
13. Persson I (2000) Documentation of GENSYS. DE solver. Östersund, Sweden
14. Kassa E, Johansson G (2006) Simulation of train–turnout interaction and plastic deformation of rail profile. Veh Syst Dyn 44:349–359
15. Nicklisch D, Kassa E, Nielsen J, Ekh M, Iwnicki S (2010) Geometry and stiffness optimization for switches and crossings, and simulation of material degradation. Proc IMechE [F] J Rail Rapid Transit 224:279–292
16. Johansson A, Pålsson B, Ekh M, Nielsen J, Brouzoulis J, Kassa E (2011) Simulation of wheel–rail contact and damage in switches & crossings. Wear 271(1–2):472–481
17. Tomas K (2009) Track irregularities for high-speed trains. KTH, Sweden Stockholm

Statistical Approach to Model Track Dynamics Towards the Monitoring of Railway Turnouts

Pegah Barkhordari and Roberto Galeazzi

Abstract This paper proposes a method for the automatic generation of statistical models that describe the track stiffness in terms of the first and second track resonance frequencies, which are associated with the dynamic behavior of the ballast and rail pad layers. The method combines the empirical mode decomposition and a subspace identification method to estimate the track resonance frequencies from track vibrations induced during train passage. The generalized extreme value distribution is found to be a robust descriptor of the estimates of the first and second track resonance frequencies across time and space. The method is demonstrated on track acceleration data collected over a period of two years at fixed locations along a turnout for three different types of trains, and models are built for the switch panel, the closure/crossing panel and a transition zone. Further, it is shown that the degradation of track components is captured through the recursive generation of such statistical models, which can then become the basis for the development of condition monitoring systems.

Executive Summary

This work offers a method to robustly estimate the first and second track resonance frequencies that are directly related to the total track stiffness. The periodic assessment of such resonance frequencies would enable the ability to closely monitor the development of degradation phenomena on some of the key components impacting infrastructure dependability, safety and ride comfort, thereby creating a stepping stone towards the actualization of predictive maintenance strategies.

Exploiting continuous measurements of loaded track vertical acceleration, a robust statistical model is built that describes the expected value of the track resonance frequency as well as the dispersion due to uncontrollable operational (train speed, train load, wheel defects, etc.) and environmental (temperature, amount of precipitations,

P. Barkhordari · R. Galeazzi (✉)
Department of Electrical Engineering, Technical University of Denmark,
2800 Kgs. Lyngby, Denmark
e-mail: rg@elektro.dtu.dk

P. Barkhordari
e-mail: pebark@elektro.dtu.dk

R. Galeazzi et al. (eds.), *Intelligent Quality Assessment of Railway Switches and Crossings*, Springer Series in Reliability Engineering,
https://doi.org/10.1007/978-3-030-62472-9_2

etc.) factors. The periodic update of such model displays changes in track resonance frequency due to degradation phenomena connected with the continuous loading of the track.

The method is extensively validated on full scale vibration data collected over a period of two years at a turnout of the Danish railway infrastructure. Statistical models have been developed for four different sections of the turnout—the switch panel, the closure panel, the crossing panel and the transition zone from turnout to straight track—utilizing vibration data associated with freight, IC3 and IR4 trains. The analysis highlights that the proposed method offers

- the automatic generation of statistical models describing the track stiffness using measurements of track acceleration
- the access to track stiffness information with higher time and spatial resolution than methods currently adopted such as indexes based on track geometry measurements
- the monitoring of track quality degradation through the periodic update of the statistical models.

The availability of such models paves the way towards the development of condition monitoring tools for the continuous supervision of the health state of key track components.

1 Introduction

A central aspect to ensure and possibly increase the capacity of a railway network is to achieve a high dependability of the infrastructure. Preventive and periodic maintenance strategies have been so far adopted by railway infrastructure managers to maximize the network throughput as well as to sustain a high safety level. These approaches naturally lead to a high operation and maintenance cost, reported as hundreds of millions Euro on a yearly basis [11]. The actualization of predictive maintenance strategies is foreseen as a turning point for infrastructure operations, which should produce significant savings in maintenance cost while preserving or even improving the dependability level of the assets. A key prerequisite for predictive maintenance is the availability of automated condition monitoring systems for the supervision of infrastructure components.

The capacity and safety of railway networks are significantly affected by the performance and state of health of the turnouts. According to the 2014 RSSB Annual Safety Performance Report [9, Sect. 8.5], 31% of the track-related derailments in the period 2009–2014 in Great Britain were caused by turnouts malfunctioning. Therefore, a considerable part of the total infrastructure maintenance expense is associated with maintenance and renewal actions of turnouts. This is estimated by the Danish railway infrastructure manager, Banedanmark, as one third of the total yearly maintenance costs [16].

Railway turnouts are a complex systems whose performance depends on the dynamic behavior and interaction of its components. The failure of a turnout can be caused by the degradation of one or more of these components, some of which are particularly challenging to monitor due to their hidden and/or distributed nature, such as the ballast and rail pad. According to the failure data recorded in the UK in 2009 [13, Chap. 7], the ballast is the third most common failed component (with a failing frequency of 7.9%) after the switch rail (45.3%) and the slide chair (30.4%). This emphasizes the necessity of developing methods able to retrieve information about the health state of hidden components based on measurements performed at track level. Such knowledge is the premise for the design of condition monitoring tools capable of providing early warnings of deterioration processes and diagnosing the occurrence of faults in railway turnout components.

1.1 Literature Review

Condition monitoring of the railway infrastructure has been a prominent research topic in recent years, and several methods and tools have been proposed. A number of direct methods have been recommended for the appraisal of the status of the infrastructure components through explicit measurement of their physical properties. Examples are the ground penetrating radar, the cone penetration test, and visual inspection [1, 6, 7, 21, 22, 29, 32]. These methods provision railway infrastructure managers with information about the condition of the asset; however the high cost, the disruption of normal operations, and the limited effectiveness are among the factors limiting their use.

Indirect approaches relying on the intelligent processing of measured data collected by either measurement vehicles or track-side measurement systems have been developed for the analysis of the ballast and subgrade condition. The track stiffness has been used as a potential indicator of railway track quality in numerous studies adopting measurements cars as source of information. Examples are the track loading vehicle [30], the moving rail car [26], the railway Portancemetre [14] and the rolling stiffness measurement vehicle [6, 29].

The assessment of the track condition through the estimation of its dynamic properties from field measurements by using model-based techniques has been proposed as another alternative to direct methods. Kaewunruen and Remennikov [17] and Remennikov et al. [28] employed a model-based technique for the analysis of structural degradation of rail pads and estimated the yearly deterioration rate. They extracted effective mass, dynamic stiffness and damping ratio of new and worn pads based on an analytical formulation, derived from a single-degree-of-freedom model and data obtained from hammer tests. The same technique was utilized in [19] to further investigate the dynamic behavior of rail pads under large preloads. Burrow et al. [8] proposed the combination of the falling weight deflectometer test with a finite element model as an inverse method for the condition evaluation of ballasted railway tracks. Suitability of a two-degrees-of-freedom model in combination with hammer excitation measurements for detecting cracked sleepers and broken fastening systems along a railway track was investigated in [18]. In a research study conducted by

[12], a practical method was proposed as a potential tool for diagnostic applications. Employing such method, the mechanical properties of the rail and the substructure were estimated by minimizing the difference between the measured frequency response function and a simulated one obtained from a discretely supported Euler beam model.

The detection of degradation of the ballast layer by employing complex mechanical models in combination with Bayesian updating algorithms has been carried out recently by [23–25]. The proposed methods were validated based on acceleration data obtained from field measurements. The large dimensionality of the mechanical model can be considered as the main drawback of these methods, as it may result in the design of highly complex diagnostic tools. Barkhordari et al. [4] proposed a low-complexity behavioural model of the railway turnout capable of predicting the track response due to dynamic behavior of the ballast and rail pad. The model parameters were estimated by applying the Eigenstructure Realization Algorithm (ERA), a subspace identification method, to vibration data collected during a receptance test. The model was validated by comparing its predicted output with the measured track vibration due to train passage. Despite the good prediction capability, the creation of such models relies on the availability of free vibration data gathered through receptance tests, an experimental campaign rarely performed by railway infrastructure managers.

Low-complexity data-driven behavioural models that contain the dynamic characteristics of main track components are industrially appealing since the low parametrization will enable easy tuning and guarantee high portability across the entire railway network despite the natural presence of uncertainties due to e.g. geographical location and physical age of the components. Moreover, simpler and more robust monitoring tools for diagnosing faults and prognosing failures could be designed by using low-complexity behavioural models.

1.2 Main Contribution

This work presents a novel method for the automatic generation of statistical models that describe the dynamic characteristics of two key turnout components, i.e. the rail pad and ballast, by exploiting measurements of track vertical acceleration collected by a track-side measurement system during train passages. The method combines the empirical mode decomposition (EMD) with numerical algorithms for subspace state space system identification (N4SID) to estimate the parameters of two second order models associated to the ballast and rail pad. Through subsequent eigenvalue analysis of each estimated model, the first (ballast) and second (rail pad) track resonance frequencies are computed.

To achieve a robust estimate of the resonance frequencies, which is insensitive to uncontrollable external factors—quality of the train's wheels, meteorological conditions, train load and speed—, a statistical characterization is performed by recursively estimating such frequencies over monthly periods and fitting a probability density function to the obtained pool of estimates. The statistical characteristics of the track resonance frequencies are well approximated by the generalized extreme value distribution, which robustly describes the estimates for different periods of time and at different locations along the turnout. Further, the periodic generation of this statistical models allows to capture variations in the first and second track resonance frequencies due to degradation processes as well as maintenance actions. Last, the effect of train load on the consistency of the estimates is also discussed by considering measured vibrations induced by freight, IC3 and IR4 trains.

The method is demonstrated on vibration data collected at a turnout of the Danish railway network, and statistical models are built for given locations in the switch panel, the closure/crossing panel, and the transition zone between the turnout and the straight track. These statistical models provide a robust basis for the future development of condition monitoring tools for the supervision of the ballast and rail pad layers in turnouts.

2 Track Side Measurement Set-Up

A turnout in the Danish railway infrastructure was instrumented with 3 wheel detectors, 12 2-axis accelerometers (measurement range: ±500 g) and 3 displacement sensors (measurement range: ±20 mm). All sensors are connected to a data acquisition unit where signals are conditioned and temporarily stored. A schematic overview of the sensor location along the turnout is shown in Fig. 1. The wheel detectors are used to automatically start/stop the data collection during the passage of trains through the turnout. The accelerometers are magnetically mounted on the rail web and measure the train-induced acceleration in the vertical and lateral directions. The displacement sensors measure the vertical motion of the sleepers with respect to a rest position. Sensor readings are sampled at $F_s = 20$ kHz. The instrumented turnout and examples of the installed instrumentation are shown in Fig. 2.

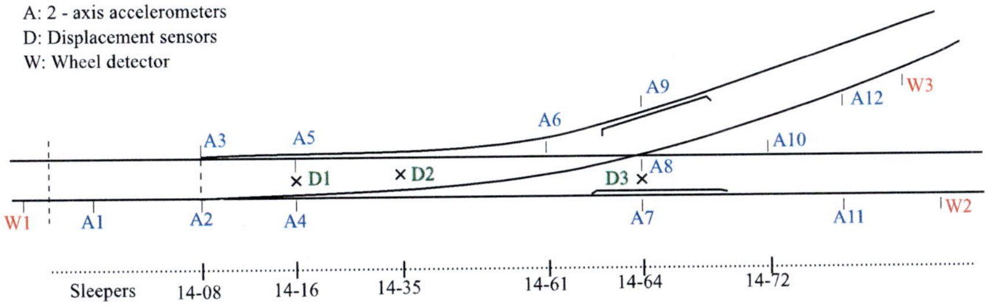

Fig. 1 Layout of the sensors location along the turnout at Tommerup station (Fyn—Denmark). The data utilized in this work refer to vertical accelerations measured by accelerometers A2, A4, A7 and A11

(a) Overview of the turnout in proximity of Tommerup station

(b) Accelerometer A5 magnetically connected to the rail web

(c) Displacement sensor resting on the sleeper

(d) Wheel detector mounted on proximity of the rail web

Fig. 2 Pictures of the instrumented turnout nearby Tommerup station (Fyn—Denmark)

A railway turnout consists of three main sections: the switch panel, the closure panel and the crossing panel. The present study is intended to describe a method to develop low-complexity statistical models capable of predicting the dynamic behavior of track components in these three sections. For the purpose of building such models the track vertical accelerations measured at locations A2, A4, A7 and A11 are considered (see Fig. 1). As an example, the normalized vertical acceleration measured at location A7 and the time-shifted signal from the wheel detector are shown in Fig. 3. The measured acceleration is normalized according to $\bar{a} = a/\max(|a|)$ to ensure data anonymity.

The time-shifted signal from the wheel detector is used to find the time instants at which train bogies are passing through the measurement locations. The time delay τ for each measurement location is obtained based on the distance from the wheel detection sensor and the train speed. The time synchronization between the wheel detector signal $WS(t)$ and the vertical acceleration $\bar{a}(t)$ allows the automatic slicing of the acceleration signals into segments corresponding to each train bogie for the model identification presented in Sect. 3.

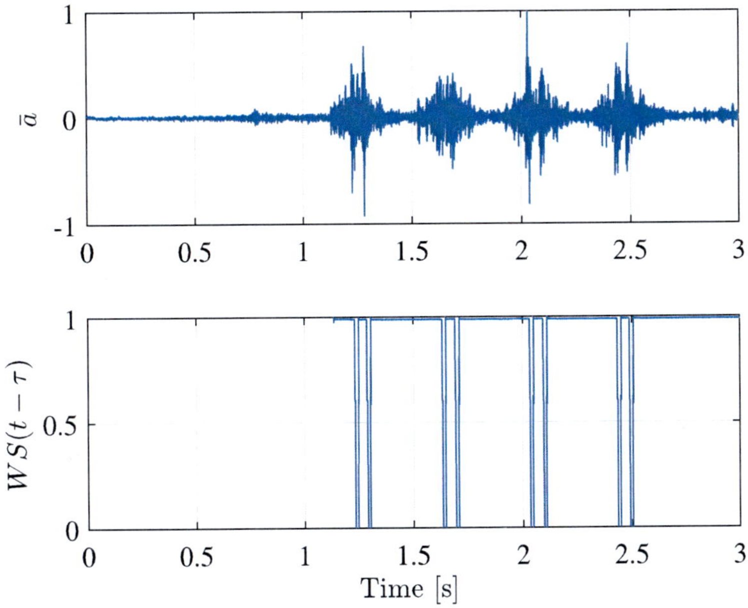

Fig. 3 The track vertical acceleration and the wheel detector signal corresponding to the passage of an IC3 passenger train at location A7

3 Robust Estimation of the Track Resonance Frequencies

The dynamic response of the track under the excitation of a passing train is function of the characteristics of its components and their physical interactions. In particular, track stiffness is a key parameter affecting both ride comfort and ride safety. Several components contribute to the stiffness of the track such as rail pad, ballast, sub-ballast and sub-grade; however the rail pad and the ballast have a prominent role since they are responsible for absorbing most of the impact energy from the train passage. Permanent changes in the dynamic behaviour of the track can be mostly attributed to degradation in at least one of these two components; hence models able to describe the dynamic characteristics of the ballast and rail pad are of clear interest to monitor the overall condition of the track.

The first and second track resonance frequencies are examined in this study as they correspond to the in-phase vibration of rail and sleepers on the ballast stiffness, and the out-of-phase vibration on the flexibility of the rail pad, respectively. A novel method for the robust estimation of the track vibration characteristics is proposed by combining a signal processing technique with a subspace identification algorithm. The schematic of the method is presented in Fig. 4.

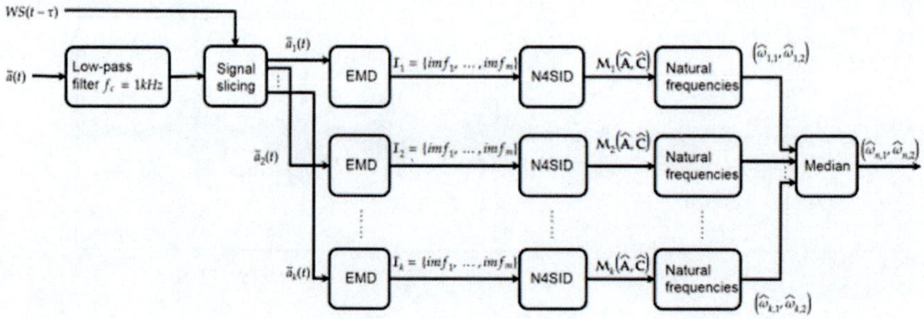

Fig. 4 Block diagram of the estimation method for ballast and rail pad resonance frequencies

3.1 Empirical Mode Decomposition

The empirical mode decomposition, first presented in [15], is a time domain technique decomposing a signal into its so-called intrinsic mode functions (IMFs). EMD is here employed to extract the fundamental oscillatory modes representing the dynamic behavior of the track in relation to the ballast and rail pad layers. The EMD method is chosen as it is an appropriate technique for processing non-stationary and non-linear signals, e.g. rail acceleration response recorded in the presence of wheel flat. Moreover, EMD is a fully data-driven algorithm where the mode decomposition is based on an automatic and adaptive time-varying filtering, and it does not requires an a-priori basis.

Let $\bar{a}(t)$ for $t \in [t_0, t_1]$, be the measured track vertical acceleration induced by the train $\mathcal{T}$ consisting of N_b bogies when it passes through the turnout between the initial time t_0 and the final time t_1. The acceleration signal is first low-pass filtered with a cut-off frequency of 1 kHz, then it is sliced at every bogie passage before being fed to the EMD algorithm. The filter cut-off frequency is chosen to limit the analysis to the frequency range that contains the first and second track resonance frequencies [10]. For the considered turnout, this was verified by performing a series of receptance tests in [5]. The slicing of the signal is performed to remove the influence of parametric frequencies in the frequency range where the ballast is active, since this could deteriorate the identification of the first track resonance frequency. The k-th slice of the acceleration signal $\bar{a}(t)$ corresponding to the passage of k-th bogie is $\bar{a}_k(t)$ for $t \in [t_{w_1,k}, t_{w_2,k} + \delta]$, where $t_{w_1,k}$ and $t_{w_2,k}$ are the time instants when the first and second wheels of the bogie pass the accelerometer and $0 < \delta < t_{w_1,k+1}$. It must be noted that the response of the track to each passing wheel is not considered individually since the signal is not completely damped in between the wheel passages and, therefore, the responses to two successive wheels are not entirely distinguishable from each other (see the example shown in Fig. 3).

By applying the EMD algorithm summarized in Algorithm 1 to each slice of the acceleration signal, the IMFs are obtained. Two stopping criteria are defined in the algorithm. The first criterion is used to stop the overall sifting process, when $Q_{residual}$ is smaller than $\epsilon_1 = 45 - 50$ dB. The second one stops the sifting process associated with the j-th IMF, when $Q_{resolution}$ (i.e. the energy ratio of the signal in each iteration to the average of the envelopes) is larger than $\epsilon_2 = 45 - 50$ dB. The IMFs extracted from the EMD algorithm are utilized as identification data sets to estimate low-complexity behavioral models of track components at different locations along the turnout.

Algorithm 1: Empirical Mode Decomposition adopted from [27]

Data: Track vertical acceleration $\bar{a}_k(t)$
Result: Set of intrinsic mode functions $\mathcal{I}_k = \{IMF_1, \ldots, IMF_m\}$
Residual $r(t) \leftarrow \bar{a}_k(t)$;
$i \leftarrow 1$;
while $Q_{residual} < \epsilon_1$ **do**
  $h_{10}(t) \leftarrow r(t)$;
  $j \leftarrow 1$;
  repeat
    Find all local extrema $h_{1(j-1)}(t)$;
    Extrapolate the extrema;
    Find the upper and lower envelopes by spline interpolation of the local extrema;
    Compute the mean value $m(t)$ between the lower and upper envelopes;
    Compute $Q_{resolution}$;

$$Q_{resolution} = 10log_{10}\left(\frac{\|h_{1j}(t)\|^2}{\|m(t)\|^2}\right)$$

    Subtract the mean value from the original signal $h_{1j}(t) = h_{1(j-1)}(t) - m(t)$;
    $j \leftarrow j + 1$,
  until $Q_{resolution} > \epsilon_2$;
  $IMF_i(t) \leftarrow h_{1(j-1)}(t)$;
  $r(t) \leftarrow r(t) - IMF_i(t)$;
  Compute $Q_{residual}$;

$$Q_{residual} = 10log_{10}\left(\frac{\|\bar{a}_k(t)\|^2}{\|r(t)\|^2}\right)$$

  $i \leftarrow i + 1$;
end

3.2 *N4SID Subspace Identification Method*

The subspace identification method N4SID [31] is adopted to estimate low-complexity behavioural models in state space form, which describe the dominant track behavior due to the rail pad and ballast dynamic characteristics. A second order model is deemed sufficient to represent the behavior of each IMF, which contains only one dominant oscillatory mode of the system. Since the excitation input, i.e. the impact force from the train wheels, is not measured and only the vertical track acceleration is available then the identification of output models is pursued.

For the j-th IMF extracted from the EMD method a canonical realization of the discrete-time state space model is given as,

$$\begin{cases} \hat{\mathbf{x}}_j(k+1) = \hat{\mathbf{A}}_j \hat{\mathbf{x}}_j(k) \\ \hat{\mathbf{y}}_j(k) = \hat{\mathbf{C}}_j \hat{\mathbf{x}}_j(k) \end{cases}. \tag{1}$$

The past and future output data in the N4SID algorithm are defined as $\mathbf{Y}_p = (IMF_j)_{0|k-1} \in \mathbb{R}^{k\times N}$ and $\mathbf{Y}_f = (IMF_j)_{k|2k-1} \in \mathbb{R}^{k\times N}$. Since the input signal (i.e., train excitation) is unknown in the current study, the past and future input vectors in the identification algorithm (i.e., $\mathbf{U}_f \in \mathbb{R}^{k\times N}$ and $\mathbf{U}_p \in \mathbb{R}^{k\times N}$) are set to zero.

The N4SID algorithm consists of the following steps [20]:

1. Define $\mathbf{W}_p$ as,

$$\mathbf{W}_p = \begin{bmatrix} \mathbf{U}_p \\ \mathbf{Y}_p \end{bmatrix} \in \mathbb{R}^{2k\times N} \tag{2}$$

2. Construct $\mathbf{Y}_p$ and $\mathbf{Y}_f$ as,

$$\mathbf{Y}_p = \begin{bmatrix} IMF_j(0) & IMF_j(1) \dots & IMF_j(N-1) \\ IMF_j(1) & IMF_j(2) \dots & IMF_j(N) \\ \vdots & \vdots \quad \ddots & \vdots \\ IMF_j(k-1) & IMF_j(k) \dots & IMF_j(k+N-2) \end{bmatrix} \in \mathbb{R}^{k\times N}$$

$$\mathbf{Y}_f = \begin{bmatrix} IMF_j(k) & IMF_j(k+1) \dots & IMF_j(k+N-1) \\ IMF_j(k+1) & IMF_j(k+2) \dots & IMF_j(k+N) \\ \vdots & \vdots \quad \ddots & \vdots \\ IMF_j(2k-1) & IMF_j(2k) \quad \dots & IMF_j(2k+N-2) \end{bmatrix} \in \mathbb{R}^{k\times N}$$

3. Compute the oblique projection of $\mathbf{Y}_f$ onto $\mathbf{W}_p$ along $\mathbf{U}_f$ (i.e., ξ) by using Eq. (3) and the LQ decomposition in Eq. (4),

$$\xi = \hat{\mathbf{E}}_{||\mathbf{U}_f}\{\mathbf{Y}_f|\mathbf{W}_p\} = \mathbf{R}_{32}\mathbf{R}_{22}^{\dagger}\mathbf{W}_p = \mathbf{O}_k\mathbf{X}_f \tag{3}$$

$$\begin{bmatrix} \mathbf{U}_f \\ \mathbf{W}_p \\ \mathbf{Y}_f \end{bmatrix} = \begin{bmatrix} \mathbf{R}_{11} & \mathbf{0} & \mathbf{0} \\ \mathbf{R}_{21} & \mathbf{R}_{22} & \mathbf{0} \\ \mathbf{R}_{31} & \mathbf{R}_{32} & \mathbf{0} \end{bmatrix} \begin{bmatrix} \bar{\mathbf{Q}_1}^T \\ \bar{\mathbf{Q}_2}^T \\ \bar{\mathbf{Q}_3}^T \end{bmatrix}, \tag{4}$$

where $\mathbf{R}_{11} \in \mathbb{R}^{k\times k}$ and $\mathbf{R}_{22} \in \mathbb{R}^{2k\times 2k}$ are upper triangular, and $\bar{\mathbf{Q}_1} \in \mathbb{R}^{N\times k}$, $\bar{\mathbf{Q}_2} \in \mathbb{R}^{N\times 2k}$ and $\bar{\mathbf{Q}_3} \in \mathbb{R}^{N\times k}$ are orthogonal matrices. $\mathbf{R}_{22}^{\dagger}$ is the pseudo-inverse of $\mathbf{R}_{22}$, $\mathbf{O}_k \in \mathbb{R}^{k\times 2}$ is the extended observability matrix and $\mathbf{X}_f$ the future state vector.

4. Compute the state vector $\mathbf{X}_f$ using the singular value decomposition of ξ,

$$\begin{aligned}\xi &= \mathbf{U}_1\Sigma_1\mathbf{V}_1^T = \mathbf{O}_k\mathbf{X}_f\\ \mathbf{O}_k &= \mathbf{U}_1\Sigma_1^{\frac{1}{2}}\mathbf{T}\\ \mathbf{X}_f &= \mathbf{T}^{-1}\Sigma_1^{\frac{1}{2}}\mathbf{V}_1^{-1},\end{aligned} \tag{5}$$

where $\mathbf{X}_f = [\mathbf{x}(k), \mathbf{x}(k+1), \ldots, \mathbf{x}(k+N-1)] \in \mathbb{R}^{2\times N}$.

5. Consider the regression equation,

$$\begin{bmatrix}\bar{\mathbf{X}}_{k+1}\\ \bar{\mathbf{Y}}_{k|k}\end{bmatrix} = \begin{bmatrix}\mathbf{A}_j & \mathbf{B}_j\\ \mathbf{C}_j & \mathbf{D}_j\end{bmatrix}\begin{bmatrix}\bar{\mathbf{X}}_k\\ \bar{\mathbf{U}}_{k|k},\end{bmatrix} \tag{6}$$

with $\bar{\mathbf{X}}_{k+1}$, $\bar{\mathbf{X}}_k$, $\bar{\mathbf{U}}_{k|k}$ and $\bar{\mathbf{Y}}_{k|k}$ defined as

$$\begin{aligned}\bar{\mathbf{X}}_{k+1} &= [\mathbf{x}_j(k+1), \ldots, \mathbf{x}_j(k+N-1)] \in \mathbb{R}^{2\times N-1},\\ \bar{\mathbf{X}}_k &= [\mathbf{x}_j(k), \ldots, \mathbf{x}_j(k+N-2)] \in \mathbb{R}^{2\times N-1},\\ \bar{\mathbf{U}}_{k|k} &= [u(k), u(k+1), \ldots, u(k+N-2)] \in \mathbb{R}^{1\times N-1},\\ \bar{\mathbf{Y}}_{k|k} &= [IMF_j(k), IMF_j(k+1), \ldots, IMF_j(k+N-2)] \in \mathbb{R}^{1\times N-1}.\end{aligned} \tag{7}$$

$\bar{\mathbf{X}}_{k+1}$ is the estimated future states, $\mathbf{Y}_{k|k}$ the future output data, and $\bar{\mathbf{U}}_{k|k}$ the future input data.

6. Compute the matrices $\hat{\mathbf{A}}, \hat{\mathbf{B}}, \hat{\mathbf{C}}, \hat{\mathbf{D}}$ through solving the regression equation by using the least-squares technique,

$$\begin{bmatrix}\hat{\mathbf{A}}_j & \hat{\mathbf{B}}_j\\ \hat{\mathbf{C}}_j & \hat{\mathbf{D}}_j\end{bmatrix} = \left(\begin{bmatrix}\bar{\mathbf{X}}_{k+1}\\ \bar{\mathbf{Y}}_{k|k}\end{bmatrix}\begin{bmatrix}\bar{\mathbf{X}}_k\\ \bar{\mathbf{U}}_{k|k}\end{bmatrix}^T\right)\left(\begin{bmatrix}\bar{\mathbf{X}}_k\\ \bar{\mathbf{U}}_{k|k}\end{bmatrix}\begin{bmatrix}\bar{\mathbf{X}}_k\\ \bar{\mathbf{U}}_{k|k}\end{bmatrix}^T\right)^{-1}. \tag{8}$$

3.3 Frequency Estimation

The system and output matrices ($\hat{\mathbf{A}}_j$ and $\hat{\mathbf{C}}_j$) of the second order model identified for the j-th IMF can be expressed in terms of natural frequency ω_{nj} and damping ratio ζ_j,

$$\hat{\mathbf{A}}_j = \exp\left(\begin{bmatrix}0 & 1\\ -\omega_{nj}^2 & -2\zeta_j\omega_{nj}\end{bmatrix} \times T_s\right), \quad \hat{\mathbf{C}}_j = \begin{bmatrix}1 & 0\end{bmatrix}. \tag{9}$$

Performing the eigenvalue analysis, the resonance frequency and the damping ratio can be extracted as,

$$\omega_{nj} = \frac{|\ln(\lambda_1(\hat{\mathbf{A}}_j))|}{2\pi T_s}, \quad \zeta_j = \frac{-\mathrm{Re}(\ln(\lambda_1(\hat{\mathbf{A}}_j))/T_s)}{|\ln(\lambda_1(\hat{\mathbf{A}}_j))/T_s|}, \tag{10}$$

in which T_s is the sampling time and $\lambda_1(\hat{\mathbf{A}}_j)$ is the first eigenvalue of the complex pair associated with the matrix $\hat{\mathbf{A}}_j$. To obtain a single estimation of the resonance frequencies for each train passage $(\hat{\omega}_{n,1}, \hat{\omega}_{n,2})$, the median is taken over all the estimations extracted from the signal slices.

4 Method Implementation

The proposed method is now applied to the measured train-induced acceleration data to develop low-complexity behavioural models describing the dynamic behavior of the track at locations A2, A4, A7 and A11 along the turnout.

As an example, the track acceleration response corresponding to the passage of a bogie through location A7, and its extracted IMFs are shown in Fig. 5. The IMFs which represent the dynamic characteristics of the track in connection with the ballast and rail pad layers can be selected based on the frequency intervals in which the first and second track resonances are contained. The first and second track resonances are found in the ranges of 50–300 and 200–700 Hz, respectively [5, 10]. By performing a frequency analysis for the extracted IMFs, it can be shown that the second and the fourth IMFs (IMF_2 and IMF_4) are the only IMFs with the dominant frequency component lying within these frequency ranges.

The low-complexity behavioral models identified for IMF_4 and IMF_2 (ballast and rail pad) can be formulated in the following general form,

$$\mathcal{M}_b : \begin{cases} \hat{\mathbf{A}}_b = \begin{bmatrix} a_{b_{11}} & a_{b_{12}} \\ a_{b_{21}} & a_{b_{22}} \end{bmatrix} \\ \hat{\mathbf{C}}_b = \begin{bmatrix} c_{b_{11}} & c_{b_{12}} \end{bmatrix} \end{cases} \qquad \mathcal{M}_p : \begin{cases} \hat{\mathbf{A}}_p = \begin{bmatrix} a_{p_{11}} & a_{p_{12}} \\ a_{p_{21}} & a_{p_{22}} \end{bmatrix} \\ \hat{\mathbf{C}}_p = \begin{bmatrix} c_{p_{11}} & c_{p_{12}} \end{bmatrix} \end{cases}. \tag{11}$$

The indices b and p in the latter equation are used to denote ballast and rail pad. $\mathcal{M}_b$ and $\mathcal{M}_p$ are combined through the output into a unified model $\mathcal{M}$ able to describe the dominant track behavior

$$\mathcal{M} : \begin{cases} \hat{\mathbf{A}} = \begin{bmatrix} \hat{\mathbf{A}}_b & \mathbf{0} \\ \mathbf{0} & \hat{\mathbf{A}}_p \end{bmatrix} \\ \hat{\mathbf{C}} = [\hat{\mathbf{C}}_b \; \hat{\mathbf{C}}_p] \end{cases}. \tag{12}$$

To evaluate the prediction capability of the identified model (12), its output must be compared with the measured data. Since the input signal is unknown it is necessary to recursively estimate the initial conditions associated with each wheel passage. The estimated initial conditions can then be used along with the estimated state and output matrices ($\hat{\mathbf{A}}$ and $\hat{\mathbf{C}}$) to find the output of the identified model,

$$\hat{\tilde{a}}_i = \hat{\mathbf{C}}\hat{\mathbf{A}}^i\hat{\mathbf{x}}_0, \quad i \in \{1, \dots, N_b\} \tag{13}$$

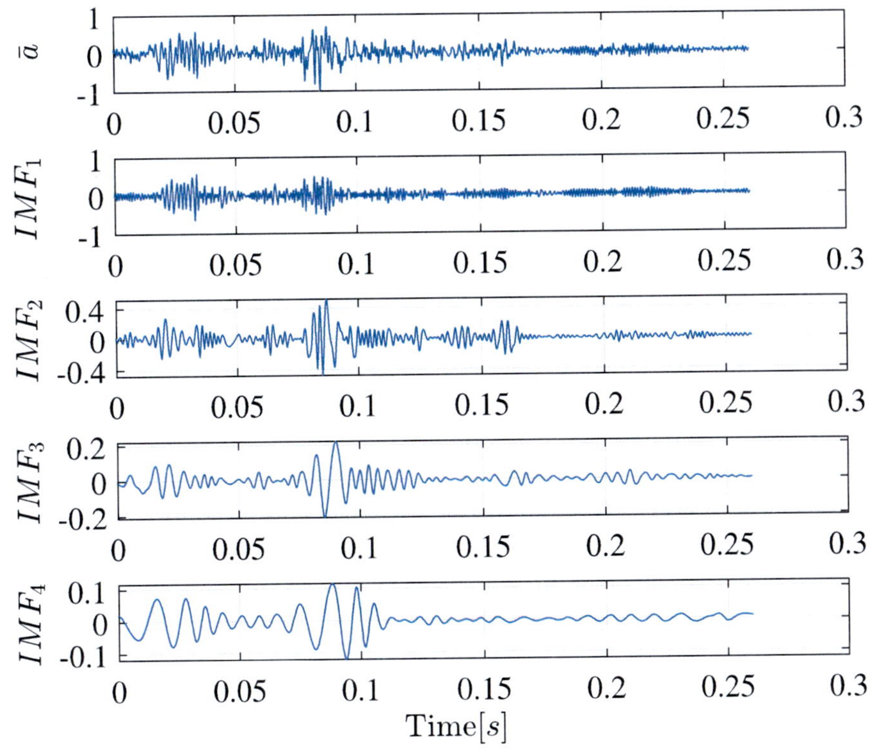

Fig. 5 Track response to a passage of two adjacent wheels (IC3 train travelling at 158 km/h) and the four extracted IMFs

4.1 Model Validation and Robustness

Considering the response of the track at different locations along the turnout (A2, A4, A7 and A11) collected during the passage of an IC3 train with the speed of 158 km/h, the discrete time state space realization of the identified models for IMF_4 and IMF_2 with their 95% confidence interval are presented in Tables 1 and 2.

The robustness and validity of the models identified for different locations are investigated by comparing their predicted output with the measured data collected during the passage of another IC3 train with the same speed. Figure 6 shows the comparison between the measured and predicted accelerations for location A7. It can be seen that the identified model is capable of predicting the dominant dynamics of the track at this location.

The dynamic characteristics of the identified models including the eigenvalues, natural frequencies and damping factors are presented in Table 3. The performance of the models in predicting the measured accelerations at different locations is evaluated through the fitting score defined as follows

Table 1 Estimated parameters of the identified models for the ballast layer

Section	$\hat{\mathbf{A}}_b$	$\hat{\mathbf{C}}_b$
A2	$\begin{bmatrix} 0.9983 \pm 0.0006 & 0.0443 \pm 0.0007 \\ -0.0443 \pm 0.0007 & 0.9983 \pm 0.0006 \end{bmatrix}$	$\begin{bmatrix} 0.7074 \pm 0.0242 & -0.0157 \pm 0.0287 \end{bmatrix}$
A4	$\begin{bmatrix} 0.9980 \pm 0.00009 & 0.0535 \pm 0.0012 \\ -0.0535 \pm 0.0012 & 0.9980 \pm 0.00009 \end{bmatrix}$	$\begin{bmatrix} 0.7073 \pm 0.0003 & -0.011 \pm 0.0004 \end{bmatrix}$
A7	$\begin{bmatrix} 0.9983 \pm 0.0003 & 0.0300 \pm 0.0009 \\ -0.0300 \pm 0.0009 & 0.9983 \pm 0.0003 \end{bmatrix}$	$\begin{bmatrix} 0.7072 \pm 0.00004 & -0.0106 \pm 0.00007 \end{bmatrix}$
A11	$\begin{bmatrix} 0.9974 \pm 0.0003 & 0.0582 \pm 0.0003 \\ -0.0582 \pm 0.0003 & 0.9974 \pm 0.0003 \end{bmatrix}$	$\begin{bmatrix} 0.7074 \pm 0.0004 & -0.0206 \pm 0.0004 \end{bmatrix}$

Table 2 Estimated parameters of the identified models for the railpad layer

Section	$\hat{\mathbf{A}}_p$	$\hat{\mathbf{C}}_p$
A2	$\begin{bmatrix} 0.9798 \pm 0.0021 & 0.1697 \pm 0.0018 \\ -0.1697 \pm 0.0018 & 0.9798 \pm 0.0021 \end{bmatrix}$	$\begin{bmatrix} 0.7092 \pm 0.0008 & -0.0600 \pm 0.0007 \end{bmatrix}$
A4	$\begin{bmatrix} 0.9851 \pm 0.0008 & 0.1539 \pm 0.0003 \\ -0.1539 \pm 0.0003 & 0.9851 \pm 0.0008 \end{bmatrix}$	$\begin{bmatrix} 0.7082 \pm 0.0001 & -0.0548 \pm 0.0001 \end{bmatrix}$
A7	$\begin{bmatrix} 0.9680 \pm 0.0063 & 0.2105 \pm 0.0037 \\ -0.2105 \pm 0.0037 & 0.9680 \pm 0.0063 \end{bmatrix}$	$\begin{bmatrix} 0.7105 \pm 0.0022 & -0.0756 \pm 0.0021 \end{bmatrix}$
A11	$\begin{bmatrix} 0.9801 \pm 0.00043 & 0.1684 \pm 0.0004 \\ -0.1684 \pm 0.0004 & 0.9801 \pm 0.00043 \end{bmatrix}$	$\begin{bmatrix} 0.7091 \pm 0.00016 & -0.0601 \pm 0.00014 \end{bmatrix}$

$$\text{fit} = 100 \times \frac{1 - \left\| \bar{a} - \hat{\bar{a}} \right\|}{\| \bar{a} - \bar{a}_m \|}, \tag{14}$$

where $\bar{a}$ and $\hat{\bar{a}}$ are the measured and estimated accelerations, and $\bar{a}_m$ is the mean value of the measured acceleration. The values obtained from the latter equation for the data recorded during the passage of a batch of 10 IC3 trains with the same speed, are averaged to calculate the overall fitting scores provided in Table 3. The results show that the identified models are capable to predict the behavior of the measured accelerations with the accuracy of more than 63% at locations A2, A4 and A11, and 48% at location A7.

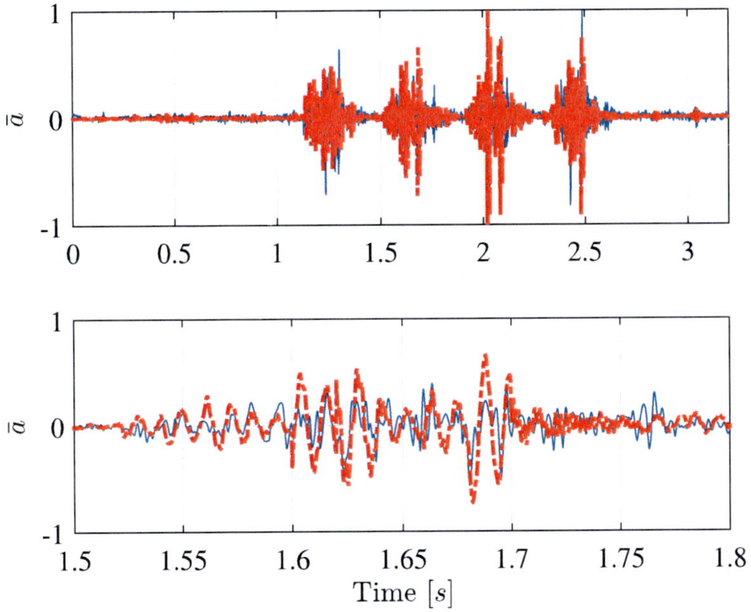

Fig. 6 Comparison of the measured (blue) and predicted accelerations (red) in correspondence to an IC3 train passage at 158 km/h. (Top) The whole train passage; (bottom) One bogie passage

Table 3 Modal characteristics of the identified models and their prediction capability

Section	λ [-]	ω_n [Hz]	ζ [-]	Fit [%]
A2	$0.9983 \pm 0.0443i$	141.3151±1.9852	0.0170	65.3
	$0.9798 \pm 0.1679i$	546.1867 ± 4.3221	0.0328	
A4	$0.9980 \pm 0.0535i$	170.6385 ± 3.6408	0.0106	68.42
	$0.9851 \pm 0.1539i$	493.5009 ± 1.600	0.0191	
A7	$0.9983 \pm 0.0300i$	95.7096 ± 2.7984	0.0416	48.35
	$0.9680 \pm 0.2105i$	682.2822 ± 6.6074	0.0441	
A11	$0.9974 \pm 0.0582i$	185.5176 ± 0.8853	0.016	63.84
	$0.9801 \pm 0.1684i$	541.9173 ± 1.0399	0.033	

5 Statistical Modelling of the Track Resonance Frequencies

To provide a robust assessment of the vibration behavior of the track along the turnout, a statistical modelling is pursued exploiting the large volume of available measurements collected over a period of approximately 2 years. The statistical model is preferred because the single estimation of the track resonance frequencies can be affected by a variety of factors that cannot be controlled, as e.g. wheel flats, weather conditions, train load and speed.

To perform the statistical modelling the measurements of track vertical acceleration induced by three different types of train (IC3, IR4 and freight) during a one-month period were utilized. To limit the impact of train speed only trains with travelling speed in the range 150–160 km/h were considered in the analysis. This speed range was chosen based on the result of a sensitivity analysis performed in [5], which indicates that high speed trains can better excite the track resonance frequencies. The proposed estimation scheme is then applied to the acceleration signals measured at locations A2, A4, A7 and A11 for each train type, and the following vectors of estimates are built for each measurement point

$$\text{Freight}: \begin{cases} \boldsymbol{\omega}_{n1}^{f} = \left[\omega_{n1,1}^{f}, \omega_{n1,2}^{f}, \ldots, \omega_{n1,N_f}^{f}\right] \\ \boldsymbol{\omega}_{n2}^{f} = \left[\omega_{n2,1}^{f}, \omega_{n2,2}^{f}, \ldots, \omega_{n2,N_f}^{f}\right] \end{cases} \tag{15}$$

$$\text{IR4}: \begin{cases} \boldsymbol{\omega}_{n1}^{r} = \left[\omega_{n1,1}^{r}, \omega_{n1,2}^{r}, \ldots, \omega_{n1,N_r}^{r}\right] \\ \boldsymbol{\omega}_{n2}^{r} = \left[\omega_{n2,1}^{r}, \omega_{n2,2}^{r}, \ldots, \omega_{n2,N_r}^{r}\right] \end{cases} \tag{16}$$

$$\text{IC3}: \begin{cases} \boldsymbol{\omega}_{n1}^{c} = \left[\omega_{n1,1}^{c}, \omega_{n1,2}^{c}, \ldots, \omega_{n1,N_c}^{c}\right] \\ \boldsymbol{\omega}_{n2}^{c} = \left[\omega_{n2,1}^{c}, \omega_{n2,2}^{c}, \ldots, \omega_{n2,N_c}^{c}\right] \end{cases} \tag{17}$$

where N_f, N_r and N_c are the number of freight, IR4 and IC3 trains in the selected speed range passing through the instrumented turnout within a month.

Figures 7 and 8 show the histograms and the probability plots for $(\boldsymbol{\omega}_{n1}^{f}, \boldsymbol{\omega}_{n1}^{r}, \boldsymbol{\omega}_{n1}^{c})$ and $(\boldsymbol{\omega}_{n2}^{f}, \boldsymbol{\omega}_{n2}^{r}, \boldsymbol{\omega}_{n2}^{c})$ at location A7. The vertical dashed lines represent the values of the first and second resonance frequencies ($\omega_{n,1} \simeq 99$ Hz, $\omega_{n,2} \simeq 616$ Hz) estimated from the receptance test performed at location A7 in the same time period [5]. It can be seen that these values lie within two standard deviations from the mean values of the histograms.

The statistical analysis of the estimated track resonance frequencies for all three types of trains at the different locations along the turnout suggests that the histograms can be well approximated by generalized extreme value (GEV) distributions, which are described by the following probability density function

$$\mathcal{G}(x) = \frac{1}{\sigma} e^{-\left(1+\xi\frac{(x-\mu)}{\sigma}\right)^{-\frac{1}{\xi}}} \left(1 + \xi\frac{(x-\mu)}{\sigma}\right)^{-1-\frac{1}{\xi}} \tag{18}$$

for $(1 + \xi(x - \mu)/\sigma) > 0$, where μ, σ and $\xi \neq 0$ are the location, scale and shape parameters, respectively. The shape parameter determines the type of the GEV distribution. $\xi = 0$, $\xi > 0$ and $\xi < 0$ represent type I, type II and type III of the GEV distribution, respectively.

Tables 4, 5 and 6 show the parameters of the estimated GEV distributions and results of the Kolmogorov-Smirnov goodness-of-fit test with a confidence level of 95%. Based on the calculated p-values, it can be concluded that for all kinds of train excitation the Type III GEV distribution well approximate the estimates of the first

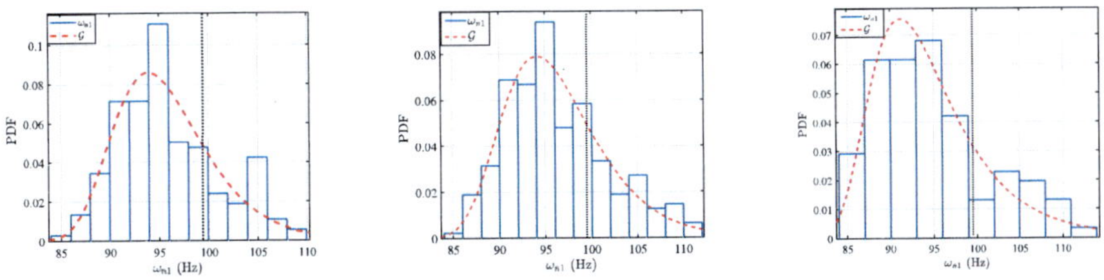

(a) Histogram freight train (b) Histogram IR4 (c) Histogram IC3

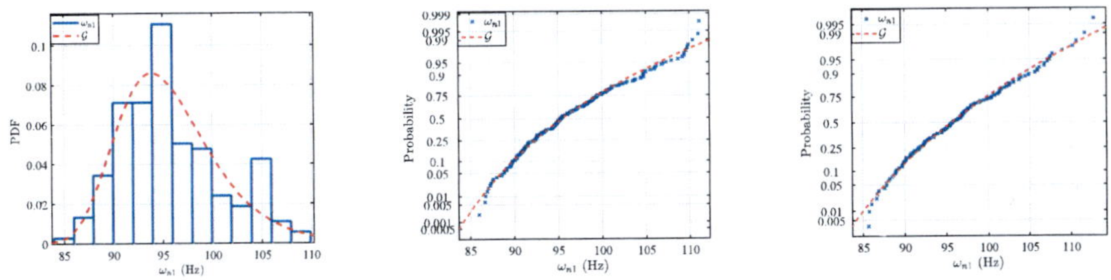

(d) Probability plot freight train (e) Probability plot IR4 (f) Probability plot IC3

Fig. 7 Statistical characterization of the first track resonance frequency at location A7 for freight, IR4 and IC3 trains

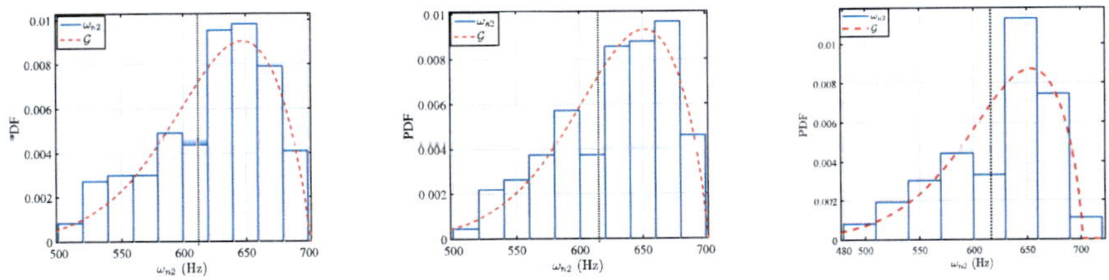

(a) Histogram freight train (b) Histogram IR4 (c) Histogram IC3

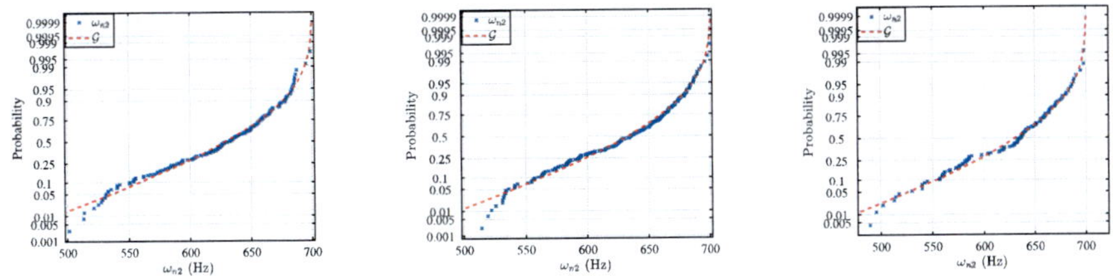

(d) Probability plot freight train (e) Probability plot IR4 (f) Probability plot IC3

Fig. 8 Statistical characterization of the second track resonance frequency at location A7 for freight, IR4 and IC3 trains

Table 4 Parameters of the GEV distribution and the corresponding p-value using a batch of freight trains

Location	Track resonances	$\mathcal{G}$-Distribution parameters	$p - value$
A_2	$\omega_{n,1}$	$\xi = -0.28, \sigma = 9.77, \mu = 158.22$	0.84
	$\omega_{n,2}$	$\xi = -0.34, \sigma = 19.55, \mu = 567.51$	0.0053
A_4	$\omega_{n,1}$	$\xi = -0.371, \sigma = 16.98, \mu = 159.92$	0.33
	$\omega_{n,2}$	$\xi = -0.307, \sigma = 42.157, \mu = 493.54$	0.35
A_7	$\omega_{n,1}$	$\xi = -0.07, \sigma = 4.31, \mu = 93.60$	0.51
	$\omega_{n,2}$	$\xi = -0.57929, \sigma = 50.567, \mu = 613.8$	0.89
A_{11}	$\omega_{n,1}$	$\xi = -0.510, \sigma = 10.99, \mu = 173.26$	0.97
	$\omega_{n,2}$	$\xi = -0.23, \sigma = 21.65, \mu = 594.03$	0.47

Table 5 Parameters of the GEV distribution and the corresponding p-value using a batch of IR4 trains

Location	Track resonances	$\mathcal{G}$-Distribution parameters	$p - value$
A_2	ω_{n1}	$\xi = -0.25, \sigma = 10.14, \mu = 157.15$	0.87
	ω_{n2}	$\xi = -0.22, \sigma = 23.63, \mu = 559.52$	0.068
A_4	ω_{n1}	$\xi = -0.37, \sigma = 15.30, \mu = 159.31$	0.51
	ω_{n2}	$\xi = -0.37, \sigma = 44.8, \mu = 502.08$	0.38
A_7	ω_{n1}	$\xi = -0.06, \sigma = 4.67, \mu = 93.81$	0.87
	ω_{n2}	$\xi = -0.63, \sigma = 49.55, \mu = 617.15$	0.45
A_{11}	ω_{n1}	$\xi = -0.42, \sigma = 10.14, \mu = 171.65$	0.62
	ω_{n2}	$\xi = -0.25, \sigma = 19.98, \mu = 598.6$	0.50

Table 6 Parameters of the GEV distribution and the corresponding p-value using a batch of IC3 trains

Location	Track resonances	$\mathcal{G}$-Distribution parameters	$p - value$
A_2	ω_{n1}	$\xi = -0.21, \sigma = 9.25, \mu = 161.77$	0.96
	ω_{n2}	$\xi = -0.36, \sigma = 26.73, \mu = 581.3$	0.60
A_4	ω_{n1}	$\xi = -0.32, \sigma = 14.11, \mu = 163.85$	0.94
	ω_{n2}	$\xi = 0.33, \sigma = 37.06, \mu = 505.16$	0.02
A_7	ω_{n1}	$\xi = -0.10, \sigma = 5.26, \mu = 92.73$	0.98
	ω_{n2}	$\xi = -0.62, \sigma = 54.89, \mu = 613.79$	0.38
A_{11}	ω_{n1}	$\xi = -0.42, \sigma = 10.97, \mu = 176.42$	0.97
	ω_{n2}	$\xi = -0.46, \sigma = 21.04, \mu = 607.77$	0.5

and second track resonance frequencies at all locations, except location A2 where the GEV distribution does not fit the estimates of the second resonance frequency when the track is excited by freight and IR4 trains.

5.1 *Towards Condition Monitoring of Railway Turnout*

To investigate the feasibility of adopting the proposed models as basis for the development of condition monitoring systems to oversee the health state of the ballast and rail pad layers over time, the statistical models of the track resonance frequencies are obtained recursively during consecutive monthly periods including a month before and a month after a maintenance event (ballast tamping). Figure 9 shows the obtained probability plots for the first and second resonance frequencies of the track when excited by freight train, IR4 and IC3 trains. After the ballast tamping event, there is a shift in the probability plot of the first track resonance frequency toward higher frequencies, which is an indication of the change in ballast stiffness. This change is apparent when analyzing the data collected during the passage of all the three types of trains. As time passes, the degradation of the ballast layer causes a gradual decrease of the ballast stiffness. This trend is more obvious when analyzing the acceleration data induced by IC3 trains and less obvious when considering freight trains. This can be explained by the fact that IC3 trains have lower axle loads with less variation compared to IR4 and freight trains [3]. Significant variation of the axle load in freight trains leads to a high level of uncertainty in the estimation of the natural frequencies. Moreover, the number of freight trains travelling with a speed in the range 150–160 km/h in a one-month period may not be sufficient to give an accurate picture of the statistical characteristics of the track resonances.

The probability plots of the second track resonance in Fig. 9d–f exhibit a different trend compared to the first track resonance. This different behavior is expected since the rail pad properties are influenced by variations of ambient temperature. In [2] an empirical model was derived to explain the effect of temperature variations on the second resonance frequency. The model showed that for low and high temperatures the natural frequency associated with the rail pad decreases as temperature increases. This relation can be observed also in Fig. 9 where the variable trend in the second resonance frequency is caused by the temperature variations over the considered monthly periods (September 2016, September 2017, October 2017 and February 2018). Table 7 presents the averaged temperature in these four months reported by the Danish meteorological institute. For freight trains, the distribution has a kind of staircase shape, which is more visible in October 2017. This behavior is attributed to the aforementioned axle load variation.

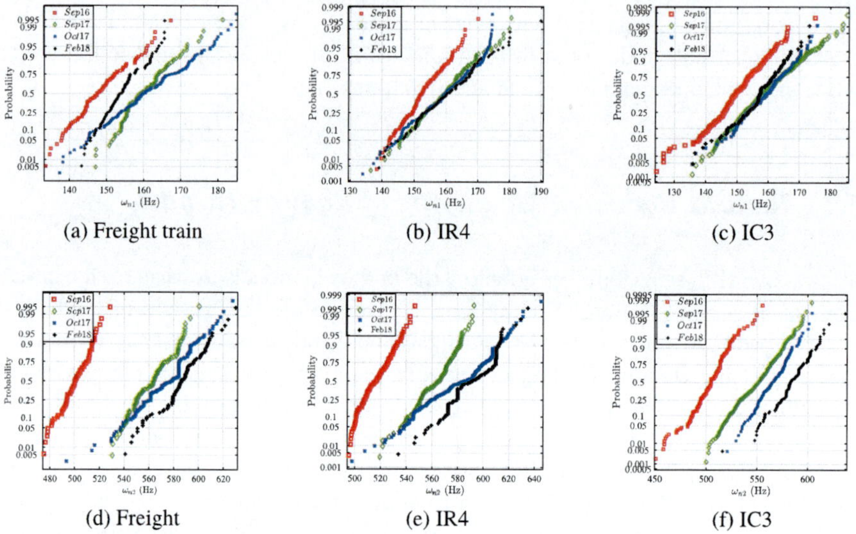

Fig. 9 Probability plots of the first and second track resonances over four different monthly periods at location A2 for freight, IR4 and IC3 trains

Table 7 Averaged temperature at Assens weather station provided by DMI

Months	Year	Temperature °C	Event information
September	2016	17	Before tamping
September	2017	14	11 months after tamping
October	2017	11.7	Receptance test
February	2018	−0.3	17 months after tamping

5.2 Discussion

The proposed method can be employed to generate statistical models of the track resonance frequencies from train-induced vibration data collected by a track-side measurement system. The statistical models are particularly beneficial since they provide a robust representation of the track infrastructure behavior despite the presence of uncontrollable parameters such as wheel/track condition, train speed, wheel load and environmental conditions.

Such statistical models provide a good basis for developing novel tools for monitoring the track infrastructure components over time, and thereby enabling the early detection of deterioration processes in the ballast and rail pad layers, which is essential for establishing efficient and cost-effective maintenance strategies. Results of the present study indicate that the type of the train-induced vibration data employed in the analysis can significantly influence the performance of the monitoring tool. Particularly, the staircase shape of the probability plots associated with the analysis of the vibration data induced by the freight trains introduces a high level of uncertainty which in turn may negatively affect the detection capability of the monitoring tool. In fact, due to the random occurrence of the stepping behavior over time the statistical characteristics of the resonances may not be approximated consistently with the GEV distribution. This can affect the accuracy of the change detection algorithm and lead to a large occurrence of false alarms. The statistical model developed based on the analysis of the data collected during the passage of IC3 trains is consistent over time and, therefore, more appropriate for designing an intelligent data-driven monitoring tool.

6 Conclusions

A novel method was proposed to generate statistical models describing the dynamic characteristics of the track infrastructure components. The method utilizes the empirical mode decomposition and the N4SID identification algorithm to first identify second order dynamical models associated to the ballast and rail pad based on the measured track vertical acceleration induced by train passage, and then compute the first and second track resonance frequencies. The statistical model is then built by fitting a parametrized probability density function to the values of the track resonance frequencies computed over a fix period of time.

Data collected at different sections of a turnout in the Danish railway infrastructure were used to successfully validate the method and create dynamical models able to predict the train induced track vibration at the points of measurement. Then, the recursive computation of track resonance frequencies allowed for the statistical analysis of such quantities over monthly periods. The analysis showed that the generalized extreme value distribution is suitable to describe these dynamical features. Further, by creating multiple statistical models over a long period of time it was possible to track significant variations in resonant frequencies due to the development of degradation processes or the occurrence of maintenance actions. Therefore, the proposed statistical models may lay the foundations for the design of condition monitoring tools for the supervision of ballast and rail pad in turnouts.

Acknowledgements This research study has been carried out as part of the INTELLISWITCH project. The financial supported by Innovation Fund Denmark, with grant number 4109-00003B, is gratefully acknowledged. We also very much appreciate the help from The Danish Meteorological Institute who made weather data freely available for this research.

References

1. Asplund M, Larsson D, Rantatalo M, Nissen A, Kumar U (2013) Inspection of railway turnouts using camera. In: Proceedings of the 10th world congress on railway research, Australia
2. Barkhordari P, Galeazzi R, Blanke M (2021) Monitoring of railpad long-term condition in turnouts using extreme value distributions. arXiv:2101.02567 [eess.SY]
3. Barkhordari P, Galeazzi R, Blanke M (2021) Monitoring of turnout ballast degradation using statistical low-complexity behavioral models. IEEE Trans Control Syst Technol 29:278–293. https://doi.org/10.1109/TCST.2020.2974697
4. Barkhordari P, Galeazzi R, de Miguel Tejada A, Santos I (2017) Low-complexity behavioral model for predictive maintenance of railway turnouts. In: Proceedings of the 2017 annual conference of the prognostics and health management society
5. Barkhordari P, Galeazzi R, de Miguel Tejada A, Santos I (2019) Identification of behavioural models for railway turnouts monitoring. arXiv:1910.06582 [eess.SY]
6. Berggren E (2009) Railway track stiffness—dynamic measurements and evaluation for efficient maintenance. PhD thesis, Royal Institute of Technology
7. Brough M, Stirling A, Ghataora G, Madelin K (2003) Evaluation of railway trackbed and formation: a case study. NDT & E Int 36(3):145–156
8. Burrow MP, Chan AH, Shein A (2007) Deflectometer-based analysis of ballasted railway tracks. Proc Inst Civil Eng Geotechn Eng 160(3):169–177
9. Clinton A (2014) Annual Safety Performance Report 2013/14. Technical report, Railway Safety and Standards Board
10. Dahlberg T (2006) Handbook of railway vehicle dynamics. CRC Press
11. EIM-EFRTC-CER Working Group (2012) Market strategies for track maintenance & renewal. Technical Report 2353 7473-11, CER—Community of European Railway and Infrastructure Companies
12. Fesharakifard R, Dequidt A, Coste O, Tison T (2014)Identification of railway track parameters for the track/train interaction analysis. In: Proceedings of the second international conference on railway technology: research, development and maintenance
13. Hassankiadeh S (2011) Failure analysis of railway switches and crossings for the purpose of preventive maintenance. MSc thesis, Royal Institute of Technology
14. Hosseingholian M, Froumentin M, Levacher D (2009) Continuous method to measure track stiffness a new tool for inspection of rail infrastructure. World Appl Sci J 6(5):579–589
15. Huang NE, Shen Z, Long SR, Wu MC, Shih HH, Zheng Q, Yen N-C, Tung CC, Liu HH (1998) The empirical mode decomposition and the Hilbert spectrum for nonlinear and non-stationary time series analysis. In: Proceedings of the Royal Society of London A: mathematical, physical and engineering sciences, vol 454, pp 903–995. The Royal Society
16. Juul Andersen K (2012) Årsrapport 2012. Technical Report 13-00144, Banedanmar
17. Kaewunruen S, Remennikov A (2005) Monitoring structural degradation of rail pads in laboratory using impact excitation technique. In: First international conference on structural condition assessment, monitoring, and improvement, pp 389–394
18. Kaewunruen S, Remennikov A (2007) Field trials for dynamic characteristics of railway track and its components using impact excitation technique. NDT & E Int 40(7):510–519
19. Kaewunruen S, Remennikov A (2008) An alternative rail pad tester for measuring dynamic properties of rail pads under large preloads. Exper Mech 48(1):55–64
20. Katayama T (2006) Subspace methods for system identification. Springer Science & Business Media
21. Kind T (2011) GPR antenna array for the inspection of railway ballast. In: Proceedings of the national seminar & exhibition on non-destructive evaluation
22. Labarile A, Stella E, Ancona N, Distante A (2004) Ballast 3D reconstruction by a matching pursuit based stereo matcher. In: Proceedings of the 2004 IEEE intelligent vehicles symposium, pp. 653–657. IEEE
23. Lam H, Alabi S, Yang J (2017) Identification of rail-sleeper-ballast system through time-domain Markov chain Monte Carlo-based Bayesian approach. Eng Struct 140:421–436

24. Lam H, Hu Q, Wong M (2014) The Bayesian methodology for the detection of railway ballast damage under a concrete sleeper. Eng Struct 81:289–301
25. Lam H, Wong M, Yang Y (2012) A feasibility study on railway ballast damage detection utilizing measured vibration of in situ concrete sleeper. Eng Struct 45:284–298
26. Norman C, Farritor S, Arnold R, Elias S, Fateh M, El-Sibaie M (2006) Design of a system to measure track modulus from a moving railcar. US Department of Transportation, Federal Railroad Administration, Office of Research and Development
27. Rato R, Ortigueira MD, Batista A (2008) On the hht, its problems, and some solutions. Mech Syst Signal Process 22(6):1374–1394
28. Remennikov A, Kaewunruen S, Ikaunieks K (2006) Deterioration of dynamic rail pad characteristics. In: Proceedings of the conference of railway engineering, pp 173–179
29. Smekal A, Berggren E, Silvast M (2006) Monitoring and substructure condition assessment of existing railway lines for upgrading to higher axle loads and speeds. In: Proceedings of the 7th world congress on railway research, Canada
30. Thompson R, Li D (2002) Automated vertical track strength testing using TTCI track loading vehicle. Technology Digest, February
31. Viberg M (1995) Subspace-based methods for the identification of linear time-invariant systems. Automatica 31(12):1835–1851
32. Yella S, Dougherty M, Gupta N (2009) Condition monitoring of wooden railway sleepers. Transp Res Part C Emer Technol 17(1):38–55

Improving Switches and Crossings Performance and Reliability

Sin Sin Hsu and Niall Fagan

Abstract Network Rail, the infrastructure manager of UK, is currently focussing on improving the reliability and performance of switches and crossings layouts. Site investigations have revealed many poorly performing switches and crossings sites that require intensive maintenance intervention with associated costs and safety implications. Analysis has shown that if the root causes are correctly identified, such as design deficiencies or wheel-rail interface issues, then applying the correct solutions can eradicate, or minimise problems. This paper presents the use of modelling and site validation to solve track maintenance issues leading to improvement in performance and reliability of switches and crossings. This is supported by practical case studies of how improvements are being achieved by a structured process which includes: undertaking site visits to identify the problem, collecting data for technical analysis and then liaising with local engineers to implement solution for site specific failures and, where applicable, working with manufacturers to optimise their products based on improved engineering requirements.

Keywords Optimisation · Switches and crossings · Failure modes

Executive Summary
The principles identified and lessons learned in this paper are applicable to all railways with turnout performance issues. This paper presents case studies of track maintenance issues which were solved using modelling and site validation leading to improvement in performance and reliability of switches and crossings saving the infrastructure manager significant costs in attending to repeat faults.

S. S. Hsu (✉)
Network Rail, London, UK
e-mail: sinsin.hsu@networkrail.co.uk

N. Fagan
Network Rail, London, UK

R. Galeazzi et al. (eds.), *Intelligent Quality Assessment of Railway Switches and Crossings*, Springer Series in Reliability Engineering,
https://doi.org/10.1007/978-3-030-62472-9_3

1 Introduction

1.1 Background

The performance of Switch & Crossing (S&C) in the UK, in terms of reliability and/or asset life, has to some extent deteriorated in recent years; this may be attributable to the general increase in both tonnage and speed, and change in rolling stock type.

However, there are sometimes underlying issues causing the poor performance. Failure modes resulting from poor performance may lead to the imposition of speed restrictions and/or operational hindrances in order to carry out emergency repairs or unduly frequent maintenance. Premature renewal of the S&C asset often follows.

Thorough site investigations, supported by subsequent detailed analysis, have proved that the majority of S&C failures could have been prevented—or at least deterioration retarded—by a better understanding of the underlying causes of the failure. Many of the causes could have been avoided, or at least minimised, through better specification of components, layout and alignment at the design stage.

Two case studies are presented herein to demonstrate the approach in analysing the failures and the proposed solutions. The novelty of this paper is the use of modelling tools that are traditionally used during the design phase.

1.2 Methodology

Recent site investigations have revealed many poorly performing switches and crossings sites that require intensive maintenance intervention. Two case studies are shown herein demonstrating solutions to site specific problems. The first is the identification of the root cause of a persistent screw failure. The second, modifications made to componentry design to prolong the service life of a layout.

The main tools used are vehicle dynamics simulations (VAMPIRE) for analysing wheel/rail interface issues and CAD and finite element modelling (SolidWorks) for evaluating component stress distribution and assembly motion and instrumentation. Matlab was used for data analysis, where recorded data such as acceleration, displacement, rail profiles are processed and input into the modelling for validation and subsequently in parametric studies for proposing improvements.

2 Case Studies

2.1 Repeat Shearing of Screws at a Crossing

This example demonstrates the effect of (normally acceptable) alignment variation through crossings that can trigger componentry failures.

2.1.1 The Problem

Repeated incidents of sheared Vossloh type screws at a crossing at the emergency crossover at Stoke Junction, on the East Coast Main Line, have been of concern to the local engineers, who have used tie bars, an extra (timber) bearer and baseplates, in attempts to secure the crossing in place. The problem area is shown in Figs. 1 and 2.

Fig. 1 Affected crossing area

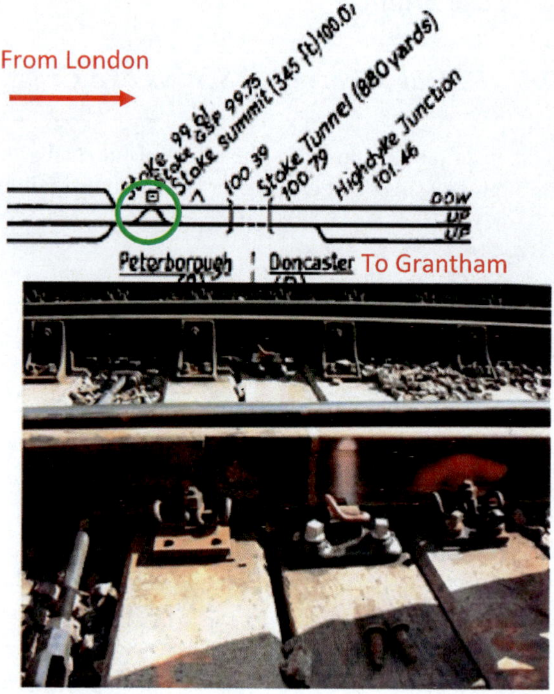

Fig. 2 Broken Vossloh type screws, an extra (timber) bearer with baseplates, and tie bars

2.1.2 Technical Analysis and the Solution

Site investigation and vehicle dynamics simulations confirmed the root cause of the screw failures to be a short (though within specification) alignment change in the crossing vicinity as shown in Fig. 3.

The track geometry measurements were taken from a measurement train. Dip angle is defined as the change in angle over a meter length and the alignment is measured over 35 m lengths.

Simulations were carried using the vehicle dynamics modelling software VAMPIRE with the latest track geometry recording and measured rail profiles as input to assess the forces acting on the rails in the vicinity of the affected crossing. Various assumptions were made:

- A laden MK4 coach vehicle model
- Standard wheel/rail friction condition of 0.3
- A set of moderately worn P8 wheel profiles.

From the simulations, the alignment change appears to be out of phase with the wheelset movements, i.e. the wheels were moving in the opposite direction to the alignment change and were therefore generating large steering and lateral forces in order to pass through the irregularity as shown in Fig. 4. This would account for the breakage of the screws on either side of the crossing baseplates.

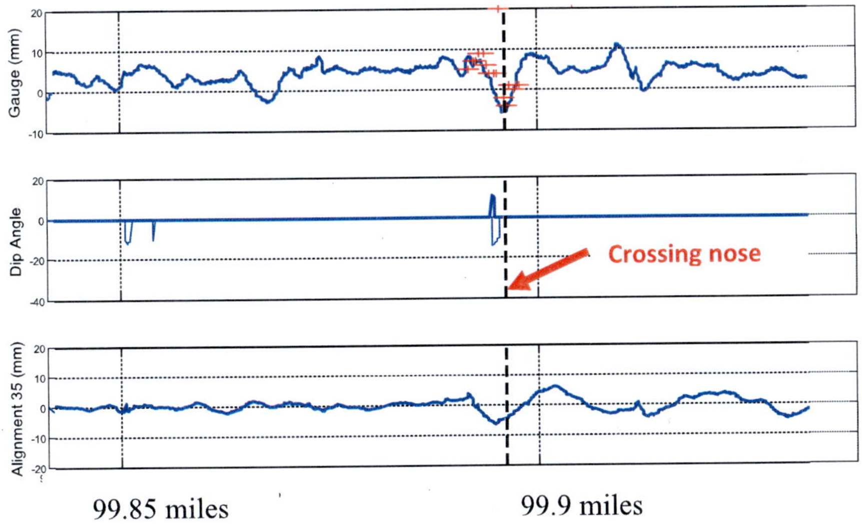

Fig. 3 Track geometry recording from a measurement train showing noticeable alignment change; and recorded dip angles at in the affected area

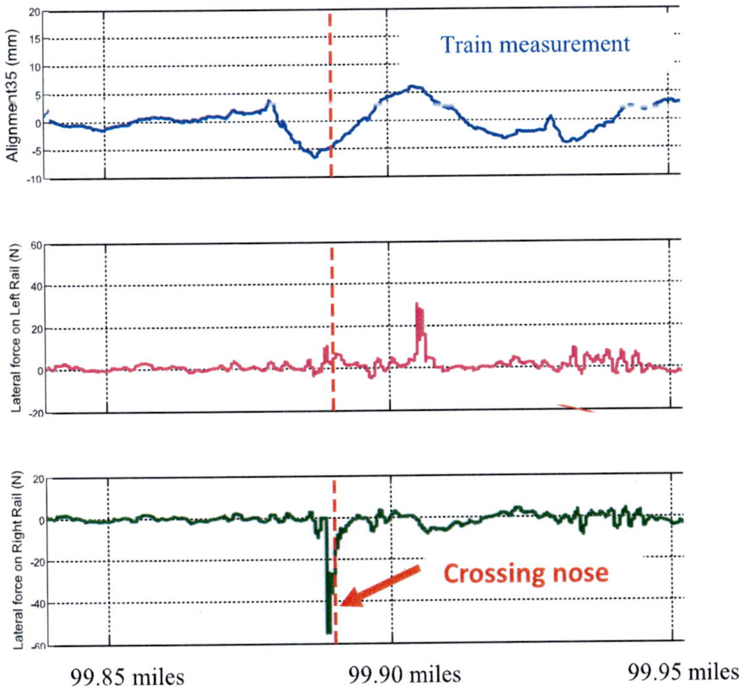

Fig. 4 Vehicle dynamics simulation results showing noticeable spike in lateral forces at the affected area

With the vehicle dynamics simulations, it was possible to assess the effects of various remedial options:

- Lowering the vehicle speed *significantly reduces* the high dynamic forces
- Correcting the alignment *eliminates* the high dynamic forces
- Correcting the gauge, however, would have little effect on reducing the forces.

It was therefore recommended to the track maintenance engineer that the alignment should be corrected as soon as was practicable; this will reduce impact loading and therefore further damage. The track support condition should be reinstated; longer chair screws should be fitted to improve engagement with the bearer.

The Track Maintenance Engineer reported that the problem has not recurred since the alignment has been corrected.

2.2 Componentry Optimisation at a Sharp Angled Double Junction

This case study shows the use of componentry optimisation to improve the crossing performance at a two-levelled layout.

2.2.1 Problem

Experiences from several sites including Newark, Great West Junction (Basingstoke), Shalford Junction, Vauxhall, Weybridge, Feltham, North Somerset junction, showed that the critical speed for sharp crossings of 1 in 6 or below appears to be no greater than 60 mph.

Great West Junction (Fig. 5) was renewed in 2006/7; since then a high number of crossing failures have occurred with five out of eight crossings replaced due to cracking from the base of the casting at the two fixed diamonds on the Down Reading line. The typical failure mode is shown in Fig. 6 and the failure history is shown in Fig. 7.

The line speed is 90 mph. With the service life of the crossings averaging about 4 to 5 years, the failure rate was very high. In particular, one of the obtuse crossings has been replaced twice in two years. The track maintainer said the situation was unsustainable.

Investigations revealed a number of design issues at the site. The most apparent was an alignment fault (no doubt made worse by the higher line speed) due to the short lengths of rails that are unavoidable with a sharp angled double junction—see photograph on the right. The other issues were deficiencies in the design of the cast crossings.

Fig. 5 Great West Junction with an alignment kink due to short sections of rail

Fig. 6 Typical cracking at the casting foot as seen on site

As a result of the investigation and finite element modelling work using Solidworks, a pair of modified obtuse crossings were made and trialled in track in 2015. To this date, the crossings are still performing well.

The next three sections detail how crossing design deficiencies have been addressed.

1. High Dynamic Forces at Wheel Transfer Area

At the wheel transfer area of a crossing, a wheel will experience a dip angle due to the differential height of the crossing nose and the wing rail. The amount of support

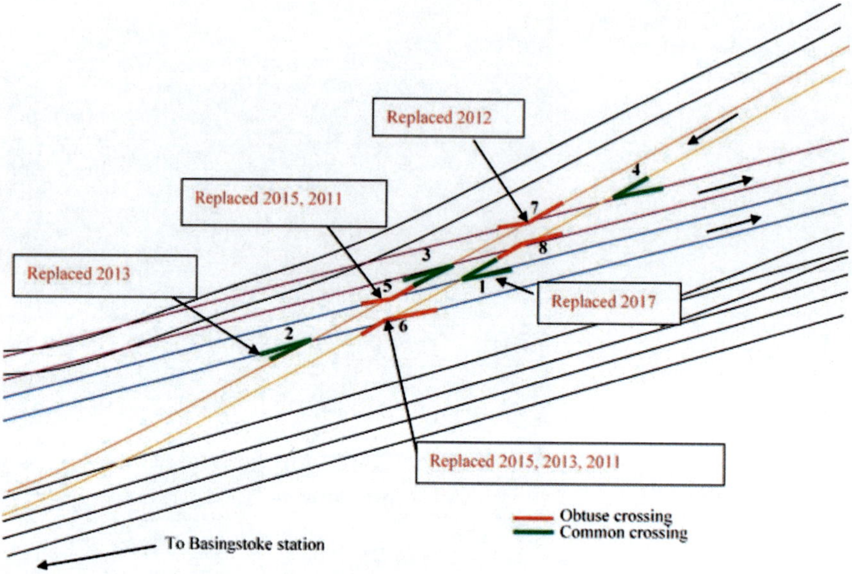

Fig. 7 Recent failure history at Great West Junction

provided by the wing rail is dictated by the width of the wheel and the angle of the crossing. Sharper angle crossings tend to result in a large dip angle as the wheel transfer generally occurs (abruptly) near to, or at, the nose tip; i.e. the lowest point of the nose, with minimal—if any—simultaneous support from the wing rail.

The effect of a wheel traversing a sharp and a flat crossing is illustrated below in Fig. 8. Note that the minimum width of the wheel tread is 127 mm, and the maximum is 140 mm.

A British Rail Research report [1] showed that the impact forces imparted during wheel transfer increased with dip angle as well as with vehicle speed. As an example, they measured a maximum force of 300% the static wheel load for a 1 in 5.5 crossing compared to only 50% greater for a 1 in 18.5 crossing angle.

The running bands observed at Great West Junction confirmed that wheels are transferring at the nose tip (see photo below in Fig. 9). With a line speed of 90 mph on the main lines, it is highly probable that the dynamic loadings on the crossings are significant, which would account for the rapid deterioration of the track support system and the premature component failure.

With a flatter crossing, the overlap between the wing rail and nose is longer, and therefore more effective in supporting the wheel, as shown by the running bands on a 1 in 10 crossing below in Fig. 10.

The simple lesson to be learnt is that sharp angled crossings should not be used for higher speed running due to the short, if not instantaneous, wheel transfer at the crossing nose. This is an inherent design issue with the layout at Great West, which cannot be rectified unless the junction is re-modelled.

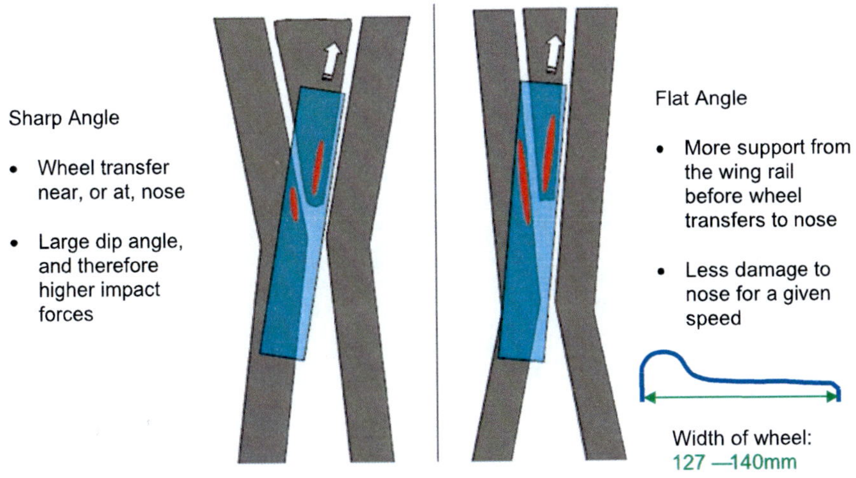

Fig. 8 Wheel transfer at sharp versus flat angled crossings

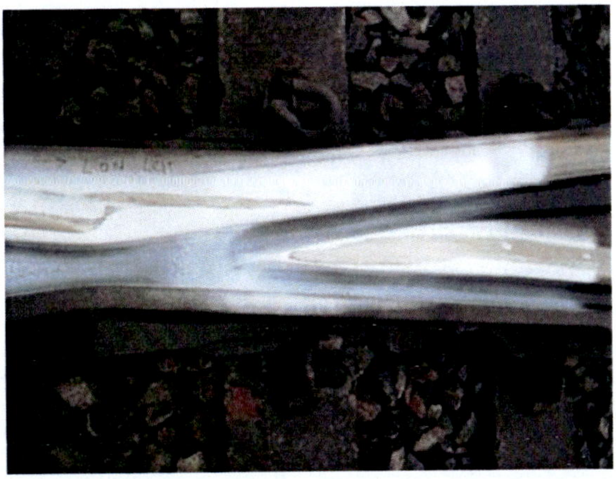

Fig. 9 Running band showing instantaneous transfer of the wheel to the crossing nose resulting in large impact forces acting on a thin section of the crossing nose

2. Stress Raisers at the Casting Foot Flange Area

Cast crossings have been designed with 'scalloped' foot flanges for the past 40 years. At the ends of the scallops are fillets with small radii, which can act as stress raisers, especially under repeated loading. In order to cope with increasing vehicle speeds and traffic tonnage, the design is in need of updating to avoid, or at least significantly reduce, dynamic stress in such locations.

Casting foot cracking is a common occurrence and accounted for 103 crossing failures in 2012/13. This failure mode has a big performance impact as the only remediation action is an (immediate) temporary speed restriction until such time as the crossing is replaced; which can take weeks—depending upon the availability of a

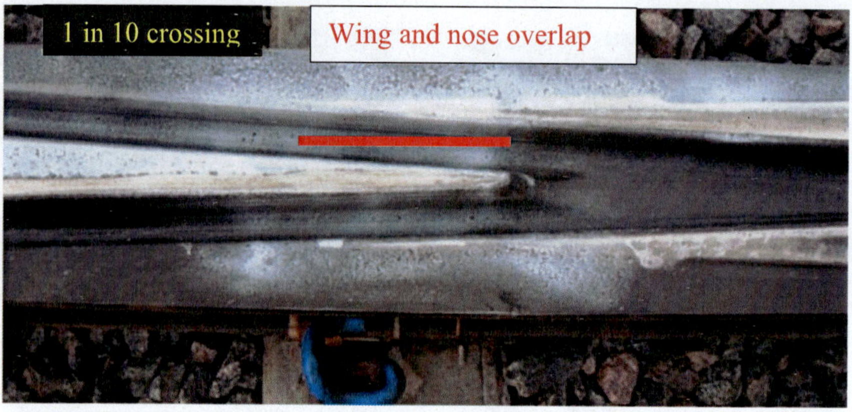

Fig. 10 Wing rail and crossing nose of a 1 in 10 crossing providing effective (overlapping) support to a passing wheel

replacement and the arranging of a possession of sufficient duration. The photographs (Fig. 11) below show examples of such failures.

There are, however, other makes of crossing that have a continuous foot flange. These appear to offer a more robust design. As examples, crossings at Finsbury Park are shown below in Fig. 12.

To improve the performance of the crossings at Great West, it was proposed that continuous foot flange design be utilised. This simple change should increase the rigidity of the section and remove any potential stress raisers.

Whilst reviewing the structural design of the crossings, it was noted that a re-design of the crossing would have to include the supporting ribs on the underside of the crossing as shown in Fig. 13; the modified design would have to minimise any sudden change in sectional stiffness, e.g. where the foot radius coincides with the hollow on the underside, as may be shown in the photograph on the right in Fig. 13.

3. Lack of Support at the Knuckle at Obtuse Crossings

The positioning of bearers under crossings appears to be a significant factor in the flexing of the casting. It was noted that the span between the supporting bearers at the wheel transfer area of the obtuse crossings at Great West Junction is relatively large which, given the impact forces discussed earlier, will have given rise to (probably considerable) flexing of the obtuse crossings. The manufacturers confirmed the span as being 650 mm, i.e. the standard bearer spacing used throughout the British RT60 design from the 1990s and the new NR60 design layouts, which have UIC 60 rail section.

Historically, smaller section 113 lb (CEN56 design) obtuse crossings have a shorter span between bearers at the wheel transfer area that gradually increases to 710 mm. This is achieved by positioning the first bearer under the knuckle and working outwards for the other bearers, each being positioned to optimise support

(a) Willesden 2014
(Crossing manufactured in 2010)

(b) Potters Bar Pts 2188 at 2014
(Crossing manufactured in 2012)

(c) Vauxhall (2009) with casting foot cracks

Fig. 11 Casting foot cracking

(a) Crossing (1999) with continuous foot flange design

(b) Crossing (2006) with continuous foot flange design

Fig. 12 Crossings at Finsbury Park

Fig. 13 Underside of crossing with reinforcing ribs but sudden change in section stiffness at the foot radius (Stock photographs from Brian Whitney)

under the likely impact area (at the noses). For example, in a 113 lb 1 in 5.5 obtuse crossing, the next bearer spacing is just 419 mm from knuckle bearer (Fig. 14).

It seems that this part of the design philosophy for 113 lb obtuse crossings was not transferred to the RT60, and subsequently NR60, obtuse crossings. It is also noted that the new NR56 (modern equivalent of the 113 lb section) obtuse crossings also adopted the larger spacing design. The performance of these crossings is expected to be worse than the NR60 because of the smaller rail section. This was seen at Shalford Junction, a lower track category line, where one of the obtuse crossing cracked at the casting foot within 6 year of installation in April 2016 (see photographs below in Fig. 15).

A simple beam bending calculation shows that by reducing the bearer spacing at the obtuse crossing noses from 650 to 419 mm, reduces the bending moment and therefore the vertical flexing (stress), by at least 40% at the crossing foot.

The main failure mode at this Great West Junction is crossing foot cracking. Optimising the bearer support position at the knuckle appears to be a simple and

Fig. 14 The narrowest (weakest) section of the obtuse crossing casting at Great West is not supported by a bearer

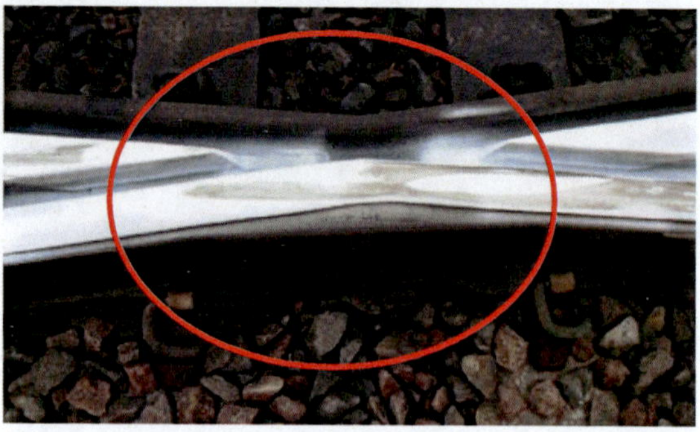

Fig. 15 Obtuse crossing at Shalford Junction cracked at the casting foot within 6 years of installation

effective solution; and one which could have the greatest benefit of stress reduction at the base of the casting.

2.2.2 Technical Analysis and Proposed Solution

Understanding the root causes of the failure modes enable potential solutions to be examined. Simple structural analysis was used as a first pass to assess that the proposed improvements were correct. The manufacturer then came out with products based on the guidelines. A finite element analysis was then carried out on the proposed designs to check that the products are adequate and that no new problems are created.

<u>Structural Analysis</u>

Results from a simple structural analysis confirmed that significant improvements can be achieved with a continuous foot flange and reduced bearer spacing.

The main findings from a simple structural assessment of the sectional properties are:

1. By simply introducing the continuous foot flange, I_{xx}, the strength of the section can be increased by 44% and the resultant casting foot stress at the knuckle reduced by 40% as shown in Fig. 16.
2. By also moving the (necessary) production hollows at the base of the casting away from the narrow section at the knuckle, the overall reduction in foot stress could be as much as 55% as illustrated in Fig. 17.
3. Assuming the crossing is simply supported, it was found that reducing the bearer spacing to 400 mm, reduces the foot stress by 36% as shown in Fig. 18.

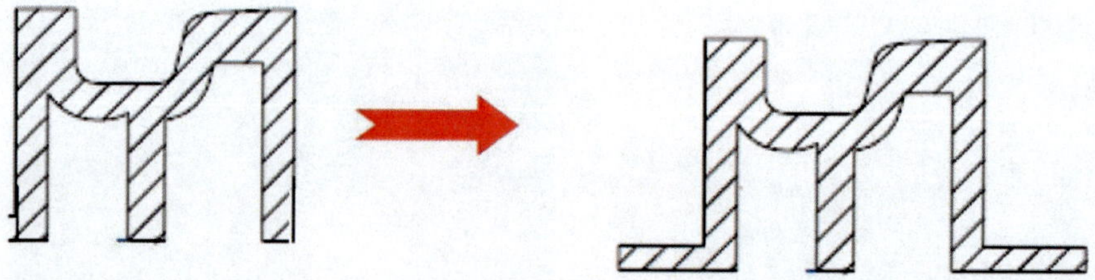

Fig. 16 Existing pocket/scallop foot design versus continuous foot flange at knuckle

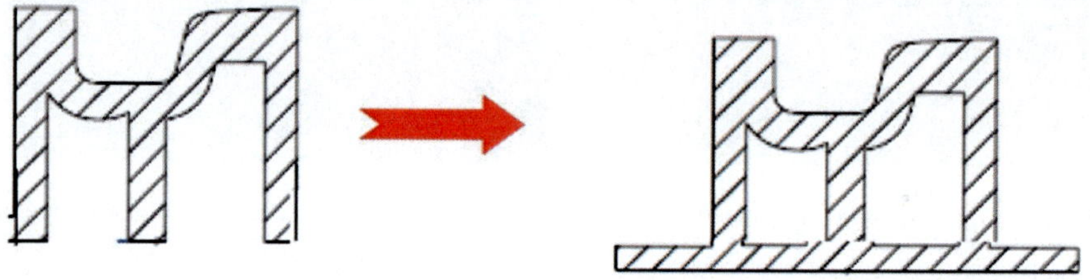

Fig. 17 Existing versus continuous foot flange and closed up the section at knuckle

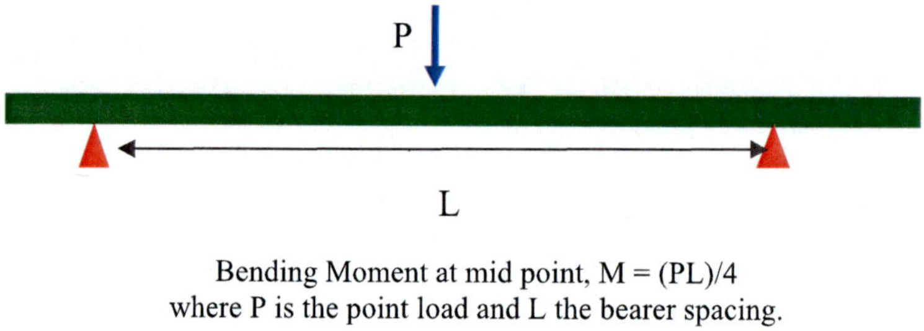

Fig. 18 Simply supported beam under a point load

Finite Element Modelling

The manufacturers presented three new crossing designs for further analysis by finite element method. All three designs shown in Fig. 19 have continuous casting foot flanges but different bearer spacings; two of the designs (nos. 2 & 3) have five supporting bearers—as opposed to the existing four (Fig. 20).

When subjected to the worst case loading, i.e. a vertical load of 500 kN, all three new crossings were predicted to experience high stresses at the same location as existing crossings, but with lower values than the fatigue stress limits. In other words, the crossings have not simply been 'beefed up' in such a way that the problem is shifted somewhere else. The new crossings would still fail at a similar location—crossing nose and the foot casting directly under the nose as shown by the coloured regions in the figure below but under a significantly higher load. The new designs could therefore be expected to have a much longer service life.

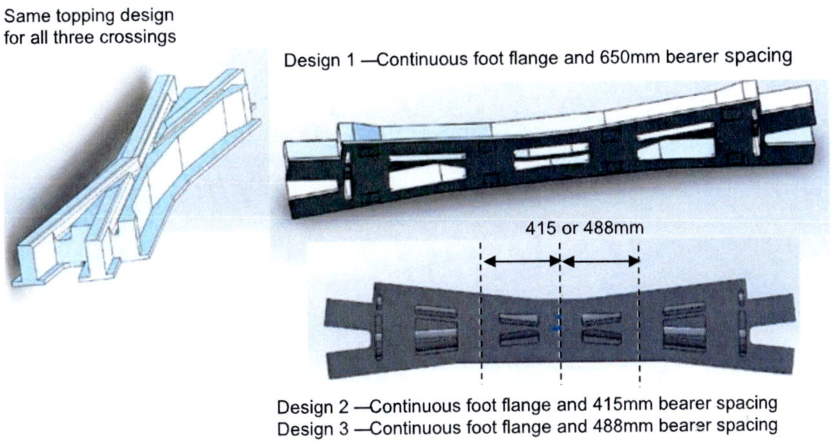

Fig. 19 Three proposed designs by manufacturer

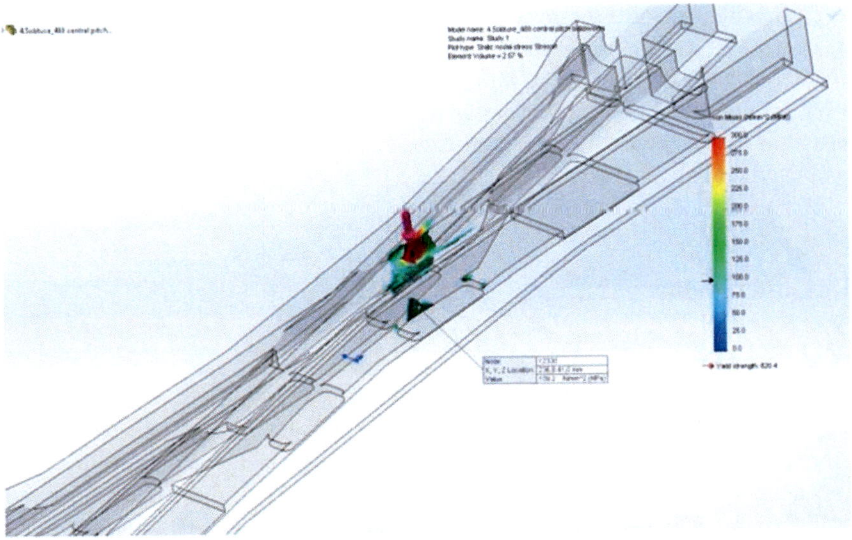

Fig. 20 The crossing is predicted to fail at the same location—crossing nose and the foot casting directly under the nose—not shifting the failure mode elsewhere

Designs 2 and 3, with improved support provided by the additional bearer, can be expected to have a longer service life than Design 1. Design 2 resulted in marginally (approx. 9%) lower maximum stresses under the nose, when considering an elastic support condition—representative of some form of voiding. However, for maintainability, the infrastructure manager has chosen Design 3, as the bigger gap between

Fig. 21 New crossings with beefed up casting and reduced bearer spacing

bearers will allow manual packing; and, should the need arise, emergency plain-lining will be simpler.

The modified obtuse crossings, shown in Fig. 21, were manufactured and installed at Great West Junction in April 2015. They have reduced bearer spacings and a more robust foot flange design. The existing crossing designs had a history of failing within two years; to date (2019) the new design has not failed. Owing to the persistent speed restriction imposed at the site, the infrastructure owner is reviewing the possible option of building a flyover.

3 Conclusions

This paper demonstrates the tangible benefits of a structured, analytical approach to identifying the root causes of the failure modes and developing appropriate solutions as part of the design process prior to the manufacture and installation of Switch and Crossing. With the confidence of modelling work validated by comprehensive site measurements, the standards, drawings and design processes have been improved, facilitated by better understanding of the technical issues by designers, manufacturers and maintainers. It was also shown that with good engagement with suppliers, it was possible to ensure better quality of product and consistent production processes to further improve asset performance. The examples demonstrated Network Rail's overarching drive to achieve safety by design as well as performance improvements. Furthermore, the principles identified and lessons learned are transferable to all railways with switches and crossings performance issues.

Reference

1. Cheesewright P (1977) Wheel/rail forces measured at crossings of cast manganese and built up types

Metallurgical Characterization

Damage and Microstructure Evolution in Cast Hadfield Steels Used in Railway Crossings

Ankit Kumar, Roumen Petrov, and Jilt Sietsma

Abstract Although the damage development in the railway steels is relatively well documented in the scientific literature, its appearance in railway crossings remains a quite complicated issue because of the extreme loading conditions during the train passage. The exact local conditions cannot be determined with sufficient accuracy, but the steel microstructure contains the traces of the mechanisms of microstructure and damage evolution. In this work, we discuss the microstructural evolution and damage in a field-loaded railway crossing made of cast austenitic Hadfield steel and subjected to impact and rolling contact fatigue (RCF) loading. A nanoscale twinning substructure surrounded by dislocation cells and a high dislocation density are identified using Electron Channeling Contrast Imaging (ECCI) and Transmission Electron Microscopy (TEM) in the deformed microstructure of Hadfield steels. The effect of these substructures as well as of the non-metallic inclusions and other casting defects on the damage development in the austenitic Hadfield steels is also discussed. Additionally, the strain-induced transformation of the austenite into martensite in the deformed crossing surface is studied by X-ray diffraction and magnetometer measurements. The results do not show evidence of strain-induced austenite-to-martensite transformation under impact and RCF loading of the railway crossings. ECCI in controlled diffraction conditions is used to study the fatigue crack growth in undeformed cast Hadfield steel specimens when subjected to laboratory scale

A. Kumar · R. Petrov · J. Sietsma (✉)
Department of Materials Science and Engineering, Delft University of Technology, Delft, The Netherlands
e-mail: J.sietsma@tudelft.nl

A. Kumar
e-mail: A.Kumar-2@tudelft.nl

R. Petrov · J. Sietsma
Department of Electrical Energy, Metals, Mechanical Constructions and Systems, Ghent University, Ghent, Belgium

R. Galeazzi et al. (eds.), *Intelligent Quality Assessment of Railway Switches and Crossings*, Springer Series in Reliability Engineering,
https://doi.org/10.1007/978-3-030-62472-9_4

fatigue testing. Furthermore, the role of twins and grain boundaries on the fatigue crack growth is discussed.

Executive Summary
In railway crossings, the understanding of microstructural damage in rail steels is important for improving the service life, defining maintenance strategies, improving quality of steel production and for designing new materials. The extreme loading conditions in railway crossings during the train passage leads to the degradation of steel microstructure which may lead to crossing failure. Hence, the steel microstructure contains the traces of the mechanisms of microstructure and damage evolution. In this work, we established an understanding of microstructural evolution and damage from macro to nano meter length scales in a field-loaded railway crossing made of cast austenitic Hadfield steel. The deformed microstructure of the Hadfield steel crossings show the formation of nanoscale twinning substructure surrounded by dislocation cells and a high dislocation density which affects the damage phenomenon. In addition, the material defects such as non-metallic inclusions, gas porosities and solidification shrinkage cavities are identified in the microstructure which adversely affect the damage development in the austenitic Hadfield steels. With this information, strict guidelines can be set for quality control and the clean production of the steels used in railway crossings. Additionally, this work also focuses on the understanding of fatigue crack growth in undeformed cast Hadfield steel in relation with the grain boundaries and the evolving twins ahead of the crack tip. This understanding can lead to ideas for designing new materials for railway crossing applications.

1 Introduction

Cast Hadfield steels are widely used in railway crossings as they exhibit high toughness, high ductility, excellent wear resistance and high work hardening ability [1–5]. Typical for the Hadfield steels is that they demonstrate their outstanding wear resistance only if the wear is combined with compressive stresses. Such combination of properties makes them suitable material for production of switches and crossings, rock crushers, crawler treads of tractors and balls for mills etc. These steels typically have high carbon content (above 1 wt% C) and high Mn content of around 13–14 wt%. After quenching they attain a fully austenitic microstructure (Face Centered Cubic crystal structure) and sometimes combined with undissolved carbides. Under external load, Hadfield steels show planar slip and Twinning Induced Plasticity (TWIP) as primary mechanisms of deformation. These mechanisms strongly depend on the Stacking Fault Energy (SFE) of the alloy. The cast Hadfield steels have a low SFE (i.e. ~23 mJ/m^2) [1, 4]. This is why the Hadfield steels strain harden rapidly due to the difficulty of cross slip and show high strain hardening capability.

Railway crossings are subjected to impact, wear and rolling contact fatigue (RCF) loading during service [6]. This leads to the formation of various surface and subsurface defects such as spalling, flaking, squats and microcracks, especially in the crossing nose. These defects can lead to fracture of the entire crossing nose during further wheel-rail contacts, and hence may pose severe safety threats to the passengers if they were not detected and repaired. Therefore, it is important to study the microstructural evolution and damage in rail crossings which can provide strategies for the sustainable material design and/or suitable maintenance of the railway crossings.

The damage initiation and propagation in polycrystalline alloys strongly depends on their microstructural features such as phases, phase and grain boundaries, twin boundaries, grain orientations and grain size etc. [7–16]. Various studies [2, 4, 5, 17–21] have been conducted to study the damage in the Hadfield steel rail crossings. However, they show different opinions regarding the damage mechanism in relation with the microstructure of the Hadfield steels. Xiao et al. [17] claim that the damage in Hadfield steels in crossing applications has no correlation with their microstructure and the cracks initiate and propagate due to the combined effect of maximum contact stresses at the crossing surface and the maximum residual stresses below the surface. Similarly, Lv et al. [4] claim that the grain boundaries, plastic deformation bands, twinning and crystal orientation of austenitic grains in the cast Hadfield steels have no correlation with the nucleation and growth of fatigue cracks. They rather propose that the high concentration of vacancy clusters distributed as layers below the crossing surface is responsible for the damage. Yong et al. [20] also suggest that the formation of micro-voids and their coalescence in the region of high stresses in the crossings are responsible for damage in Hadfield steel and the damage has no correlation with the crystal orientations. However, Harzallah et al. [5] show that fine twinning substructures formed under impact loading in the Hadfield steel crossings are responsible for the crack initiation. Hence, there is no unambiguous understanding of damage in Hadfield steels in crossings applications, in relation with their microstructure.

In this study, the microstructural evolution and microstructural features (deformation twinning, casting defects, non-metallic inclusions and possible martensite phase formed from austenite upon loading) which can be responsible for damage in the Hadfield steel in a field-loaded crossing are discussed. Additionally, the effect of the microstructural features on the damage in Hadfield steels is studied in laboratory scale fatigue testing.

2 Material and Experimental Methods

The cast Hadfield steel used in this study has a chemical composition of Fe-1.1C-13Mn-0.4Si (in wt%). The steel was first cast in the shape of a rail crossing. Afterwards, the crossing was solution annealed at 1050 °C for 3 h and subsequently quenched to form a fully austenitic microstructure. The crossing was installed in the Dutch rail network and removed after an approximate load passage of 400 MGT in service. The region of the maximum impact and RCF loading from this rail crossing was cut using a diamond saw (Fig. 1). A specimen was subsequently extracted from

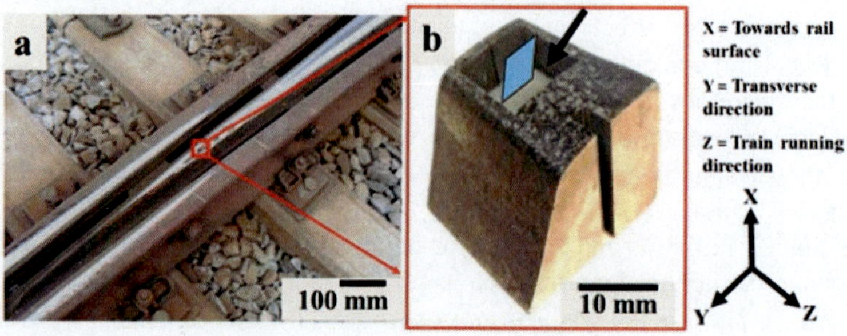

Fig. 1 **a** Photograph of a field-loaded railway crossing, **b** Region of the crossing nose after cutting from the maximum impact and RCF loading zone (black arrow shows the specimen cut from this region and rectangle shows the cross-sectional/XY plane where microstructural observations were carried out)

this region using wire cutting (shown by black arrow in Fig. 1b) and the microstructure was investigated on the XY plane (shown in blue in Fig. 1b), which is perpendicular to the train running direction. Furthermore, features such as casting defects and non-metallic inclusions are also investigated which can affect the damage in a cast Hadfield steel crossings.

Various characterization methods such as Optical Microscopy (OM), Scanning Electron Microscopy (SEM), Electron Channeling Contrast Imaging (ECCI) were used to study the microstructural evolution of Hadfield steel after the impact and RCF loading. The specimens for OM and SEM were prepared by a routine metallographic preparation sequence with final polishing using 1 μm diamond paste and etching with 2 vol.% Nital solution. An Olympus BX60M^{TM} light optical microscope was used for the OM. The Vickers microhardness measurements were performed on a Dura-scan 70 (Struers) hardness tester, using a load of 1 N for 10 s. A JEOL JSM 6500F SEM with a field emission gun was used for Secondary Electron (SE) imaging and Energy Dispersive X-ray Spectroscopy (EDS) analysis. For ECCI investigation, the specimen was polished using an OPS colloidal silica solution containing 40 nm size particles after polishing with 1 μm diamond paste. A Zeiss Cross-beam SEM with a Gemini-type field emission gun was used for ECCI investigations, using an accelerating voltage of 30 kV, a probe current of 2 nA and a working distance of 7 mm. The specimens for Transmission Electron Microscopy (TEM) were prepared using a Focused Ion Beam (FIB) milling in a FEI Helios Nanolab 600i dual beam FIB/SEM. The TEM analysis was performed in a Jeol JEM-2200FS FEG-TEM microscope operated at 200 kV.

Specimens with dimensions of $10 \times 10 \times 1$ mm^3 were taken from the crossing surface and measured for the phase analysis using X-Ray Diffraction (XRD) in a Bruker D8-advance diffractometer operating at 45 kV and 40 mA, using Cu Kα radiation ($\lambda = 0.15406$ nm) and a scan rate of 0.002° 2θ s^{-1}. Cubic specimens with $2 \times 2 \times 2$ mm^3 dimensions were also cut from region shown by black arrow in Fig. 1b and measured in a Lake Shore 7307 Vibrating Sample Magnetometer (VSM) at room temperature for determination of the martensite fraction. Before the magnetic measurements, the magnetometer was calibrated using a standard nickel specimen (NIST). The martensite fraction ($f(\alpha')$) in the specimen is calculated as proposed in [22] using

$$f(\alpha') = \frac{M_s}{x_{Fe} M_{s,Fe}} \quad (1)$$

where, M_s is the specimen saturation magnetization which was calculated from the measured magnetization of the specimen minus the contribution from paramagnetic austenite. $M_{s,Fe}$ is the saturation magnetization of pure iron (i.e. 215 Am^2/kg [23]) and x_{Fe} represents the atomic fraction of iron in the material.

Furthermore, fatigue crack growth experiments were performed on notched rectangular specimens (Fig. 2) from an undeformed cast Hadfield steel. The tests were performed in situ on a fatigue testing machine (Instron 8800 Electro plus), having a 10 kN load cell. An optical microscope is placed in a horizontal configuration to observe the crack growth. A stress ratio of 0.1 (maximum load of 1400 N and minimum of 140 N) was used in load controlled tensile-tensile fatigue testing. The specimens were also investigated in an interrupted manner before and after different crack growth stages to study the crack growth in relation with microstructural features such as the pre-existing grain boundaries and the twin boundaries (evolving during the crack growth due to plasticity ahead the crack). A controlled ECCI (c-ECCI) in combination with EBSD is used for this investigation [24–29]. The specimens were first mapped using EBSD in a Zeiss Cross-beam SEM. The orientation data from EBSD is used to simulate the Kikuchi diffraction pattern with respect to the electron beam direction using TOCA software [24, 30]. The software enables the calculation of specimen tilt

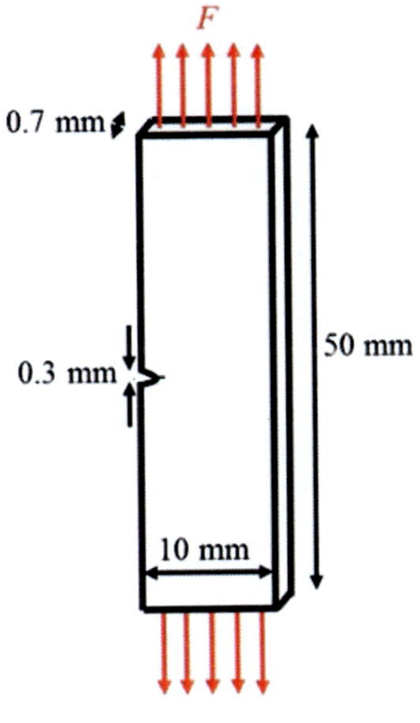

Fig. 2 Schematic drawing of a notched specimen for fatigue crack growth experiments (cut from an undeformed cast Hadfield steel)

and rotation values, required to achieve a two-beam diffraction/electron channeling condition for a well-defined lattice plane. This allows the imaging of crystal defects such as stacking faults, dislocations and twinning in the microstructure.

3 Results and Discussion

3.1 Evolution of Deformation Twinning Under Impact and Rolling Contact Fatigue Loading

Railway crossings are subjected to high impact, wear and RCF loading which leads to the changes in the steel microstructure due to the plastic deformation. Fig. shows the microhardness distribution and microstructural evolution in a field-loaded cast Hadfield steel crossing. A maximum microhardness of 760 ± 18 HV is reported close to the surface of this crossing and the hardness gradually decreases along the depth and reaches a minimum value of 330 ± 5 HV. The high hardness close to the crossing surface is attributed to the high density of deformation twins evolved due to the plastic deformation during the wheel-crossing contacts (Fig. 3b). We observe that multiple twinning systems have been activated in the grains during the plastic deformation under wheel-crossing contact (Fig. 3b–d). The density of twins decreases along the depth of the crossing (Fig. 3b–d). Figure 3e shows no indication of deformation twinning at the depth of 3.7 mm below the crossing surface. This indicates that the critical stress for deformation twinning is reached only up to a depth of approximately 3.7 mm in this crossing.

The deformation twinning is further characterized using ECCI and TEM on the surface of the field- loaded cast Hadfield steel crossing (Fig. 4). This enables to study the microstructural evolution and deformation mechanism of the cast Hadfield steel under the impact and RCF loading conditions. Figure 4a depicts the ECCI micrograph showing the nano-sized twins and equiaxed dislocation cells (encircled) in the vicinity of these twins. The size of these dislocation cells varies from 100 to 200 nm (Fig. 4a). The formation of dislocation cells is commonly observed in low, medium and high stacking fault alloys [25, 31, 32]. It is also observed that multiple twin systems are present within a single grain (Fig. 4a). Figure 4b is a magnified ECCI micrograph from Fig. 4a, showing the twin size of the order of 10 nm. Figure 4c shows the TEM image of the deformed microstructure of Hadfield steel under the impact and RCF loading with the corresponding Selected Area Electron Diffraction (SAED) Pattern. It is observed that five twinned zones are present within a width of 20 nm, which indicates a twin size of less than 5 nm in the deformed Hadfield steel microstructure (Fig. 4c). Such fine twins in the microstructure can lead to high strain hardening in the Hadfield steels [33, 34]. Figure 4d shows a TEM image of the deformed Hadfield steel indicating the accumulation of a high density of dislocations in the vicinity of the twins. Twin boundaries act as barriers to the dislocation motion, leading to high strength and strain hardening in Hadfield steels.

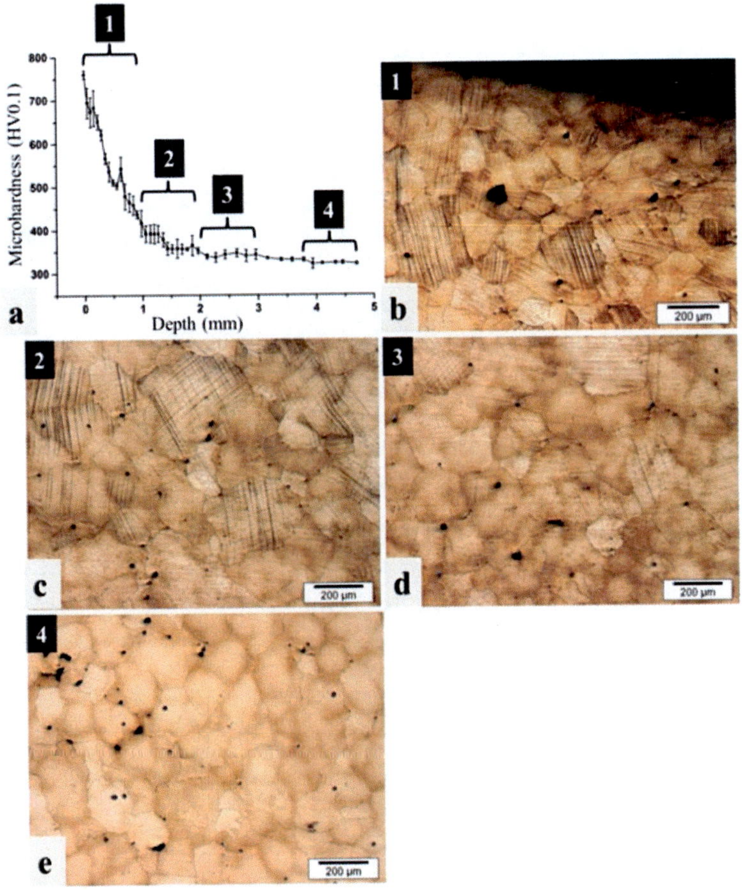

Fig. 3 Microhardness distribution from the surface of a field-loaded cast Hadfield steel crossing into the depth, (**b–e**) Optical micrographs showing the microstructural evolution in this crossing from the surface into the depth, **b** Region 1 close to the crossing surface showing high density of deformation twins, **c** Region 2 (≈ 1–2 mm below the crossing surface) showing lower density of deformation twinning than region 1, **d** Region 3 (≈ 2–3 mm below the crossing surface) showing lower density of deformation twinning than region 1 and 2, **e** Region 4 (≈3.7–4.7 mm below the crossing surface) showing no deformation twinning

3.2 *Solidification Porosity and Non-Metallic Inclusions and Their Role in Damage*

The Hadfield steel used in railway crossing application is produced by conventional casting as explained in Sect. 2. The casting process leads to solidification defects such as gas porosity and solidification shrinkage porosities [35, 36]. The gas porosities generally form due to the entrapment of atmospheric gases such as oxygen and nitrogen in the casting mold and in the liquid metal especially due to turbulence

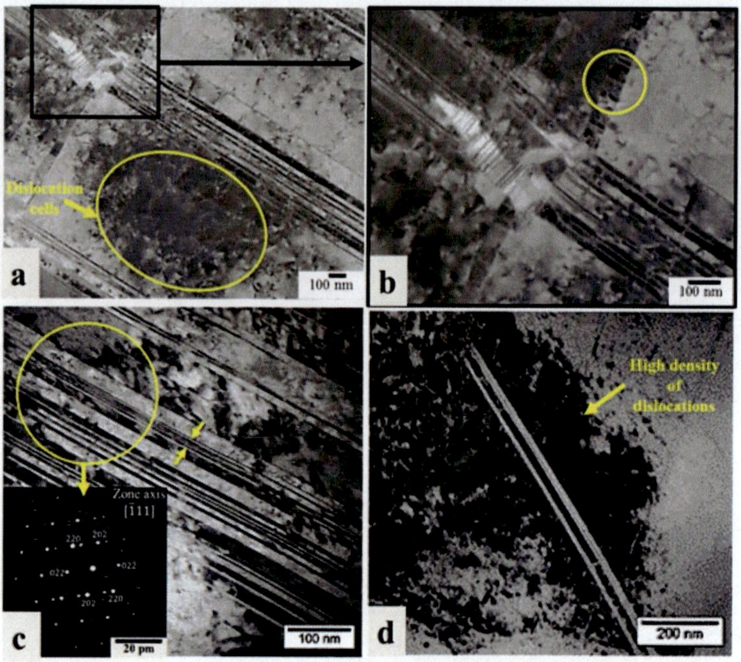

Fig. 4 **a** ECCI image showing the formation of dislocation cells (encircled) in the vicinity of twins, on the surface of a field-loaded Hadfield steel crossing, **b** Magnified ECCI micrograph from the region in (**a**), showing the twin size of the order of 10 nm, **c** TEM image with corresponding Selected Area Electron Diffraction (SAED) pattern, showing the nano-sized deformation twinning (arrows show the presence of 5 twins in a width of approximately 20 nm), **d** TEM image showing the high density of dislocations (in black contrast) accumulated at the twin boundaries

in the liquid flow during casting. However, the solidification shrinkage porosities are formed due to volume changes caused by solidification and thermal contraction [37]. A poor riser design also can lead to formation of shrinkage porosities during casting when it provides insufficient liquid feeding (misrun). The casting defects in the cast Hadfield steel crossing can be in the range of few micrometers to few mm in size. These defects are considered to be detrimental for the fatigue life of the Hadfield steels depending on their size, shape, position, and stress concentration during loading and can be responsible for the crack initiation in crossings [38]. Figure 5a shows a gas porosity size of around 220 μm and solidification shrinkages at the grain boundaries. Figure 5b depicts a magnified optical micrograph of cast Hadfield steel showing that the solidification shrinkages are mainly found at the grain boundaries. This can reduce the damage tolerance of the Hadfield steels due to intergranular crack propagation under the impact and RCF loading. Peters et al. [39] reported such intergranular cracking in the cast Hadfield steels crossings due to the presence of shrinkage porosities at the grain boundaries. Figure 5c represents a SEM image showing intergranular microcracking from the solidification shrinkage

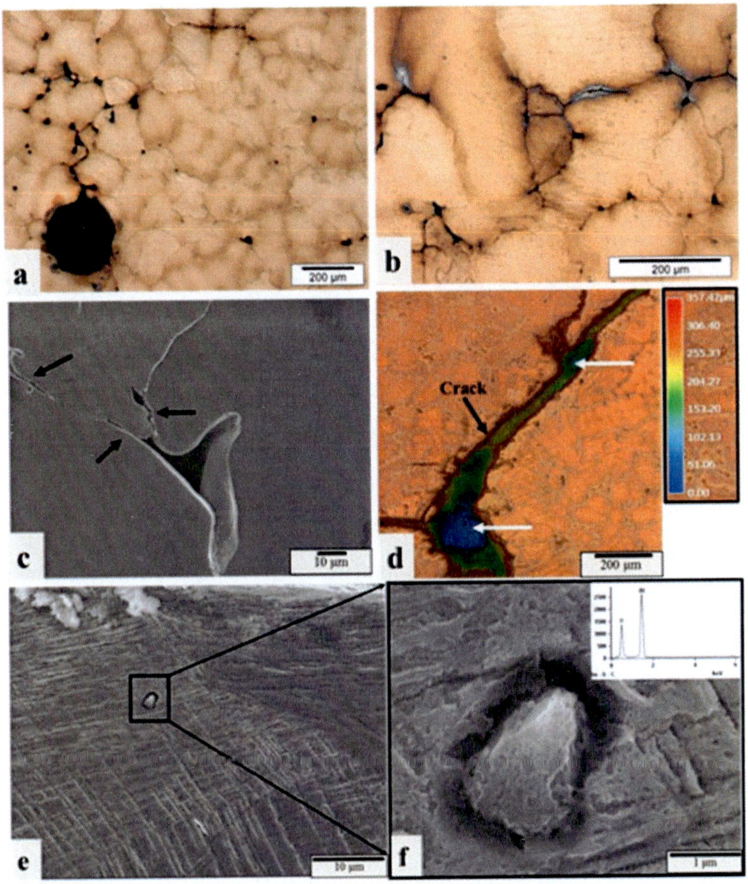

Fig. 5 **a** Optical micrograph showing gas porosities and solidification shrinkages in the cast Hadfield steel; **b** Magnified optical micrograph showing solidification shrinkages at the grain boundaries, **c** SEM image showing the intergranular microcracking from a solidification shrinkage at 1 mm below the crossing surface (black arrows show the microcracks), **d** 3D optical micrograph of a failed Hadfield steel crossings showing a crack following the casting porosities and the crack branching at the porosities (color coded scale shows the height distribution with respect to the deepest point visible in the image), **e** SEM image showing the presence of aluminium oxide inclusion at 20 µm below crossing surface, **f** Magnified SEM image of the aluminium oxide inclusion (with EDS result) showing interface decohesion

porosity at 1 mm below the crossing surface. Figure 5d shows a 3D optical micrograph from the cross-section of a failed Hadfield steel crossing (color coded scale shows the height distribution with respect to the deepest point visible in the image). The crack is observed to follow the casting porosities, which are represented by deeper areas in blue in Fig. 5d. Crack branching is also observed at the casting porosities (Fig. 5d).

Figure 5e shows the presence of an aluminum oxide inclusion at 20 μm below the surface of the field- loaded Hadfield steel crossing. Such inclusions can affect the fatigue life of the crossings due to the fact that they can act as stress raisers in the steel microstructure during the wheel-rail contact. Studies show that the presence of such inclusions promotes decohesion at the inclusion/matrix interface due high strain localization and can lead to void nucleation in the material [40, 41]. Such interface decohesion between an aluminum oxide inclusion and the deformed Hadfield steel matrix is observed in the magnified SEM image in Fig. 5f.

3.3 Strain-Induced Austenite-to-Martensite Transformation Under the Impact and RCF Loading and Its Effect on the Damage in Hadfield Steel

Some early studies show that one of the strain hardening mechanisms in the Hadfield steels is the strain-induced martensite transformation from the metastable austenite [42, 43]. This transformation occurs by a mechanism involving the dislocation motion on an invariant set of planes, which leads to the change in the crystal structure of parent austenite phase into ε or α' martensite due to the change in the stacking sequence of atomic planes [44]. However, Adler et al. [1] proposed that such transformation cannot occur in Hadfield steel (with a composition of 13 wt% Mn and 1.1 wt% C) due to its stacking fault energy being higher than 20 mJ/m^2. It is well known that after casting a heterogeneous distribution of some alloying elements (Mn, Cr, Mo, Ni etc.) is often observed in the as cast structure of steels [45]. The cast Hadfield steel studied in this work shows Mn segregation up to 18 wt% in the vicinity of the grain boundary, whereas the Mn concentration varies in between 10 and 13 wt% in the grain interiors (Fig. 6). This means that the time and temperature during the solutionizing (i.e. 3 h at 1050 °C) of Hadfield steels after casting, are not sufficient for the homogenization of

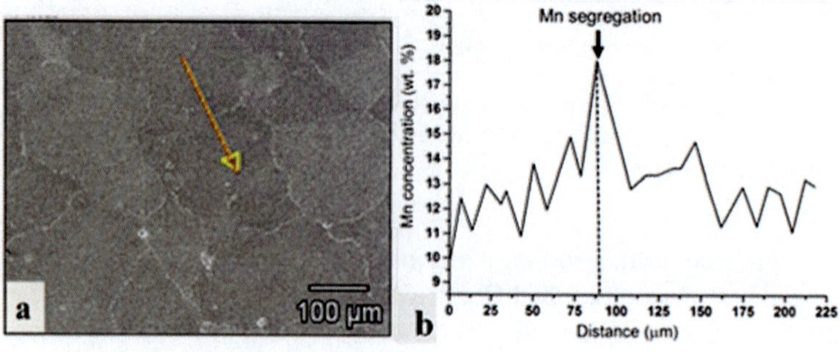

Fig. 6 SEM image of the cast Hadfield steel showing a line (in yellow) across a grain boundary (in white contrast) along which EDS analysis was performed; (b) EDS analysis showing the Mn segregation in the vicinity of the grain boundary (Dotted line show the position of the grain boundary)

manganese. The Mn atoms can only diffuse up to a distance of approximately 3.5 μm during such solutionizing treatment. The Mn diffusion distance calculations were performed according to the procedure mentioned in [46]. Therefore, the variation in Mn content still exists in the studied cast Hadfield microstructure. This variation in the Mn content affects the stacking fault energy of the alloy locally. A locally high Mn content leads to a high stacking fault energy and minimizes the possibility of strain-induced martensite transformation [47, 48]. However, regions with low Mn content (10–11 wt%) can be potential sites for such transformation due to the low stacking fault energy in these regions [47].

Jost et al. [49] studied the strain-induced austenite-to-martensite transformation in Hadfield steels under dry sliding wear conditions. They showed that such transformation on the surface of the Hadfield steel can lead to crack initiation due to the brittle nature of the martensite. Adler et al. [1] later proposed that this transformation was primarily due to the decarburization at the Hadfield steel surface which leads to a decrease in the mechanical stability of austenite. Furthermore, some studies propose that such transformation in low stacking fault alloys leads to void nucleation at the interface of the martensite and austenite and negatively affects the fatigue life [23, 50, 51]. Therefore, it is important to study the strain-induced austenite-to-martensite transformation in the field-loaded cast Hadfield steel crossing in order to know if such critical regions containing martensite and austenite exist in the microstructure.

Figure 7a shows the results of the XRD measurements from the surface of the field-loaded Hadfield steel crossing and 2 mm below the surface. Only the presence of austenite peaks is observed in these measurements, which indicates that no transformation to martensite has taken place under the impact/RCF loading in the present Hadfield steels. Furthermore, this transformation is also investigated using magnetometer measurements at the crossing surface, a representative result of which is shown in Fig. 7b. The result shows a saturation magnetization of 0.16 Am^2/kg, which is equivalent to a very low fraction of martensitic phase of around 0.0009. In addition to this, the martensitic phase is also not observed in the TEM analysis of multiple specimens taken from the crossing surface Fig. 7(c, d). Therefore, it can be concluded that no strain-induced austenite-to-martensite transformation took place in this Hadfield steel under the impact and RCF loading conditions in present case of crossing. Das et al. [52] proposed that the probability of strain-induced martensite transformation decreases with an increase in strain rate in the low stacking fault alloys. Therefore, the high strain rates in the case of railway crossings can be responsible for the absence of strain-induced martensite transformation in the Hadfield steels crossings.

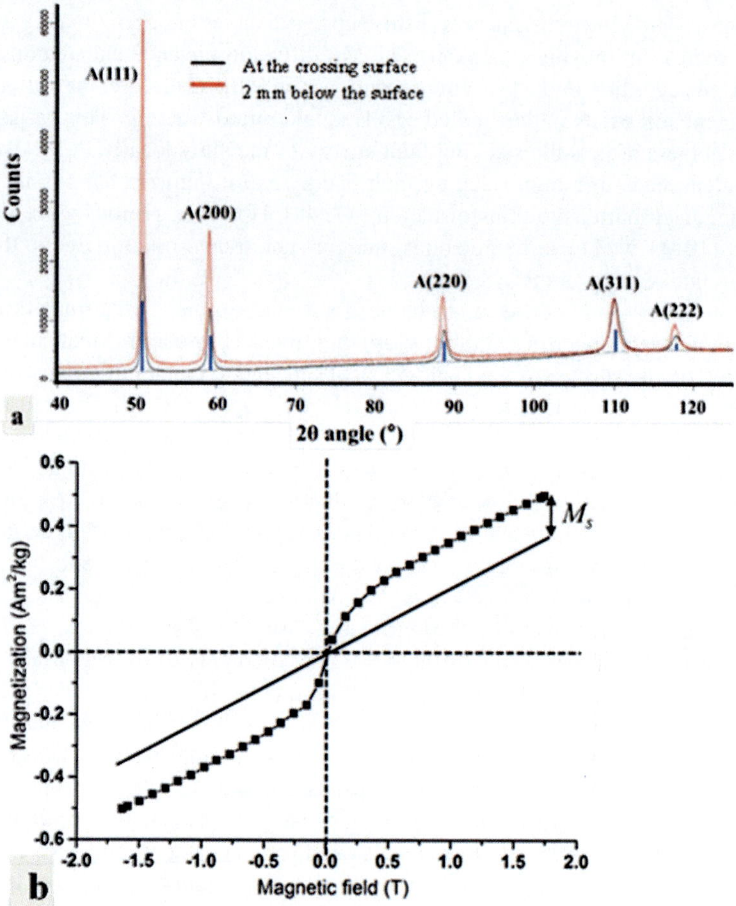

Fig. 7 X-Ray Diffraction (XRD) measurements from the surface (in black) and 2 mm below the surface (in red) of the field-loaded Hadfield steel crossing showing only the peaks of the austenite phase, **b** Results of magnetometer measurements from the surface of the field-loaded Hadfield steel crossing (Inclined solid straight line shows the contribution from the paramagnetic austenite (0.36 Am^2/kg at 1.75 T))

3.4 *Effect of the Grain and Twin Boundaries on Fatigue Crack Growth in an Undeformed Cast Hadfield Steel*

Crack growth in polycrystalline metals involves complex phenomena such as atomistic bond breaking, plastic deformation ahead of the crack tip, dislocation motion near the crack and dislocation pile ups at microstructural obstacles etc. [8, 13, 53–56]. In metals, the microstructural features such as grain orientations, grain boundaries, twin boundaries, stacking faults and different phases in the microstructure are known to have a significant effect on their fatigue crack propagation characteristics [12,

15, 57–60]. Additionally, the microstructural evolution based on the stress intensity ahead of the crack tip also affects the fatigue crack growth behavior.

Figure 8a–f show the microstructure and crack evolution in the undeformed cast Hadfield steel during the laboratory fatigue crack growth experiments. Figure 8a, b show the EBSD Inverse Pole Figure (IPF) map and c-ECCI image of the undeformed, cast Hadfield steel displaying the deflection of the fatigue crack at a grain boundary. The grain boundary refers to a high angle grain boundary with an approximate misorientation of 37° between the grains. Additionally, crack branching is observed near the grain boundary. This is in agreement with the observations of Chen et al. [57]. They showed that the grain boundaries resist the fatigue crack growth due to dislocation pile ups at the boundaries and cause large angle crack deflections at the boundaries. Zhai et al. [58] proposed a crystallographic model for the propagation of short fatigue cracks, which states that the crack plane twist and tilt angles are main factors responsible for crack retardation at the grain boundaries. These tilt and twist angles of the cracks depend on the orientations of the grains.

The microstructural evolution in the vicinity of the crack tip is shown in Fig. 8c–f. Figure 8c shows the c-ECCI image displaying the formation of nano-sized deformation twins in the vicinity of the fatigue crack. Fatigue crack propagation is observed along these twins. Figure 8d shows the formation of nano-sized deformation twins (shown by arrows) which are rather inclined to the fatigue crack growth direction. Repeated crack branching is observed in the primary fatigue crack in Fig. 8d. The crack branching retards the fatigue crack growth by lowering the stress intensity ahead of the primary crack [54]. It is observed that the crack planes in the branched cracks are parallel to the deformation twins shown in Fig. 8d. This indicates that the crack branching in cast Hadfield steels is associated with the deformation twins inclined to the fatigue crack growth direction. Figure 8e shows the formation of deformation twins perpendicular to the direction of crack growth. Figure 8f shows the magnified c-ECCI image from sub-region in Fig. 8e (shown by rectangle). We observe the formation of a high density of dislocations (in white contrast) and dislocation cells in the vicinity of the deformation twins. The twin boundaries act as obstacles to the dislocation motion ahead of the crack tip and cause the retardation in the fatigue crack growth [15]. Based on the observations in Figs. 8(c–d), it can be concluded that the angle between the crack growth direction and the deformation twins evolving ahead of the crack, is one of the factors affecting fatigue crack growth behavior in the cast Hadfield steels.

4 Conclusions

The understanding of the deformation and damage phenomena in cast Hadfield steels is important for sustainable design of steel microstructures for railway crossings application. Detailed study of the microstructure of in-field and laboratory loaded samples from a Hadfield steel allow drawing the following conclusions from the current study:

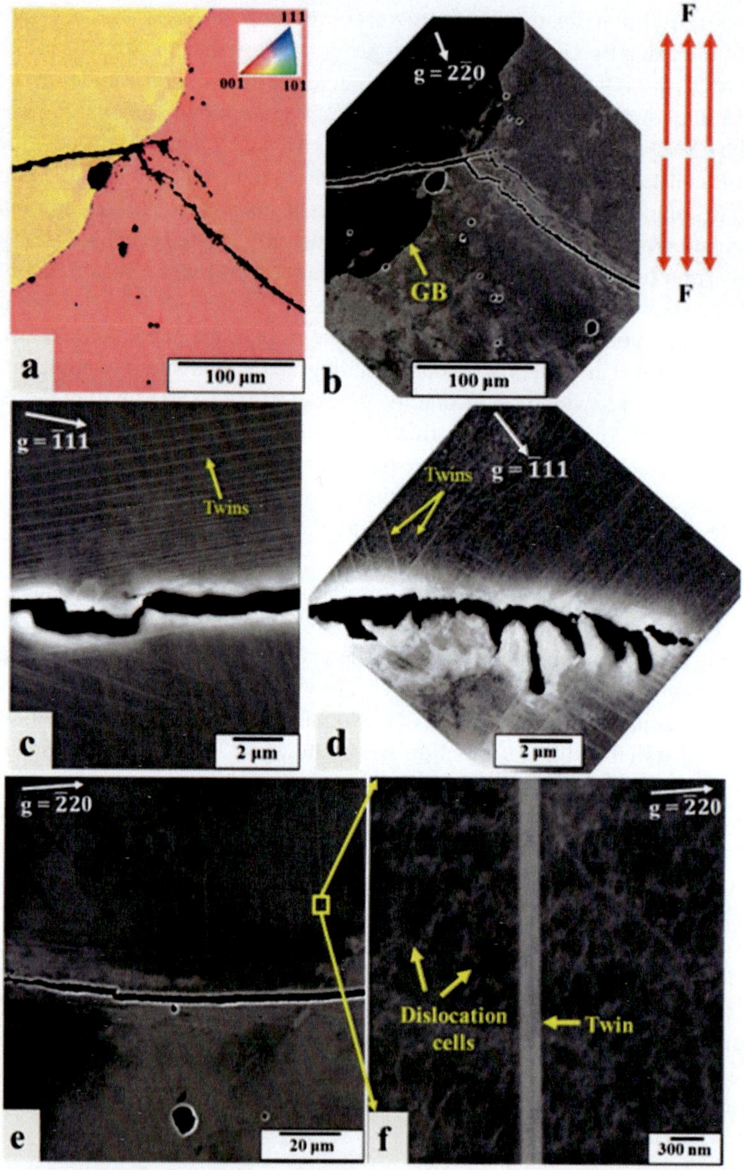

Fig. 8 Microstructural observations after the fatigue crack growth experiments on undeformed cast Hadfield steel in laboratory settings, **a**, **b** EBSD IPF map in normal direction and c-ECCI image showing the deflection of fatigue crack at the grain boundary, **c** c-ECCI image showing the formation of deformation twins ahead the crack and fatigue crack growing along the twins, **d** c-ECCI image showing the crack branching in the direction parallel to the deformation twins, **e** c-ECCI image showing the formation of deformation twins perpendicular to the fatigue crack growth direction, **f** Magnified c-ECCI image from the sub-region in **e** showing the formation of dislocation cells and high density of dislocations in the vicinity of deformation twins

1. Deformation twinning is the main deformation mechanism in the cast Hadfield steels under impact and RCF loading due to their low stacking fault energy. Heavy plastic deformation is responsible for evolution of nano-sized deformation twins near the surface of the cast Hadfield steel rail crossing.
2. Formation of nano-sized deformation twinning at the crossing surface leads to accumulation of a high density of dislocations and formation of dislocation cells in the vicinity of twins. The nano-twins and the high density of dislocation structures are responsible for the strong strain hardening response of the cast Hadfield steels under studied conditions of impact and RCF loading.
3. It is required to control the casting defects such as gas porosities/solidification shrinkages and non-metallic inclusions in cast Hadfield steels as these defects contribute to the crack initiation and propagation in crossings.
4. No evidence was found for strain-induced austenite-to-martensite transformation in the field-loaded cast Hadfield steels crossing. Therefore, the damage initiation cannot be correlated with such transformation in cast Hadfield steels in crossing application.
5. Fatigue crack growth in cast Hadfield is affected by the evolution of dislocations ahead the crack tip and the dislocation pile ups at the microstructural obstacles such as twin boundaries and grain boundaries. In addition, the grain boundaries can deflect the fatigue crack due to crystallographic misorientation between the grains and it can lead to crack retardation. Furthermore, the angle between the fatigue crack growth direction and the deformation twins evolving ahead of the crack, also affects the fatigue crack growth behavior.

Acknowledgements This research was carried out under project number F91.10.12475b in the framework of the Partnership Program of the Materials innovation institute M2i (www.m2i.nl) and the Foundation for Fundamental Research on Matter (FOM) (www.fom.nl), which is part of the Netherlands Organisation for Scientific Research (www.nwo.nl). Authors would like to thank Dr. Michael Herbig, Max-Planck-Institut für Eisenforschung GmbH, for providing necessary experimental facilities for this research. We would like to thank ProRail for its financial support and arranging the specimens for this research.

References

1. Adler PH, Olson GB, Owen WS (1986) Strain hardening of Hadfield manganese steel. Metall Mater Trans A 17:1725–1737
2. Yan W, Fang L, Sun K, Xu Y (2007) Effect of surface work hardening on wear behavior of Hadfield steel. Mater Sci Eng A 460:542–549
3. Srivastava AK, Das K (2008) Microstructural characterization of Hadfield austenitic manganese steel. J Mater Sci 43:5654–5658
4. Lv B, Zhang M, Zhang FC, Zheng CL, Feng XY, Qian LH, Qin XB (2012) Micro-mechanism of rolling contact fatigue in Hadfield steel crossing. Int J Fatigue 44:273–278
5. Harzallah R, Mouftiez A, Felder E, Hariri S, Maujean J-P (2010) Rolling contact fatigue of Hadfield steel X120Mn12. Wear 269:647–654. https://doi.org/10.1016/J.WEAR.2010.07.001

6. Kumar A, Makineni SK, Dutta A, Goulas C, Steenbergen M, Petrov RH, Sietsma J (2019) Design of high-strength and damage-resistant carbide-free fine bainitic steels for railway crossing applications. Mater Sci Eng A 759:210–223. https://doi.org/10.1016/J.MSEA.2019.05.043
7. Garnham JE, Davis CL (2008) The role of deformed rail microstructure on rolling contact fatigue initiation. Wear 265:1363–1372. https://doi.org/10.1016/J.WEAR.2008.02.042
8. Singh A, Dao M, Lu L, Suresh S (2011) Deformation, structural changes and damage evolution in nanotwinned copper under repeated frictional contact sliding. Acta Mater 59:7311–7324. https://doi.org/10.1016/J.ACTAMAT.2011.08.014
9. Morris WL (1980) The noncontinuum crack tip deformation behavior of surface microcracks. Metall Trans A 11:1117–1123. https://doi.org/10.1007/BF02668135
10. King A, Ludwig W, Herbig M, Buffière JY, Khan AA, Stevens N, Marrow TJ (2011) Three-dimensional in situ observations of short fatigue crack growth in magnesium. Acta Mater 59:6761–6771. https://doi.org/10.1016/J.ACTAMAT.2011.07.034
11. Güngör S, Edwards L (1993) Effect of surface texture on the initiation and propagation of small fatigue cracks in a forged 6082 aluminium alloy. Mater Sci Eng A 160:17–24. https://doi.org/10.1016/0921-5093(93)90493-X
12. Herbig M, King A, Reischig P, Proudhon H, Lauridsen EM, Marrow J, Buffière JY, Ludwig W (2011) 3-D growth of a short fatigue crack within a polycrystalline microstructure studied using combined diffraction and phase-contrast X-ray tomography. Acta Mater 59:590–601. https://doi.org/10.1016/J.ACTAMAT.2010.09.063
13. Suresh S (1983) Crack deflection: implications for the growth of long and short fatigue cracks. Metall Trans A 14:2375–2385. https://doi.org/10.1007/BF02663313
14. Navarro A, De Los Rios ER (1992) Fatigue crack growth modelling by successive blocking of dislocations. Proc R Soc A Math Phys Eng Sci 437:375–390. https://doi.org/10.1098/rspa.1992.0067
15. Liu L, Wang J, Gong SK, Mao SX (2014) Atomistic observation of a crack tip approaching coherent twin boundaries. Sci Rep 4:4397
16. Sinha T, Kulkarni Y (2014) Alternating brittle and ductile response of coherent twin boundaries in nanotwinned metals. J Appl Phys 116:183505
17. Xiao JH, Zhang FC, Qian LH (2013) Contact stress and residual stress in the nose rail of a high manganese steel crossing due to wheel contact loading. Fatigue Fract Eng Mater Struct 37:219–226. https://doi.org/10.1111/ffe.12108
18. Yan W, Fang L, Zheng Z, Sun K, Xu Y (2009) Effect of surface nanocrystallization on abrasive wear properties in Hadfield steel. Tribol Int 42:634–641
19. Kang J, Zhang FC, Long XY, Lv B (2014) Cyclic deformation and fatigue behaviors of Hadfield manganese steel. Mater Sci Eng A 591:59–68. https://doi.org/10.1016/J.MSEA.2013.10.072
20. Wei-yong W, Bing L, Venkatesh K (2013) Effect of temperature on strength and elastic modulus of high-strength steel. J Mater Civ Eng 25:174–182. https://doi.org/10.1061/(ASCE)MT.1943-5533.0000600
21. Qian L, Feng X, Zhang F (2011) Deformed microstructure and hardness of Hadfield high manganese steel. Mater Trans 52:1623–1628
22. Zhao L, van Dijk N, Brück E, Sietsma J, van der Zwaag S (2001) Magnetic and X-ray diffraction measurements for the determination of retained austenite in TRIP steels. Mater Sci Eng A 313:145–152. https://doi.org/10.1016/S0921-5093(01)00965-0
23. Van Tol RT (2014) Microstructure evolution in deformed austenitic Twinning Induced Plasticity steels, Delft University of Technology. https://doi.org/10.4233/uuid:b2a057e2-c545-4bcc-854c-fe4b9344e486
24. Zaefferer S, Elhami N-N (2014) Theory and application of electron channelling contrast imaging under controlled diffraction conditions. Acta Mater 75:20–50. https://doi.org/10.1016/J.ACTAMAT.2014.04.018
25. Gutierrez-Urrutia I, Raabe D (2012) Grain size effect on strain hardening in twinning-induced plasticity steels. Scr Mater 66:992–996. https://doi.org/10.1016/J.SCRIPTAMAT.2012.01.037

26. Gutierrez-Urrutia I, Zaefferer S, Raabe D (2010) The effect of grain size and grain orientation on deformation twinning in a Fe–22 wt.% Mn–0.6 wt.% C TWIP steel. Mater Sci Eng A 527:3552–3560. https://doi.org/10.1016/j.msea.2010.02.041
27. Gutierrez-Urrutia I, Zaefferer S, Raabe D (2009) Electron channeling contrast imaging of twins and dislocations in twinning-induced plasticity steels under controlled diffraction conditions in a scanning electron microscope. Scr Mater 61:737–740. https://doi.org/10.1016/J.SCRIPTAMAT.2009.06.018
28. Gutierrez-Urrutia I, Zaefferer S, Raabe D (2013) Coupling of electron channeling with EBSD: toward the quantitative characterization of deformation structures in the sem. JOM 65:1229–1236. https://doi.org/10.1007/s11837-013-0678-0
29. Kumar A, Agarwal G, Petrov R, Goto S, Sietsma J, Herbig M (2019) Microstructural evolution of white and brown etching layers in pearlitic rail steels. Acta Mater. https://doi.org/10.1016/J.ACTAMAT.2019.04.012
30. Makineni SK, Kumar A, Lenz M, Kontis P, Meiners T, Zenk C, Zaefferer S, Eggeler G, Neumeier S, Spiecker E, Raabe D, Gault B (2018) On the diffusive phase transformation mechanism assisted by extended dislocations during creep of a single crystal CoNi-based superalloy. Acta Mater 155. https://doi.org/10.1016/j.actamat.2018.05.074
31. Bay B, Hansen N, Hughes DA, Kuhlmann-Wilsdorf D (1992) Overview no. 96 evolution of f.c.c. deformation structures in polyslip. Acta Metall Mater 40: 205–219. https://doi.org/10.1016/0956-7151(92)90296-q
32. Hansen N, Jensen DJ (1997) Development of microstructure in FCC metals during cold work. Philos Trans R Soc London A Math Phys Eng Sci 357:1447–1469
33. Beladi H, Timokhina IB, Estrin Y, Kim J, De Cooman BC, Kim SK (2011) Orientation dependence of twinning and strain hardening behaviour of a high manganese twinning induced plasticity steel with polycrystalline structure. Acta Mater 59:7787–7799. https://doi.org/10.1016/J.ACTAMAT.2011.08.031
34. Steinmetz DR, Jäpel T, Wietbrock B, Eisenlohr P, Gutierrez-Urrutia I, Saeed-Akbari A, Hickel T, Roters F, Raabe D (2013) Revealing the strain-hardening behavior of twinning-induced plasticity steels: theory, simulations, experiments. Acta Mater 61:494–510. https://doi.org/10.1016/J.ACTAMAT.2012.09.064
35. Agarwal G, Kumar A, Gao H, Amirthalingam M, Moon SC, Dippenaar RJ, Richardson IM, Hermans MJM (2018) Study of solidification cracking in a transformation-induced plasticity-aided steel. Metall Mater Trans A Phys Metall Mater Sci. https://doi.org/10.1007/s11661-018-4505-7
36. Agarwal G, Kumar A, Richardson IM, Hermans MJM (2019) Evaluation of solidification cracking susceptibility during laser welding in advanced high strength automotive steels. Mater Des 183:108104. https://doi.org/10.1016/j.matdes.2019.108104
37. Sun D, Garimella SV (2007) Numerical and experimental investigation of solidification shrinkage. Numer Heat Transf Part A Appl 52:145–162. https://doi.org/10.1080/10407780601115079
38. Hardin RA, Beckermann C (2013) Effect of porosity on deformation, damage, and fracture of cast steel. Metall Mater Trans A 44:5316–5332. https://doi.org/10.1007/s11661-013-1669-z
39. Peters NW, Eng P (2005) The performance of Hadfield's manganese steel as it relates to manufacture, Dostupné Na Internetu. Accessed May 11, 2015. http://www.arema.org/files/library/2005_Conference_Proceedings/00040.pdf
40. Grabulov A, Petrov R, Zandbergen HW (2010) EBSD investigation of the crack initiation and TEM/FIB analyses of the microstructural changes around the cracks formed under Rolling Contact Fatigue (RCF). Int J Fatigue 32:576–583. https://doi.org/10.1016/J.IJFATIGUE.2009.07.002
41. Kumar A, Dutta A, Makineni SK, Herbig M, Petrov RH, Sietsma J (2019) In-situ observation of strain partitioning and damage development in continuously cooled carbide-free bainitic steels using micro digital image correlation. Mater Sci Eng A 757:107–116. https://doi.org/10.1016/J.MSEA.2019.04.098

42. Collette G, Crussard C, Kohn A, Plateau J, Pomey G, Weisz M (1957) Contribution à l'étude des transformations des austénites à 12% Mn. Rev Met Paris 54:433–486. https://doi.org/10.1051/metal/195754060433
43. Spreadborough J (2018) Stacking faults in iron–manganese and cobalt–nickel. Acta Crystallogr 13:603–605. https://doi.org/10.1107/S0365110X60001448
44. Olson GB, Cohen M (1972) A mechanism for the strain-induced nucleation of martensitic transformations. J Less Common Met 28:107–118
45. Pickering EJ (2013) Macrosegregation in steel ingots: the applicability of modelling and characterisation techniques. ISIJ Int 53:935–949
46. Takahashi J, Kawakami K, Ueda M (2010) Atom probe tomography analysis of the white etching layer in a rail track surface. Acta Mater 58:3602–3612. https://doi.org/10.1016/J.ACTAMAT.2010.02.030
47. Mazancová E, Mazanec K (2009) Stacking fault energy in high manganese alloys. Mater Eng 16:26–31
48. Pierce DT, Jiménez JA, Bentley J, Raabe D, Wittig JE (2015) The influence of stacking fault energy on the microstructural and strain-hardening evolution of Fe–Mn–Al–Si steels during tensile deformation. Acta Mater 100:178–190. https://doi.org/10.1016/J.ACTAMAT.2015.08.030
49. Jost N, Schmidt I (1986) Friction-induced martensitic transformation in austenitic manganese steels. Wear 111:377–389. https://doi.org/10.1016/0043-1648(86)90134-1
50. Pineau AG, Pelloux RM (1974) Influence of strain-induced martensitic transformations on fatigue crack growth rates in stainless steels. Metall Trans 5:1103–1112. https://doi.org/10.1007/BF02644322
51. Baudry G, Pineau A (1977) Influence of strain-induced martensitic transformation on the low-cycle fatigue behavior of a stainless steel. Mater Sci Eng 28:229–242. https://doi.org/10.1016/0025-5416(77)90176-8
52. Das A, Tarafder S (2009) Experimental investigation on martensitic transformation and fracture morphologies of austenitic stainless steel. Int J Plast 25:2222–2247. https://doi.org/10.1016/J.IJPLAS.2009.03.003
53. Pippan R, Hohenwarter A (2016) The importance of fracture toughness in ultrafine and nanocrystalline bulk materials. Mater Res Lett 4:127–136
54. Suresh S, Ritchie RO (1984) Propagation of short fatigue cracks. Int Met Rev 29:445–475. https://doi.org/10.1179/imtr.1984.29.1.445
55. Kumar A, Saxena AK, Kirchlechner C, Herbig M, Brinkmann S, Petrov RH, Sietsma J (2019) In situ study on fracture behaviour of white etching layers formed on rails. Acta Mater. https://doi.org/10.1016/J.ACTAMAT.2019.08.060
56. Goulas C, Kumar A, Mecozzi MG, Castro-Cerda FM, Herbig M, Petrov RH, Sietsma J (2019) Atomic-scale investigations of isothermally formed bainite microstructures in 51CrV4 spring steel. Mater Charact. https://doi.org/10.1016/j.matchar.2019.03.038
57. Chen YQ, Pan SP, Zhou MZ, Yi DQ, Xu DZ, Xu YF (2013) Effects of inclusions, grain boundaries and grain orientations on the fatigue crack initiation and propagation behavior of 2524-T3 Al alloy. Mater Sci Eng A 580:150–158. https://doi.org/10.1016/J.MSEA.2013.05.053
58. Zhai T, Wilkinson AJ, Martin JW (2000) A crystallographic mechanism for fatigue crack propagation through grain boundaries. Acta Mater 48:4917–4927. https://doi.org/10.1016/S1359-6454(00)00214-7
59. Taylor D, Knott JF (1981) Fatigue crack propagation behaviour of short cracks; the effect of microstructure. Fatigue Fract Eng Mater Struct 4:147–155
60. Saxena AK, Kumar A, Brinckmann S, Herbig M, Dehm G, Kirchlechner C (2019) Micro fracture investigations of white etching layers. Mater Des 180:107892. https://doi.org/10.1016/j.matdes.2019.107892

Metallurgical Investigation of Crossing Noses

Hilmar Kjartansson Danielsen, Somrita Dhar, Carsten Jørn Rasmussen, and Dorte Juul Jensen

Abstract Worn crossing noses extracted from the track were studied through extensive metallographic examinations, including optical microscopy and X-ray tomography used for three-dimensional mapping of fatigue crack networks within the S&C nose. The gradient in the residual strain/stress profile from the running surface originating from the rail wheel contact was analyzed using synchrotron as well as laboratory X-ray diffraction.

Executive summary

This chapter looks at specific novel characterization techniques for railway components, and it is illustrated how they can provide insight into the properties and degradation mechanisms of railway components. X-ray computerized tomography can effectively map out large crack networks under the running surface, providing an insight into how cracks propagate in 3D and if they are surface or subsurface cracks. Other defects inside the rail component can also be mapped out, for example mapping porosities in repair welds, providing a measure of the quality of the welding. Synchrotron X-ray diffraction is an effective technique to determine the residual micro-strains non-destructively in bulk specimens. As such residual strains extend to significant depth below the running surface and can be large in magnitude, they can affect the failure mechanisms and should be taken into account when calculating the component lifetime.

H. K. Danielsen (✉) · S. Dhar
Department of Wind Energy, Technical University of Denmark, 4000 Roskilde, Denmark
e-mail: hkda@dtu.dk

C. J. Rasmussen
Banedanmark, 1577 Copenhagen, Denmark

D. Juul Jensen
Department of Mechanical Engineering, Technical University of Denmark, 2800 Kgs Lyngby, Denmark

R. Galeazzi et al. (eds.), *Intelligent Quality Assessment of Railway Switches and Crossings*, Springer Series in Reliability Engineering,
https://doi.org/10.1007/978-3-030-62472-9_5

1 Introduction

Switches and crossings (S&C) are an integral part of any rail network, allowing for necessary flexibility of directing trains from one track to another at junctions. They are also the most vulnerable part of the network due to the complex shape and the severity of the loading on them. During service, the crossings are exposed to rolling contact fatigue and impact loads, eventually causing deterioration of the mechanical properties of the steel. Therefore, understanding the damage and deformation mechanisms of the different crossing steels are important in order to prevent failure for safety and reliability. This includes residual stresses, material degradation, plastic deformation, crack formation and propagation. This chapter looks at some new characterization techniques for investigating the microstructure and properties of the most commonly used crossing steels, namely head hardened pearlitic steels and austenitic manganese steel. Here the focus is on the crossing nose, but the methods used may equally well be used for characterization of other S&C components.

2 Experimental

Two crossings made from different steels were extracted from the rail network, one made from head hardened pearlitic steel and one from austenitic manganese steel, see Fig. 1. In both cases only the noses were investigated, as these were deemed to be the most vulnerable part of the crossing, although this also depends on traffic direction.

For in-depth metallurgical investigations of crossing noses, a range of experimental techniques are used including optical and electron microscopy, absorption

Fig. 1 Two crossings of different steel types were extracted from the Danish rail network, showing the normal direction (ND), transverse direction (TD) and longitudinal (running) direction (LD). In **a** the extracted head hardened pearlitic steel crossing and in **b** is shown the extracted austenitic manganese steel crossing

contrast probe tomography, X-ray diffraction, hardness measurements, and mechanical testing. Most of these techniques are well known and will not be described in this section. Only two techniques which so far rarely have been used for characterization of rail components are described in this section, namely laboratory X-ray absorption contrast tomography for non-destructive 3D mapping of cracks or other defects and synchrotron X-ray diffraction for measurements of local residual stresses.

Absorption contrast X-ray tomography can characterize non-destructively the three-dimensional (3D) structure of inner defects of an object which has absorption difference to the bulk of the specimen with a spatial resolution down to 1 μm. In a computerized tomography (CT) system, the specimen is placed on a stage in between the radiation source and an imaging system. The specimen is then rotated up to a full angle of 360° during the scan, and the digital detector registers thousands of individual 2D images from many angles. An algorithm uses these 2D images to reconstruct the structure into a complete 3D representation. This helps in the characterization of the internal structure of the specimen including dimensions, shape, defects, and density from the whole volume of the specimen. In the present work, the 3D X-Ray tomography scans were done on a ZEISS Xradia 510 Versa machine. A polychromatic X-ray beam was used with energies ranging up to 160 kV from a tungsten target. For all the present experiments, a total of 1601 projections were acquired during a full 360° specimen rotation. 3D maps were reconstructed by a Feldkamp algorithm for cone beam reconstruction [1] to 2 k × 2 k × 2 k voxel volumes with a voxel size of around 10 μm, depending on the exact specimen size. The crack network and other defects were segmented and visualized using the Avizo 3D software. X-ray tomography reconstructions are 3D density maps, which allow detection of cracks due to the difference in density by applying an appropriate intensity threshold value.

Residual stresses in rails are usually determined by destructive methods [2]. Such measurements, however, are typically not very reliable as the specimen sectioning releases the residual stresses near the surface. Non-destructive methods are more reliable and include X-ray and neutron diffraction as well as ultrasonic and magnetic methods [3]. The latter two methods are indirect in nature, neutron diffraction methods are limited by low intensity meaning that the stresses can only be determined as an average over large specimen volumes and laboratory X-ray methods are limited in terms of intensity and penetration depth [4]. As something new, diffraction of *synchrotron* X-ray offers good penetration depth and high intensity, meaning that residual stresses within small volumes below the surface can be measured non-destructively [5, 6]. It has to be noted that strains and stresses determined in this way are *micro* strains/stresses which represent the *local* stress/strain state in the selected grain. The micro stresses cannot directly be compared to the *macro* stresses measured by laboratory X-rays which average over far larger volumes.

In the present work, synchrotron residual strain measurements were conducted at beam line 34-ID-E at the Advanced Photon Source (APS), Argonne National Laboratory [7, 8]. With this method, a polychromatic beam was used first to determine the orientations of the grains within a selected volume of the nose. The focusing of the beam was done with the help of non-dispersive Kirkpatrick-Baez (K-B) focusing mirrors. The resulting microbeam had a full-width half maximum of ~0.5 μm, which represent the spatial resolution of this method. Diffraction patterns from different depths below the specimen surface were recorded. For austenitic manganese steel, a penetration depth of 75 to 100 μm is possible. Diffraction patterns at each depth were indexed by the use of the LaueGo software [9]. A diffraction spot with high intensity and high energy as well as the diffraction vector normal to the specimen surface was then selected for monochromatic energy scans to determine the lattice spacings, d, at the selected depths.

The residual strain was calculated as

$$\varepsilon = (d - d_0)/d_0 \tag{1}$$

where is the lattice spacing at the measured position and $_0$ is the lattice spacing of the specimen in stress-free state and a precise determination of d_0 is critical for the strain determination. In the present work, this spacing ($_0$) was obtained from laboratory X-ray diffraction measurements using a specimen taken from the base of the nose at a depth of 20 mm from the running surface. Optical microscopy suggests that at this depth there is no plastic deformation from wheel rail interaction. This interaction may have introduced residual stresses even at 20 mm depth which of course would affect the $_0$ value. As the $_0$ value obtained here matches well with what is expected from the chemical composition of the steel, therefore it is a reasonable estimate for the stress-free state.

The magnitude of the residual stress may to a first approximation be estimated from the strain by multiplying with E/(1+), where E is the Young's Modulus and is the Poison's ratio. For the present material $\nu = 0.28$ and E = 203 GPa. For further details see [10].

3 Results and Discussion

3.1 Head Hardened Pearlitic Steel Nose (Welded)—Microscopy

As the pearlitic head hardened crossing nose was cut up, it became clear that it had been repair welded. Many large porosities could be found in the welded layer, see Fig. 2a, and cracks could typically be seen extending out from the porosities, indicated by arrows. The distinct layers formed due to etching and difference in microstructure caused by the welding on the pearlitic steel nose can be seen in Fig. 2b. The bright zone at the contact surface is the weld material, indicated by the white arrow, followed by the heat affected zone (HAZ) indicated by a black arrow and finally the base metal. The transitions zone between the bright weld metal and the fine grained HAZ can be seen in Fig. 3, as well as the transition zone between the fine grained HAZ and coarse grained base metal.

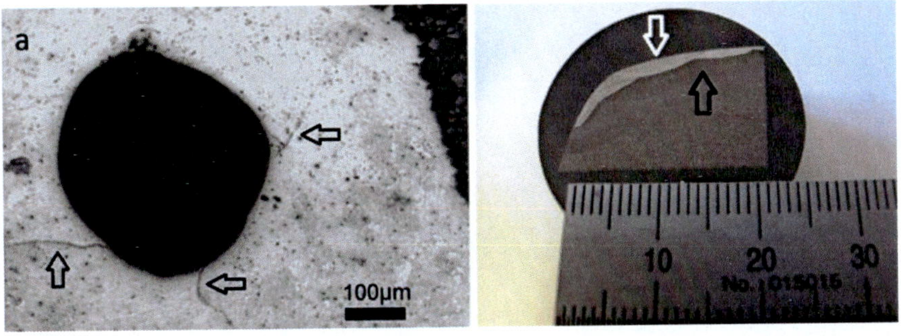

Fig. 2 **a** Optical microscopy image of a void within the weld metal. Cracks can be seen extending from the porosity. **b** An etched mounted specimen from the head hardened pearlitic steel crossing nose

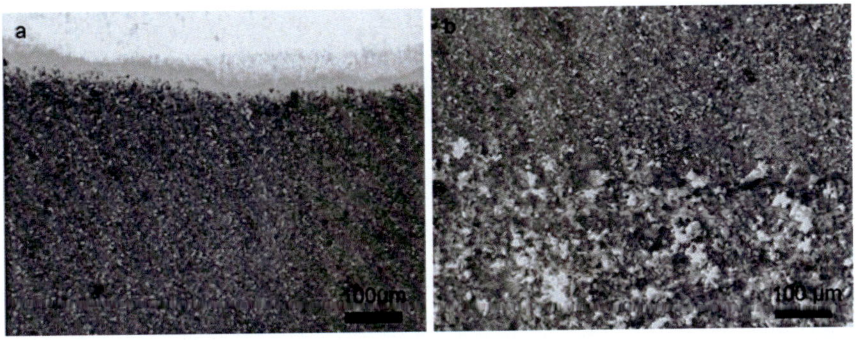

Fig. 3 Optical microscopy imaging from Fig. 2b (perpendicular plane to LD) of the welded head hardened pearlite steel crossing nose. **a** Transition from weld metal to heat affected zone. **b** Transition from heat affected zone to base metal

3.2 *Head Hardened Pearlitic Steel Nose (Welded)—Tomography*

As multiple porosities were evident from the initial sectioning of the repair welded pearlitic steel crossing nose, it was decided to use 3D mapping to get a better overview of the porosity size and distribution. Figure 4 shows the 3D tomography images of defects beneath the running surface of the head hardened pearlitic steel crossing nose. In this figure, the steel is represented as transparent and the volumes with lighter density (such as air) are represented with a blue color, making the extent of porosities visible. Multiple porosities of various shapes and dimensions are present, with a porosity size up to 1 mm. Some of the porosities have a smooth surface whereas some have rough surface. The porosities with rough surfaces are probably corroded and oxidized due to access to air and water, for example through cracks which have propagated to the running surface. Many of the porosities with smooth surfaces are located several mm inside the material and are not connected to the running surface.

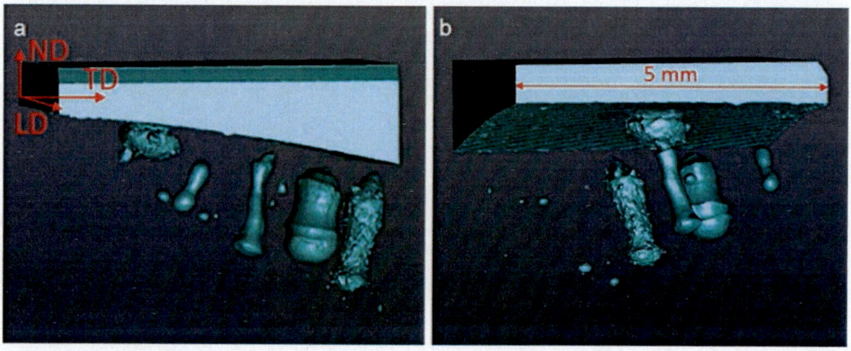

Fig. 4 3D tomographic representations of a 5 × 5 ×5 mm volume where the porosities found in the weld metal of the head hardened pearlitic steel crossing nose are seen from different directions. The steel is represented as transparent and the blue colored area represent air, including the porosities and running surface

The porosities act as sites of crack nucleation, with many cracks extending through the weld metal, which will most likely lead to accelerated failure of the material.

3.3 *Austenitic Manganese Steel Nose—Microscopy*

The deformed microstructure close to the running surface of the austenitic manganese steel crossing nose is manifested in the form of deformation bands inside the grains, shown in the optical micrographs in Fig. 5, and not by the usual elongation of grains as seen in pearlitic steels. The deformation bands are inhomogeneously distributed and have a wavy appearance, with the orientation and direction of the bands differing from grain to grain. At the contact surface, the density of deformation bands is highest, with

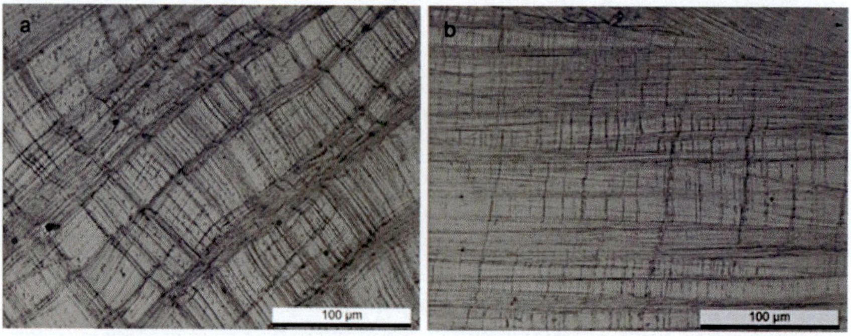

Fig. 5 Optical microscopy imaging of the austenitic manganese steel crossing nose (Perpendicular plane to LD) in different grains just below the running bands. Deformation bands crossing each other are apparent. Same magnification of both images

a large number of bands crossing each other within the grains. At deeper locations the grains mostly have deformation bands in a single direction. With increasing depth from the running surface the density of the deformation bands decreases, and at around 10–12 mm in depth the microstructure is almost free from deformation bands [11]. The microstructure of undeformed austenitic manganese steel is composed of very large equiaxed grains with an average grain size of 500 μm and some inclusions in the microstructure along with porosities, which could act as nucleation sites for cracks.

The deformation bands can either be deformation induced mechanical twins or dislocation boundaries, which the optical microscopy is not capable of differentiating between. In some austenitic steels, martensitic transformation also imparts additional hardening, but in this investigation no martensite was found. In general, there are only few available studies of the relations between the deformation microstructure and the crack propagation within the nose of austenitic manganese steel crossings published in the open literature [12, 13].

3.4 Austenitic Manganese Steel Nose—Tomography

The investigated austenitic manganese steel crossing nose was heavily damaged and 3D X-ray computerized tomography was used to get a 3D representation of the crack morphology beneath the surface. The surface morphology of the crack is much more easily characterized in 3D imaging [14–16] compared to 2D imaging [17, 18], although it is hard to show in a non-3D media. Figure 6 shows a volume at the top of the crossing nose, which illustrates that it is possible to reconstruct a crack network with good spatial resolution and sufficient detail using X-ray tomography, even though the voxel size in reconstructions of this size are around 10 μm. The steel and air are represented as being transparent and the yellow colored area is the interface between them, making the crack network as well as the crossing nose surface visible. Smaller cracks and crack tips may not be visible. The figure shows the extent and severity of cracks beneath the surface of the crossing nose, also giving an overview of the spallation and the cracks related to it. At higher magnification, the cracks seem to be rough with ridges, which are typical for rolling contact fatigue. Apart from the larger surface crack, the 3D scans show that some smaller subsurface cracks are located at various depths. Within the nose, many of the surface and subsurface cracks are interconnected, whereas some are propagating individually. Most of the cracks have a definite direction of propagation. Scans were performed on five adjacent specimens, which showed the crack network to be extensive and continuous through 50 mm beneath the surface, extending up to a depth of 3 mm along the running surface. For further information see [11, 19].

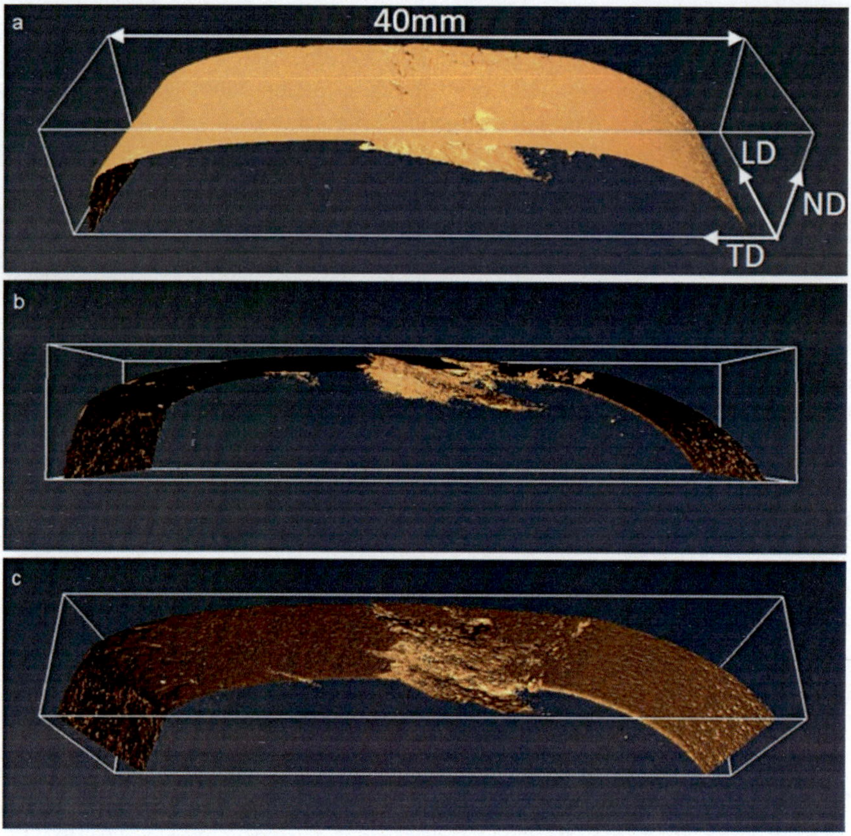

Fig. 6 X-ray tomography of a 10 × 10 ×40 mm volume containing the crack network within the austenitic manganese steel crossing nose. The 3D imaging reveals the complete morphology of the crack network. The steel and air are represented as transparent and the interface between them in yellow, which makes the crack and rail surface visible

3.5 *Austenitic Manganese Steel Nose—Residual Stresses*

The residual stresses within various parts of the investigated austenitic manganese steel crossing nose was characterized using both conventional laboratory as well as synchrotron X-rays [10, 20]. In the following, only results of the synchrotron measurements will be reported.

A 5 mm thick specimen was cut from a position near the tip of the crossing nose and the strains were measured at four depths: 3, 4, 6.5 and 15 mm below the running surface. The cutting out of the sample will relax residual stresses near the new free surface. As shall be shown below, for the present sample, the results suggest that this is mainly a problem for the outer 10 μm. The specimen and the measurement location at 6.5 mm below the running surface are sketched in Fig. 7.

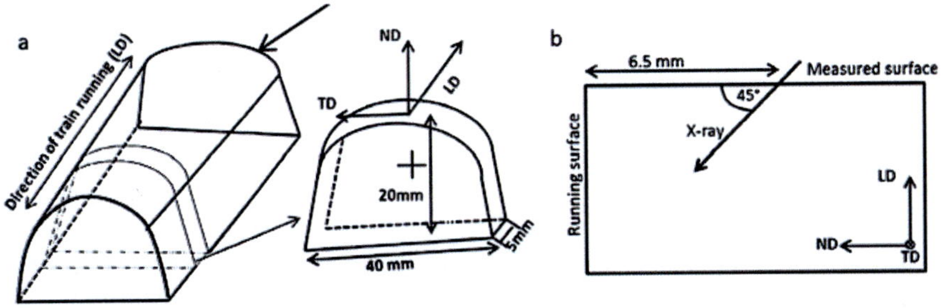

Fig. 7 Nose specimen used for the synchrotron strain measurement: **a** Sampling slice for strain measurements showing the normal direction (ND), transverse direction (TD) and longitudinal (running) direction (LD). The black cross marks a location for synchrotron measurement. **b** Sketch showing the experimental set-up for the synchrotron measurement [20]

The synchrotron measurements were conducted on the transverse specimen section, see Fig. 7a. Monochromatic X-ray scans were performed for all the four selected locations (see an example in Fig. 7b) and for penetrations down to 75 μm below the specimen surface. The measurements reveal that the strains are relaxed at the outermost approximately 10 μm from the specimen surface. This is because of the cutting out of the sample. These values are therefore not included in the results given below.

Table 1 gives an overview of all the measured strains. The strains were found to be fairly constant as a function of penetration from 10 to 75 μm below the free transverse surface. The table reveals that there are significant compressive residual strains at all depths. The results thus document that the residual stresses are not only at the running surface but extends to significant depths.

It is noted that the compressive residual strain at 15 mm from the running surface is about four times as large as the strains at 3, 4 and 6.5 mm (see Table 1). This match nicely to the well known fact that impact related residual stress is typically not maximum at the surface, but somewhere below [21]. At 15 mm the residual strain values are around 0.001 (see Table 1). Using a simple conversion from strain to stress, the stress value at 15 mm depth is around 158 MPa. This is indeed a high value, but it is well below the yield point of 363 MPa of the un-deformed austenitic manganese steel. The residual strain measurements may be related to the microstructural investigations

Table 1 Overview of residual strains measured at different depths below the running surface. The 'range of strains' reports observed scatter in strain values measured for penetrations from 10 to 75 μm below the specimen surface

Depth below the running surface (mm)	Average compressive residual strain	Range of compressive residual strains
3.0	0.00020	0–0.00026
4.0	0.00025	0–0.00070
6.5	0.00026	0–0.00045
15.0	0.00100	0.00080–0.00120

showing that the microstructure near the running surface is heavily deformed while at 15 mm, only few deformation induced boundaries are observed [11]. For the present nose, it is thus suggested that the residual micro-strain profile is related to the complex interplay between impact forces and plastic deformation originating from the rail-wheel interaction.

The depth profiles of residual stresses are of importance when interpreting the crack propagation and vice versa [10]. The measured micro strains are compressive. This means they are actually beneficial and contribute to crack resistance. That the compressive strains are largest at significant depth may contribute to explaining why cracks formed near the surface of the nose, due to a combination of the external loading, surface condition and plastic deformation, typically propagate to a certain depth and then move parallel to the running surface instead of propagating deeper into the material. However, it has to be underlined that the present micro strain measurements are for individual grains and in the direction normal to the rail surface, *only*. Many more measurements are needed if the aim is to get a complete map of the local strain distribution, and thus being able to fully relate to the crack propagation path.

4 Conclusions

X-ray computerized tomography can effectively map cracks and porosities inside the crossing nose in 3D. Information can be gained in 3D which would not be readily evident from 2D imaging, for example determining if cracks are connected to the surface or not. The present results showed extensive formations of large porosities in a head hardened repair welded nose, which could likely lead to accelerated failure of the nose. Extensive crack formation in an austenitic manganese steel nose was observed, with a continuous crack network extending more than 50 mm along the surface, however the depth of the cracks only extended down to 3 mm, possibly due to residual compressive strains. Synchrotron X-ray diffraction is an effective technique to determine the residual micro-strains non-destructively in bulk specimens. It is shown that residual strains extend to significant depth below the running surface and are large in magnitude. It is therefore suggested that residual stresses are considered both in the design of crossings, while interpreting failure mechanisms, and in lifetime predictions.

Acknowledgements The authors gratefully acknowledge support from the Innovation Fund Denmark through the project "INTELLISWITCH – Intelligent Quality Assessment of Railway Switches and Crossings" (Grant no 4109-00003B).

References

1. Feldkamp LA, Davis LC, Kress JW (1984) Practical cone-beam algorithm. J Opt Soc Am A 1(6):612–619
2. Totten G, Howes M, Inoue T (eds) (2002) Handbook of residual stress and deformation of steel. ASM International
3. Withers PJ, Bhadeshia HKDH (2001) Residual stress part 1 mesurement techniques. J Mat Sci Technol 17:355–365
4. Webster PJ, Wang X, Mills G, Webster GA (1992) Residual stress changes in railway rails. Phys B Cond Matter 180–181:1029–1031
5. Zhang Y, Andriollo T, Fæster S, Liu W, Hattel J, Barabash RI (2016) Three-dimensional local residual stress and orientation gradients near graphite nodules in ductile cast iron. Acta Matter 121:173–180
6. Webster PJ, Hughes DJ, Mills G, Vaughan GBM (2002) Synchrotron X-Ray measurements of residual stress in worn railway rail. Mat Sci Forum 4004–4007:767–772
7. Larson BC, Yang W, Ice GE, Budai JD, Tischeler JZ (2002) Three dimensional X-ray structural microscopy with submicrometre resolution. Nature 415:887–890
8. Liu W, Zschack P, Tischler JZ, Ice GE, Larson BC (2010) X-ray laue diffraction microscopy in 3D at the advanced photon source. In AIP Conference Proceedigs, vol 1365:108–111
9. Tischler JZ (2014) Reconstructing 2D and 3D X-ray orientation map from white beam Laue. In: Barabash R, Ice GE (eds) Strain and dislocation gradients from diffraction chapter 10. Imperial College Press, London
10. Dhar S, Danielsen HK, Ruqing R, Zhang Y, Grumsen F, Rasmussen C, Juul Jensen D (n.d) Residual strain and stress in a railway crossing of manganese steel determined by synchrotron and laboratory X-ray measurements, accepted for publication in Materials Science & Technology
11. Dhar S, Danielsen HK, Fæster S, Rasmussen CJ, Zhang Y, Juul Jensen D (2019) Crack formation within a Hadfield manganese steel crossing nose, 438–439:203049
12. Qian L, Feng X, Zhang F (2011) Deformed microstructure and hardness of Hadfield high manganese steel. Mater Trans 52(8):1623–1628
13. Zhang FC, Lv B, Wang TS, Zheng CL, Li M, Zhang M (2009) Microstructure in worn surface of Hadfield steel crossing. Int J Mod Phys B 23(6 & 7):1185–1190
14. Jessop C, Ahlström J, Hammar L, Fæster S, Danielsen HK (2016) 3D characterization of rolling contact fatigue crack networks. Wear 366–367:392–400
15. Zhou Y, Zheng X, Jiang J, Kuang D (2017) Modeling of rail head checks by X-ray computed tomography scan technology. Int J Fatigue 100:21–31
16. Naeimi M, Li Z, Qian Z, Zhou Y, Wu J, Petrov RH, Sietsma J, Dollevoet R (2017) Reconstruction of the rolling contact fatigue cracks in rails using X-ray computed tomography. NDT E Int 92:199–212
17. Schilke M, Larijani N, Persson C (2014) Interaction between cracks and microstructure in three dimensions for rolling contact fatigue. Fatigue Fract Eng Mater Struct 37:280–289
18. Garnham JE, Fletcher DI, Davis CL, Franklin FJ (2016) Visualization and modelling to understand rail rolling contact fatigue cracks in three dimensions. Proc Instit Mech Eng [F] J Rail Rapid Transit 225:165–178
19. Dhar S, Danielsen HK, Fæster S, Rasmussen CJ, Juul Jensen D (2019) 2D and 3D characterization of rolling contact fatigue cracks in a manganese steel wing rails from a crossing, 438–439:203049
20. Dhar S, Zhang Y, Xu R, Danielsen HK, Juul Jensen D (2017) Synchrotron X-Ray measurement of residual strain within the nose of a worn manganese steel railway crossing. In: IOP conference series: materials science and engineering, vol 219, 012016 p
21. Liu H, Zhang X, Hansen N (2014) Numerical investigation of a shot peening process by a finite element approach. In: Proceedings of the Risø international symposium on materials science, vol 35, pp 381–388

Analysis of Sleeper Screw Failures

Xiaodan Zhang, Yubin Zhang, Carsten Jørn Rasmussen, Martin Tandrup Lange, and Dorte Juul Jensen

A small but critical component in the railway is the sleeper screws fastening the baseplates with the rails to the sleepers. High loads from the passing trains, are transferred into the screws. Flexing and displacement of the rails can cause cyclic tensile and shear loads on the screws which together with corrosion may lead to cracks and failures. Due to the complex conditions, the failure mechanisms of screws are neither well characterised nor understood in spite of their importance for safety and reliability of the railway network. In the present study, more than 20 cracked and failed screws were collected at Switches and Crossings on concrete sleepers close to the Central Station of Copenhagen and characterized by four complementary techniques: 3D X-ray tomography, optical microscopy (OP), scanning electron microscopy (SEM) and electron backscatter diffraction (EBSD). 3D X-ray tomography was used for three-dimensional mapping of changes in screw shapes and of cracks. OP and SEM were applied to study the characteristics of the fractured surfaces. SEM and EBSD were combined to investigate the sub-surface deformation and corrosion. All this information was put together for a comprehensive analysis of failure. The failure

X. Zhang (✉) · Y. Zhang · D. Juul Jensen
Department of Mechanical Engineering, Technical University of Denmark, 2800 Kgs Lyngby, Lyngby, Denmark
e-mail: xzha@mek.dtu.dk

C. J. Rasmussen · M. T. Lange
Banedanmark, 1577 Copenhagen, Denmark

R. Galeazzi et al. (eds.), *Intelligent Quality Assessment of Railway Switches and Crossings*, Springer Series in Reliability Engineering,
https://doi.org/10.1007/978-3-030-62472-9_6

cause and mechanisms for the present screws are discussed, and the type of investigation may be used as an example for screw failure characterization and analysis in general.

An Executive Summary

This chapter presents a state-of-the-art combination of advanced complementary laboratory techniques suited for a comprehensive multiscale analysis of failure of sleeper screws on scales from meters to nanometres.

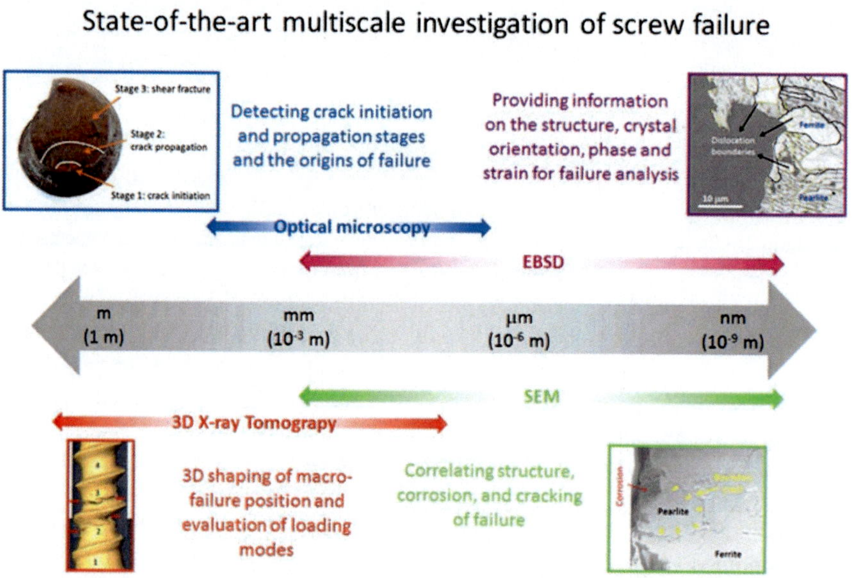

The key findings for screws mounted in dowels into concrete sleepers are:

- Three stages of crack evolution including crack initiation, propagation and fracture have been identified from the surface of failed screws.
- The crack initiation was found to be at thread 2 and 3 indicating localized cyclic bending deformation at this position.
- The cyclic bending leads to the ferrite grain subdivision into dislocation cells with a size ~1 μm. The bending may lead to the evaluation of local stress and strain modes important for the crack evolution.
- The crack propagation is enhanced by corrosion and related processes such as hydrogen-enhanced localized plasticity.

1 Introduction

Sleeper screws are of critical importance ensuring the baseplates and rails are fastened safely to the sleepers. Besides poor fastening, broken screws can lead to baseplates and blocks becoming loose and obstructing for example wheel passage or movement of switches. The possible causes for screw failure may be large dynamic forces/vibrations due to insufficient support stiffness, incorrect torque applied when fastening the screw and localised environmental conditions leading to corrosion of components [1]. However, which cause is dominant, is still unclear. This inspired the present study in which we investigate failed and broken screws with the aim of performing a comprehensive analysis of the failure. Although the present work do not attempt to give a complete analysis of the general failure mechanisms in sleeper screws under all loading conditions, the methods applied may serve as inspiration for future investigation of more screw failures.

2 Experiment

Failed and broken screws were collected at Switches and Crossings (S&Cs) on concrete sleepers close to the Central Station of Copenhagen, where the screw failure rate is high and screws need to be changed often. The yearly cost due to broken sleeper screws is estimated to 150.000 Euros. In addition, the broken screws have traffic as well as safety related consequences. The broken-off piece of a screw left deep down in a dowel is difficult and time consuming to remove. In addition it is hard to get possession time at the central station in Copenhagen. The life time of the screws should follow the life time of the sleepers and this is normally more than 30 years corresponding to more than total 750 MGT (million gross tons). At some S&Cs at Copenhagen station, the life time of the screws is reduced to 1 year. The infrastructure provider, Banedanmerk require the steel quality of the screws to be standard medium carbon steel with a carbon content around 0.35 wt%, with a tensile strength of minimum 500 N/mm^2 and an elongation of minimum 22%. The technical specifications of the screws must be in accordance to UIC 864-1. Technical details of the screws are shown in Fig. 1. More than twenty screws were visually inspected and five screws were investigated in detail in the present work.

As the screws were fastened in concrete sleepers, dowels were used. Details of the screw and the dowel are also given in Fig. 1. The 3D shape of the failed screws as well as the 3D crack network were characterized by X-ray tomography using a Zeiss Xradia 520 Versa micro-CT (contrast tomography) system. For the CT scans, a polychromatic conical beam with X-ray energies up to 140 keV and 1401 image projections over a rotation of 360° were used. The voxel size was 29.2 μm. The four screws were characterized and numbered as screw No. 1–4 according to the damage extent, i.e. No. 1 is the least damaged and No. 4 is the most severely damaged. The CT data were visualized and processed using the software Avizo®.

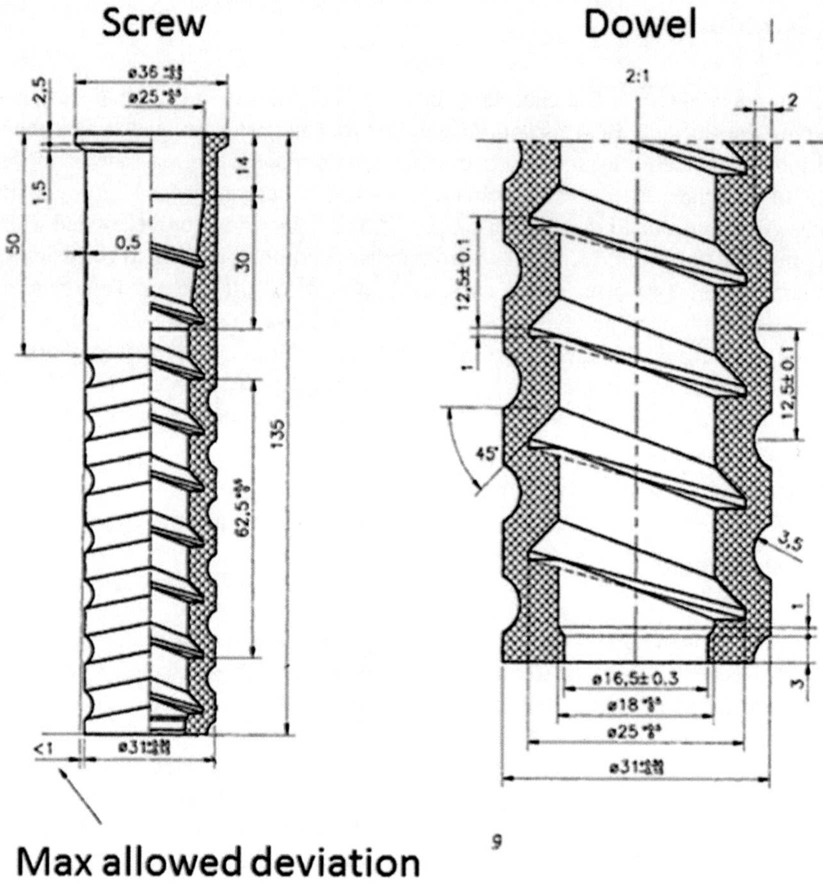

Fig. 1 Technical details of the screw and the dowel

For analysis of the fracture surface, a broken screw No. 5 was carefully cleaned in acetone using an ultrasonic cleaner and observed using an optical microscope and a scanning electron microscope. For characterization of the microstructure, subsurface deformation and corrosion beneath the fracture surface, the broken screw was cut in half along the centre line in the screw axis direction, and the cross section was mounted in conductive polymer, ground and mechanical-polished using a standard metallurgical sample preparation procedure.

3 Results and Discussion

3.1 Microstructure

The screws have a two-phase microstructure consisting of ferrite + pearlite, as shown in Fig. 2, which is for typical medium carbon steel. During the cooling process in the production of screws, proeutectoid ferrite grains precipitate and grow from the austenite grain boundaries at first, and then the lamellar pearlite with alternating layered ferrite and cementite is formed by the decomposition of the remaining eutectoid austenite [2, 3].

3.2 Three-Dimensional Mapping of Screw Shapes and Cracks

The failed screws show a similar failure mode with corrosion and cracks at directly opposite sides at the second or third thread. 3D views of the four screws No. 1–4 are shown in Fig. 3. The screws are significantly corroded which can be seen by the naked eye, and marked by the red arrows in Fig. 3. Cracks also develop from these positions, and it is observed that the crack at one side is larger than the other side. The distance between two threads is elongated about 1.0 mm for screw No.

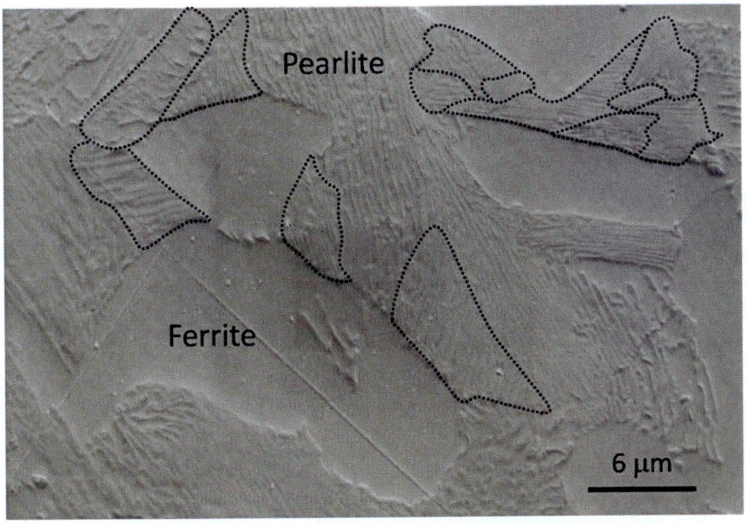

Fig. 2 Ferrite + pearlite dual-phase microstructure observed in the bulk of sleeper screws. The dashed black lines represent the pearlite colony boundaries. Only some of the pearlite colony boundaries are shown

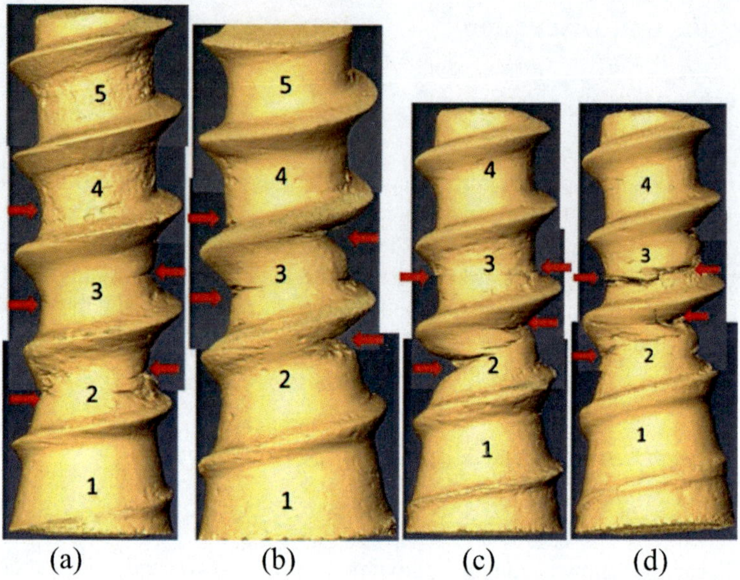

Fig. 3 3D shapes of the screws characterized by X-ray tomography. **a–d** are for screw No. 1–4, respectively. Numbers are used to mark different thread, while red arrows mark the corroded parts and cracks. Thread No. 1 is the one closest to the head of the screw

4, which is mainly attributed to the crack opening, which has a width in the range 0.77–1.1 mm (see Fig. 4). This result suggests that bending of the screws lead to the tensile-compressive stress at the two opposite sides of the screws although the detailed fatigue conditions of the screws are unknown to Banedanmark. As tensile and compressive stresses have different effects on the stress-assisted corrosion, the cracks may therefore be different at the two sides. When a crack, progresses from

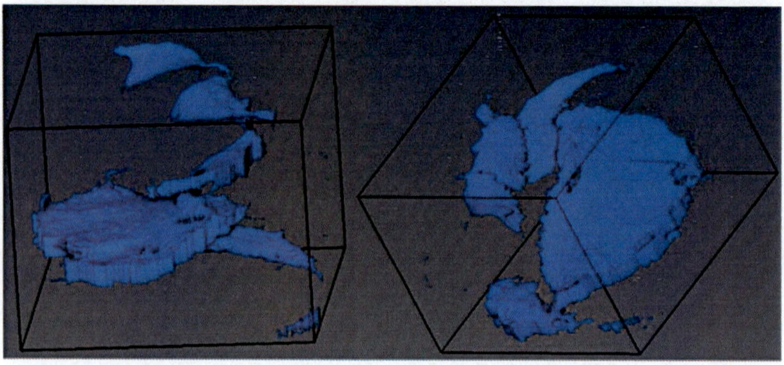

Fig. 4 Two 3D views showing the crack network in screw No. 4. For a better 3D visualization, cracks are shown in blue while the steel is chosen to be shown as transparent

each side, it may meet in the middle and combine into a large crack. When the cracks from both sides meet in the middle, it does not mean that the screw will break. This is because of the spiral propagation routes shown in Fig. 4. Eventually, however, when such a large crack further develops, the screw will break (see Fig. 4).

3.3 Fracture Surface Analysis

Figure 5 shows the fracture surface of a broken screw viewed from three directions. The fracture surface shows the classical three stages: crack initiation, propagation and fracture failure. The first two stages evolve perpendicular to the screw axis. The third stage is inclined and the fracture surface consists of two planes with angles of ~43° and ~25° to the vertical plane of the first two stages, as shown in Fig. 5b. The start of the fracture is along a ~43° plane which is the maximum shear stress

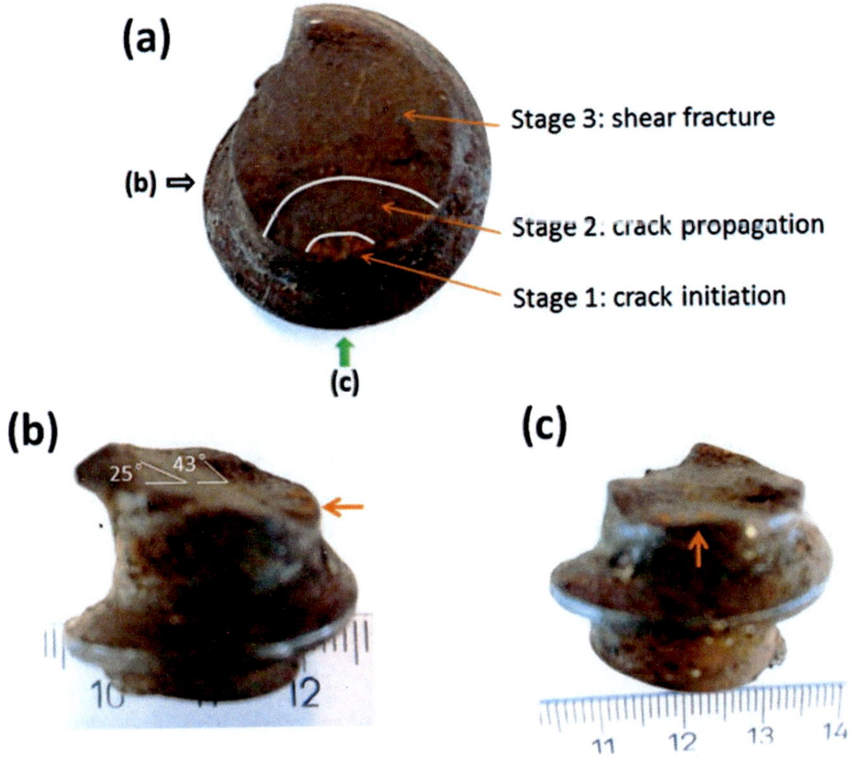

Fig. 5 Optical micrographs showing the fracture surface of a broken screw viewed from three directions: top view (**a**) and two side views ((**b**) and (**c**)). The orange arrows in (**b**) and (**c**) point out the crack initiation (stage 1 in a)

direction. As the screw at that stage is still mounted in the sleepers, there is only limited space for the facture to propagate in the maximum shear stress direction. As a result, the fracture path changes its propagation direction and forms the ~25° plane.

3.4 Sub-Surface Deformation, Cracks and Corrosion

Fresh fracture surfaces have often been used to analyse the failure mechanisms for metals [4]. In the present case, the fracture surface of the broken screw is covered by the oxides from corrosion, which inhibits such an analysis. The broken screw has therefore been cut into two pieces from the central line of the crack propagation as indicated by the green arrow in Fig. 5a. Thereby we could investigate the sub-surface area corresponding to stage 1 and 2, i.e., crack initiation and propagation. These two stages actually represent the major life of the screw after failure begins. Different representative areas were chosen to analyse the failure cause and mechanisms. Scanning electron micrographs were combined with EBSD mapping from the same area to investigate the correspondence between cracks and crystallographic orientations.

Figure 6 shows that the bending of the screw has initiated deformation induced dislocation boundaries in both the ferrite grains and pearlite colonies. Low angle dislocation cell boundaries with misorientation lower than 15° are observed in the ferrite grains, compared with the relative 'clean' ferrite without dislocation boundaries in the matrix at a position of ~2 mm from the fracture surface. Dislocation cells in the ferrite grains with the cell size around 1 μm are also observed close to the fracture surface. It is interesting that low angle dislocation boundaries are also found in the pearlite colonies. This is because there normally are no dislocation boundaries in the as-transformed state from austenite when the interlamellar spacing is less than 100 nm [2, 5, 6]. The formation of the dislocation boundaries in the present pearlite colonies during fatigue may be due to the relatively large interlamellar spacing (around 1 μm see Fig. 4) of the pearlite in the screw.

Figure 7 shows cracks along local shear bands in the pearlite colony and along the lamellar direction in the neighbour pearlite colony. There exists a stress/strain gradient from the surface on the left side to the bulk, which is documented by the much smaller cells with a size around 0.8 μm (shown by the white arrow) compared with a cell size of 1.6 μm (indicated by the black arrow) at a position of 30 μm into the matrix.

The crack may also follow pearlite colony boundaries. An example is shown in Fig. 8. This crack is quite different from the cracks in Fig. 7, which are related to the local shear deformation. The formation and propagation of the colony boundary crack in Fig. 8 may be assisted by the local chemical environment in the sleeper.

When focusing on the corrosion, it is found that its direction follows the lamellar direction, see Figs. 6, 7 and 8. It is well-known that the existence of cementite is of benefit to the mechanical properties, but is harmful to the corrosion resistance of steels, as: (i) cementite with a low hydrogen over-voltage would act as cathodic sites to promote the corrosion of the ferrite phase of the steel in acidic solution

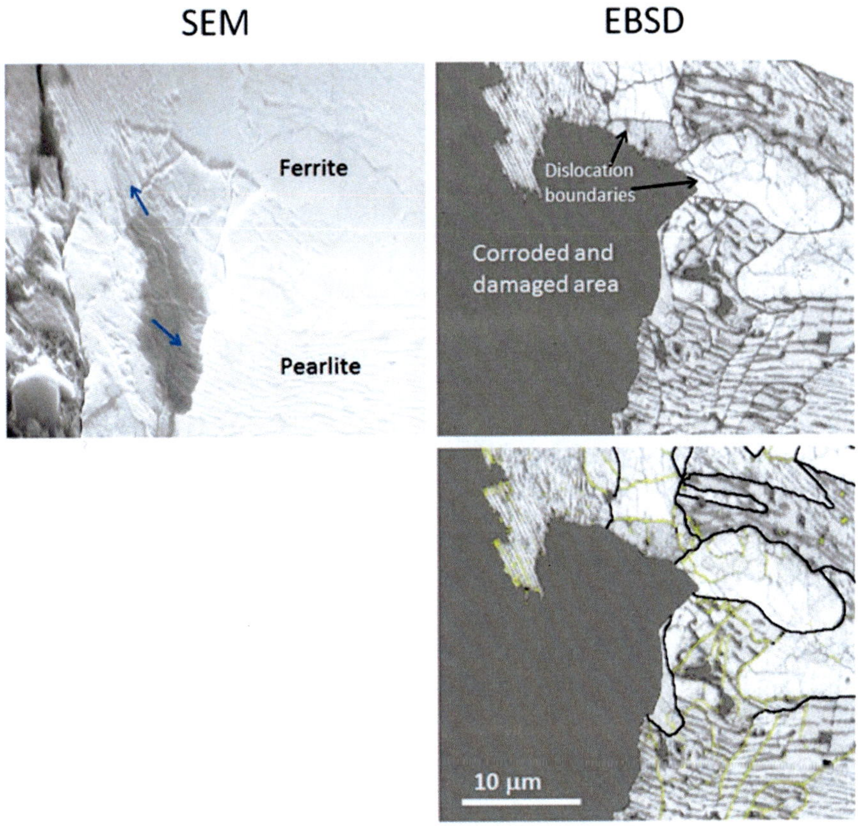

Fig. 6 SEM (left) and EBSD (right) micrographs showing the same sub-surface area in the cross section. The dark-grey area in the EBSD micrograph represents the un-indexed zone due to the oxidation and cracks. The blue arrows point to the corrosion following the lamellar direction of pearlite colonies. The black arrows point to the low angle dislocation boundaries formed in the ferrite grains. The two EBSD micrographs are identical, and only different coloring schemes are used. The top picture uses the EBSD band slope map, and the lower one uses black lines to show the high angle boundaries with misorientation angles $\theta > 15°$ and yellow lines to show the low angle boundaries with misorientation angles $1.5° < \theta < 15°$. Within the non-cracked metal, the cementite lamellae are also shown in dark grey, while the ferrite lamellae in light grey

[7], and (ii) the area of exposed cementite increases with time and forms a porous layer causing a continuous acceleration in the ferrite dissolution rate in the corrosion evolution process [8–10]. Moreover, the alternating ferrite and cementite lamellae microstructure in the pearlite colony acts like a series of circuits, and this causes the corrosion of several ferrite lamellae at the same time.

To check the combined effect of mechanical and chemical processes on the early stages of crack initiation and propagation, a short crack with a length of ~400 μm from the side surface of the broken screw was investigated. The results are shown in Fig. 9. It is interesting to note that dislocation cells are formed with the cell

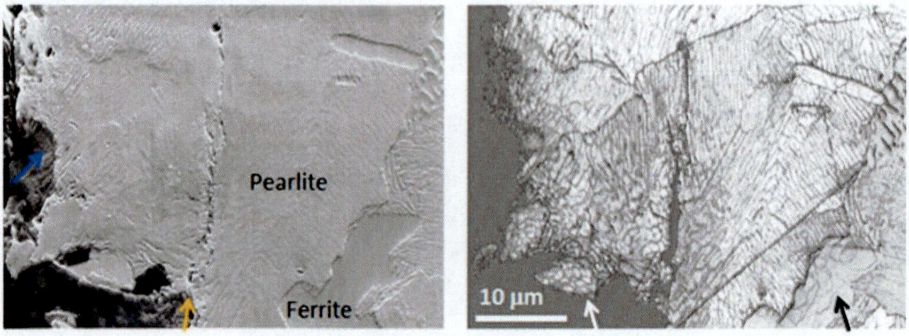

Fig. 7 SEM and EBSD micrographs of the screw showing the same sub-surface area in the cross section. The blue arrow points to the corrosion following the lamellar direction of a pearlite colony. The orange arrow points to the cracks along a shear band in one pearlite colony. The EBSD band slope micrograph is shown. The dark-grey area in the EBSD micrograph represents the un-indexed zone due to the oxidation and cracks. The cementite lamellae in the non-cracked metal are also shown in dark grey, while the ferrite lamellae are in light grey

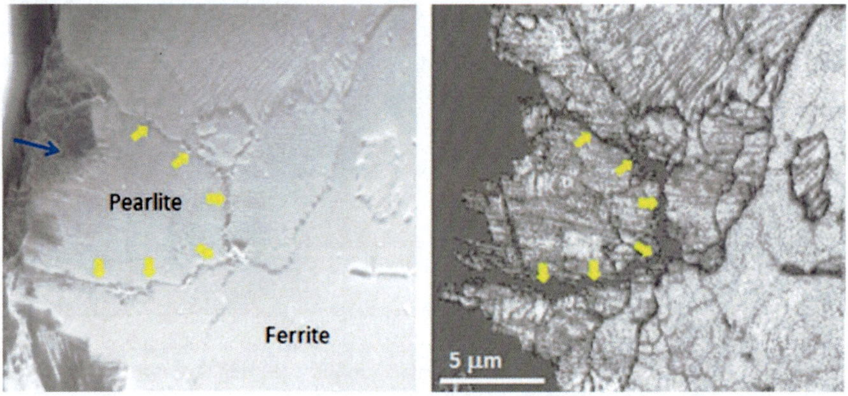

Fig. 8 SEM and EBSD micrographs showing the same sub-surface area in the cross section of the screw. The blue arrow points to the corrosion following the lamellar direction of a pearlite colony. The yellow arrows point out a crack along the pearlite colony boundaries. The coloring of the EBSD micrograph, the un-indexed zone and ferrite and cementite lamellae is the same as Fig. 7

size around ~1 μm, which is similar to the size observed from the subsurface areas (e.g. as seen in Figs. 6 and 7). The cell morphology is quite different from the lamellar morphology of dislocation boundaries corresponding to high strains formed by wear and friction processes. This indicates that the friction between the two surfaces of the present crack is limited and this main crack may have its origin in the tension caused by the bending of the screw. At the same time, corrosion holes are only observed near the small cracks in areas with remaining cementite lamellae (shown by the blue arrows), confirming that the ferrite dissolves as anodes. The round

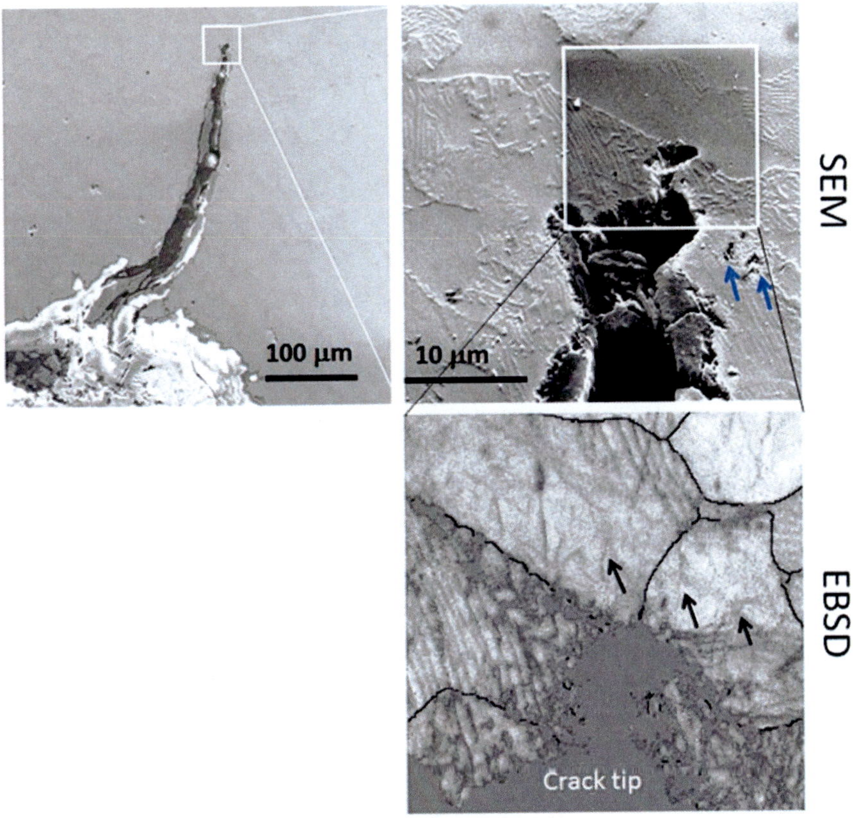

Fig. 9 SEM (top) and EBSD (bottom) micrographs showing the early stage of crack propagation. The blue arrows point at corrosion holes where only cementite lamellae are present. The black arrows show the dark grey lines in the ferrite grains representing low angle dislocation boundaries. The black lines show high angle boundaries with misorientation angles $\theta > 15°$ in ferrite matrix. The coloring of the EBSD micrograph, un-indexed zones, ferrite and cementite lamellae is the same as in Figs. 7 and 8

crack tip with corrosion products, shown by a white square in Fig. 9, also indicates corrosion-assisted cracking.

To sum up: cyclic deformation is the dominant mechanism leading to crack initiation and propagation, while corrosion and other related processes such as hydrogen-enhanced localized plasticity [11, 12] also play a role.

4 Conclusion

Cracked and failed screws from concrete sleeper S&Cs were collected near the Central Station of Copenhagen and characterized by four complementary techniques: 3D X-ray tomography, optical microscopy, scanning electron microscopy and electron backscatter diffraction. The following conclusions were found:

- The screws were elongated with cracks at thread 2 and 3 which indicates localized deformation of these sites.
- Three stages of crack evolution including crack initiation, propagation and fracture can be identified from the surface of failed screws.
- The cyclic bending leads to subdivision of the ferrite grain into dislocation cells with a size ~1 μm around the crack tip in the initial stage of crack evolution, as well as below the final fracture surface.
- Corrosion observed in the failed screw surface and at the crack tips, indicates its importance for crack initiation and propagation.
- For the present screws, cyclic bending is suggested to be the prime cause of crack formation and evolution but it is enhanced by corrosion and related processes such as hydrogen-enhanced localized plasticity may play a role too.
- The present combination of experimental techniques are found to be well suited, when the goal is to analyse failure and failure mechanisms.

Acknowledgements The authors gratefully acknowledge support from the Innovation Fund Denmark through the project "INTELLISWITCH—Intelligent Quality Assessment of Railway Switches and Crossings" (Grant no 4109-00003B). Dorte Juul Jensen, Xiaodan Zhang and Yubin Zhang further wish to acknowledge support from the European Research Council (ERC) under the European Union's Horizon 2020 research and innovation programme (grant agreement No 788567-M4D).

References

1. Capacity for rails (2015) Operational failure modes of Switches and Crossings. Public deliverable D 1.3.1. p 53. http://www.capacity4rail.eu/IMG/pdf/c4r_-_d131_-_operational_failure_modes_of_scs_public_.pdf
2. Zhang X, Hansen N, Godfrey A et al (2018) Structure and strength of sub-100 nm lamellar structures in cold-drawn pearlitic steel wire. Mater Sci Technol 34:794–808
3. Zhang X (2019) Exploring high strength metallic materials: a lesson from pearlitic steel wire. IOP Conference Series. Mater Sci Eng 580(1):012058
4. Sun C, Song Q, Zhou L et al (2019) The formation of discontinuous gradient regimes during crack initiation in high strength steels under very high cycle fatigue. Int J Fatigue 124:483–492
5. Zhang X, Godfrey A, Huang X et al (2009) Characterization of the microstructure in drawn pearlitic steel wires. In: Proceedings of the 30th Risø international symposium on materials science: nanostructured metals—fundamentals to applications. DTU, Risø, pp 409–416
6. Zhang X, Hansen N, Godfrey A et al (2014) Hirarchical structures and strength in cold-drawn pearlitic steel wire. In: Proceedings of the 35th Risø international symposium on materials science: nanostructured metals—new frontiers of nanometals. DTU, Risø, pp 153–170

7. Staicopolus DN (1963) The rate of cementite in the acidic corrosion of steel. J Electrochem Soc 110:1121–1124
8. Mora-Mendoza JL, Turgoose S (2002) Fe_3C influence on the corrosion rate of mild steel in aqueous CO_2 systems under turbulent flow conditions. Corros Sci 44:1223–1246
9. Ochoa N, Vega C, Pébère N et al (2015) CO_2 corrosion resistance of carbon steel in relation with microstructure changes. Mater Chem Phys 156:198–205
10. Hao X, Dong J, Nabuk Etim I-I et al (2016) Sustained effect of remaining cementite on the corrosion behavior of ferrite-pearlite steel under the simulated bottom plate environment of cargo oil tank. Corros Sci 110:296–304
11. Birnbaum HK, Sofronis P (1994) Hydrogen-enhanced localized plasticity—a mechanism for hydrogen-related fracture. Mater Sci Eng A 176:191–202
12. Robertson IM, Birnbaum HK, Sofronis P (2009) Hydrogen effects on plasticity. Dislocation Solid 15:249–293

Condition Monitoring and Asset Management

Data Sources and Research Models for Turnouts

Petra A. Wilfling, Michael Fellinger, and Peter Veit

Abstract The current inspection and measurement methodologies for turnouts are time driven and manual and thus based on unloaded measurements. However, these measurements are scarcely usable for behaviour prognosis. Analysis of technologies available on the market shows a possibility for a loaded and automatic inspection by combining already existing technologies. Standard Elements are currently in use as a common tool for assessing the life cycle of turnouts. Those Standard Elements include all maintenance tasks within the entire service life of turnouts, generated and based on experiences as well as on statistical data over an entire network and a long-time span. The economic justification of a maintenance task needs both, the initial investment and the remaining life span to be considered. However, the reasons for the end of service life and therefore lifetime limiting components must be analysed separately. This was done by numerous questionnaires and expert interviews, which have shown that the ballast and the sleepers are the overall reasons for the necessity of a reinvestment.

Executive Summary

Turnouts are very important assets within the railway infrastructure. As a poor condition could lead to a derailment, turnouts are also safety relevant assets. The infrastructure manager has to keep the condition of a turnout on a high status to guarantee a safe crossing. Hence, the inspection staff inspects the turnouts visually and measures different parameters. In the case of a substantial level of wear, maintenance tasks have to be executed. As the opinions of different people in everyday life differ, the description of the condition as well as the estimation of the degree of wear severity

P. A. Wilfling (✉) · M. Fellinger · P. Veit
Institute of Railway Engineering and Transport Economy, Graz University of Technology, 8010 Graz, Austria
e-mail: petra.wilfling@outlook.com

M. Fellinger
e-mail: michael.fellinger@tugraz.at

P. Veit
e-mail: peter.veit@tugraz.at

R. Galeazzi et al. (eds.), *Intelligent Quality Assessment of Railway Switches and Crossings*, Springer Series in Reliability Engineering,
https://doi.org/10.1007/978-3-030-62472-9_7

varies. This leads to a slightly subjective description process. Furthermore, no train can pass the turnout while the inspection staff is in this area. The rail section must thus be locked. As the current inspection is also done according to a fixed time interval and cannot give any information on the turnout behaviour during the passage of a vehicle, other ways to inspect turnouts are considered in this paper. To gain an objective report of the condition, different technologies were checked if they are able to measure and inspect a turnout. The analysis of these technologies has shown that they are able to measure and inspect turnouts automatically. Unfortunately, a combination of all systems would be necessary, which is hardly feasible. As the infrastructure manager has to know the service life to decide if a maintenance task is economically justifiable, the life time limiting components have to be found. As the full combination of all systems is still not available for the infrastructure manager, other ways to consider the life cycle must be used. In this case, Standard Elements give information on maintenance tasks and the life time. These elements can, generally speaking, be seen as a conclusion of all considered turnouts, provided that they have the same parameters, and their maintenance tasks within their life cycle. By referring the maintenance-costs to the point in time of re-investment, it is possible to decide if the maintenance task is economically justifiable or if the end of economic service life is reached. To confirm the statements of this approach, questionnaires and interviews with international railway managers were done. Additionally, a reinvestment list of the Austrian Federal Railways was analysed to answer the question why turnouts have to be reinvested. At least, all three approaches mutually supported the reasons and therefore lifetime limiting components.

1 Introduction

Turnouts are essential for a running train operation as they enable the change from one track to another without any interruptions of the train passing. The turnout itself consists of several individual parts and movable elements. Those assets have to be maintained differently or completely reinvested during service life. Furthermore, this infrastructure component is safety relevant and complex as well as an expensive asset. In order to transfer the high investment costs into a long service life, continuous monitoring, status recording and carrying out the required maintenance demands at the right time are prerequisites. According to a decision-making process for a turnout reinvestment, the primary necessity is the ability to describe the current condition and to predict its development. This is needed to implement a Life Cycle Management System for turnouts and therefore the prediction has to embed at least one planning period of the infrastructure manager. Furthermore, it is essential to know about the critical components leading to a reinvestment, in the event of excessive wear and tear or poor and incorrect performance.

13,592 turnouts are situated within the 9,782 kilometres railway network of Austrian Federal Railways [1]. Due to this high amount and facts as mentioned, the infrastructure manager must give priority to these assets.

2 Current Inspection Method

The current inspection method for turnouts, done by Austrian Federal Railways, is based on fixed time intervals. Depending on the daily load and maximum speed, the network is categorized into different numbered categories, with the exception of the most important track lines which are categorized with the letter "S". The higher the category number the lower the gross tons per day and the longer the periodicity of the inspection interval. Hence, the periodicity of inspections varies from 3 up to 12 months.

Nevertheless, the inspection is done manually over the whole Austrian Federal Railways network by the inspection staff. This work contains many different inspection procedures. One is the measurement of different gauges within a turnout. Beginning at front and ending with both rails at the rear, the gauge of 10 exact defined points is measured. These measurements are done with specially designed gauging tools, but do not include any further load impact, except the weight of the tools. Nevertheless, the measurements are currently used to ensure a safe crossing of vehicles as they define the geometrical limits and threshold for gauges divergence. The other inspection procedure comprises the visually observation of the whole turnout itself as well as of each individual asset. Variations to the nominal condition are estimated by the staff during this procedure. However, both the results of the gauge measurements and distances as well as the assessment of the current condition are written down, stored in reports and are currently used for the maintenance planning process.

It is obvious that during an inspection or while the inspection staff is located within the danger zone, this section may not be passed by any train. The turnout or the entire section, which is currently being inspected, must thus be locked by the infrastructure manager, or the inspection must take place at times, when there is no train traffic in the considered area. As these times are very rare in the course of the day, especially on high-load routes, the inspections will be partly carried out at night or on weekends. Due to the increasing traffic volume and shorter headways more and more inspections will need to be done during the night or at weekends in the future. If one considers the necessary tasks for the inspection and compares these with the difficulties that are already present, it is even more critical to move the inspections into the night. By contrast the carrying out the inspection during daytime requires the locking of the track sections, which has negative effects on the availability and capacity.

3 Analysis of the Current Inspection Data

As mentioned above, it is essential to know the current condition and to predict the development of turnouts. With this information the maintenance activities can on the one hand be planed and optimized and on the other hand a forecast of the reinvestment can be made. Investigation of the current inspection data has shown that the manual inspection describes the actual condition of turnouts well, but scarcely allows forecasting and time series building.

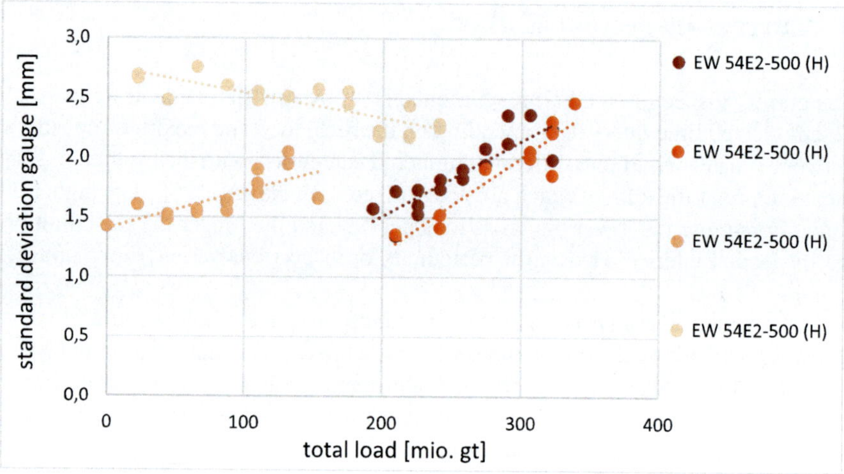

Fig. 1 Development of the standard deviation gauge of different turnouts

Figure 1 represents an excerpt of this analysis and illustrates the standard deviation of gauge measurements of four single straight turnouts with the same geometrical properties.

The increase of the standard deviation with the increase of the total load is not a surprising fact, as the condition over time cannot be equal like it is at the time of reinvestment. Surprisingly this is only visible at three of the four considered turnouts. The deviation of the gauge measurements of one turnout, depicted on the top left corner, became better and better over time with an increasing load. Since this cannot represent a realistic situation, the data used would appear to be incomplete.

In addition, the geometric features measured without load impact never describe the real situation and only allow static and unloaded viewing on these assets. In order to be able to make predictions about the behaviour of turnouts, it is necessary to know both, how a turnout behaves during the passage of a train and where the critical areas are located. This question can definitely not be answered by a manual inspection and measurements at ten different points. The manual inspection is necessary to adhere the safety critical limits whereas loaded measurements are needed to describe the real behaviour while a rail vehicle passes [2].

4 Methods of Inspecting Turnouts Automatically

Considering the effects of the increasing time limitation, the time driven and manual inspection methodology, the lack of the load impact and subjective estimation as well as the malfunction of forecasting, other possibilities must be looked for to automate and to objectify the inspection process as far as possible.

Deutschl et al. [3] introduces a visual inspection, based on two different light sources, mounted in front and after a color line scan camera. Those light sources radiate red and blue light separately. After the integrated process of shading, differencing the function and adoption, this system is able to inspect three-dimensional rail surface failures.

Another system, which is based on picture analysis is the Switch Inspection and Measurement System [4]. This system includes ten cameras, which guarantee the visualization of failures within different perspectives. The panorama cameras, mounted at the end and in front of the vehicle, view the whole turnout without the need for staff. A creation of rail profiles is possible due to the application of four line-cameras. Two further cameras of this type guarantee the visualization of the rail in colour. The last of the ten mounted cameras is one more line-camera, which records especially the rails-top according to anomalies and cracking in black-and-white images.

Auer [5] also mentioned the use of visual technologies due to different camera types. The contact-less geometry measurement system combines cameras and lasers in one technology. This system is able to measure the rail profile as well as the lateral wear. The camera itself can provide an image of the entire rail-profile, including the top and the foot of the rail. Another technology which is mentioned by Auer is the track component monitoring-system. This system uses high-resolution line-cameras combined with an illumination system. Hence, it is possible to record the side and plan-view of the rails as well as the fasteners including the detection of rail-wear.

A special kind of a track vision system, known as "V-Cube", uses the technology of laser and cameras for the inspection of the railway infrastructure [6]. The whole system consists out of three subsystems:

(1) Track Inspection System
(2) Rail Head Inspection System
(3) Track Measurement System.

The first subsystem uses a single line-camera, which scans the rails including the surrounding area and guarantees the inspection of the whole rail profile. In contrast with this system, the Rail Head Inspection System inspects only the top of the rail with a high-resolution camera. The last subsystem, the Track Measurement System, uses the images of the camera and acquires laser scans for a determination of the objects height. With the reconstruction of various images it is thus possible to illustrate the components according to a pre-defined colour-scheme.

Another system which is based on visual technology is mentioned by Brightwell [7]: the Rail Infrastructure Alignment Acquisition System. This system enables a scanning of the rails via cameras and laser followed by an export in coordinates of these data. With the combined software it is possible to correlate the data with GPS and inertial measurements of the customers rail network. By combining these technologies with a tape-recorded video, geo-referenced images are possible.

Zarembski [8, 9] has shown an inspection methodology, which uses high-image-acquisition-rate lasers mounted on a road-rail vehicle. With this system it is possible to create a contour-plot of the rail every 25 millimetres. As these images can be reprocessed with the integrated software, the measured asset could be illustrated in a tree-dimensional profile and thus deterioration can be calculated.

Cameras and lasers are also used for the Turnout & Crossing Measurement System [10]. The visual unit of this system consists out of three lasers and five cameras. With this optical system it is possible to permit a contact-less recording of the turnout within a sampling rate of one centimetre. Furthermore, this system measures the gauges every 25 cm of the through and diverging direction of turnouts.

A combination of the mentioned visual systems is combined in the SwissTrolley [11]. This system uses two laser scans for recording the surrounding within an angle of 310°, mounted at the top of the rail vehicle. Another two lasers are situated an the vehicle floor and are used for the rail scans, which are combined with two high-record cameras. Hence, while each rail is recorded by one of those cameras, the superstructure of the railway track is measured by an additional line-camera. Furthermore, this system also includes a ground penetrating radar.

However, the possibility is also provided to monitor turnouts with fix-installed wayside applications, mentioned by Fritz [12], Papacek and Pfleger [13]. The ROAD-MASTER 2000 is built for surveying the point machine by a transformer, based on measurement of the voltage during the switching process [12]. This system is also able to measure the check rail gauge and wear of the crossing nose with an installed frog sensor, which delivers an electromagnetic impulse in case of a contact between flange and sensor. Furthermore, the screw connection with the pre-stress is measured by special sensors [13]. The IS 2000 is also a sensor-based measurement system, which consists out of four inductive sensors. Those permit the monitoring of the position of the switch rails, dividing in open and closed switch rails. If one sensor notices a status outside the predefined availability window, an error message is generated [12, 13].

Stoll and Bollrath [14] examine the diagnostic of switching machines with the measurement of the switching force. This system surveys the force during the switching process and calculates the difference between the measured and reference force. As a higher force is needed in cases of wear, it is possible to predict the functionality of the switching machine and therefore its condition using this measurement.

The literature review has shown that there are different possibilities for an automatic inspection method. Most of the mentioned systems use various visual techniques for the inspection and measurement of the railway infrastructure. The fact, that some of these systems are specifically designed for turnouts while other systems are built for the open track, leads to the necessity for a verification for the potential of an automated turnout inspection.

5 Automated Turnout Inspection

Due to the widely scattered possibilities of inspection and measurement technologies, the systems of the following international companies were finally used for the analysis:

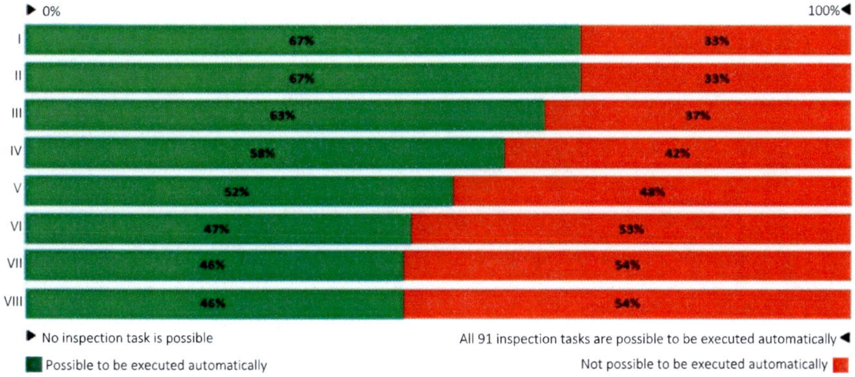

Fig. 2 Possibility for an automated turnout inspection

- DMA Srl, Torino (Italy)
- Eurailscout Inspection & Analysis B.V., Amersfoort (Netherlands)
- Fugro RailData, Utrecht (Netherlands)
- MER MEC S.p.A., Monopoli (Italy)
- Plasser & Theurer - Export von Bahnbaumaschinen GmbH, Vienna (Austria)
- Protran Technology LLC, New Jersey (USA)
- Terra Vermessungen AG, Zurich (Switzerland)
- VAE GmbH and voestalpine SIGNALING Zeltweg GmbH, Zeltweg (Austria).

This selection [15] represents a range of systems including wayside applications, on-board monitoring systems and self-propelled diagnostic vehicles in order to automate turnout inspection. Mentioned that all systems guarantee an inspection including the load-impact of the rail-vehicle and therefore without the need of inspection staff in the danger area. For data protection reasons and to avoid a ranking between the companies, the following evaluations are presented anonymously.

To be able to compare the elected technologies, it was first necessary to gather and classify all single inspection tasks for turnouts, mentioned in the national technical regulations of the Austrian and Swiss Federal Railways. In sum, 91 individual inspection tasks [15] were compared with the possibilities of all the chosen technologies and systems, illustrated in Fig. 2.

The results have shown that from nearly 50 % up to 70 % of all necessary 91 inspection tasks for turnouts could be done automatically by the systems of the different companies. Nevertheless, some parts of the inspection tasks still have to be done manually, as the systems cannot guarantee a totally automatic inspection. It should also be mentioned that the feasibility of each system should not be seen as an indicator of concordant inspection tasks between the different technologies.

Nevertheless, the results have clearly shown the strength of the different modes of operation of the selected technologies and systems. While optical systems guarantee the inspection of geometrical specifications and wear condition, fix-installed wayside application (WSA) are largely unable to do this. The possibility to inspect the

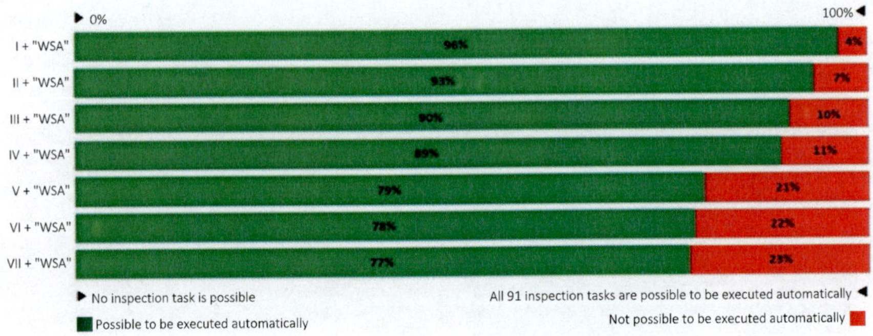

Fig. 3 Results of an automated turnout inspection by combining the technologies with WSA

signalling components of turnouts is contrary. In this case the fix-installed way-side application is the only technology that is able to execute these tasks automatically. Due to the different mode of operation they can be combined well with the optical systems. A theoretical combination of the different technologies was thus made, which shows an increase of approximately 30 % for each system combining it with "WSA" (Fig. 3).

At least 96 % of all 91 considered inspections tasks could be executed automatically, taking the load-impact of the rail vehicle into consideration and providing an inspection method without the necessity of inspection staff in the danger area. Furthermore, the combined system would allow the collecting of reproducible inspection data and ensure potential time series, for predicting the optimal maintenance tasks in the required time frequencies. This would have a positive influence on the lifetime of a turnout.

These considerations underline the demand for loaded measurements within inspections to support behaviour prediction. However, there are other tools in existence to deal with life cycle costs of turnouts.

6 The Standard Element Approach

Turnouts consist of various components such as the switch toes and the switching machine, the crossing nose, the closure rails and the wing rails. Furthermore, turnouts have different basic conditions such as the turnout type, the daily traffic load, the operating area or the sleeper material. Exactly one specific combination of these parameters defines a Standard Element.

With those parameters the standardised (average) behaviour of similar turnout elements, used in the railway network (Fig. 4—headlines), can be indicated. Furthermore, the necessary maintenance activities were determined during the entire life cycle for each Standard Element. These are named standard working cycles and they are available for each frequently occurring parameter combination in Austria (Fig. 4—4th row downwards). Each of the cycles is based on statistical analyses of

"Südbahn", v <200km/h	standard turnout	R = 500 m	double track																					
total load per day and track	profile	sleeper	crossing nose			substructure			fastenings															
55000 gross tons	60E1	concrete	Mn / rigid			good			indirect															
service life	years	36,0	0	1	2	3	4	5	6	7	8		24	25	26	27	28	29	30	31	32	33	34	35
renewal		1,0	1																					
tamping	# in service life	8,0					1				1				1			1			1			
grinding	# in service life	5,0							1				1					1						
half set of switches exchange	# in service life	1,0																						
crossing nose exchange	# in service life	2,0											1											
check rail exchange	# in service life	1,0												1										
overlay welding	# in service life	2,0																						
lipping removal	# in service life	17,0			1		1		1		1		1		1		1		1		1		1	
small maintenenace	# in service life	36,0	0,5	0,5	0,5	0,5	0,5	0,5	0,5	0,5	0	1	1,5	1,5	1,5	1,5	1,5	1,5	1,5	1,5	1,5	1,5	1,5	1,5
ballast cleaning	# in service life	1,0																						
rail pad exchange	# in service life	0,0																						
screw hole rehabilitation	# in service life	0,0																						
set of slepper exchange	# in service life	0,0																						
single sleeper exchange	# in service life	0,0																						
inspection	# in service life	36,0	1	1	1	1	1	1	1	1		1	1	1	1	1	1	1	1	1	1	1	1	1

"Südbahn", v <200km/h	standard turnout	R = 500 m	double track																					
total load per day and track	profile	sleeper	crossing nose			substructure			fastenings															
55000 gross tons	54E2	wooden	Mn / rigid			good			indirect															
service life	years	28,0	0	1	2	3	4	5	6	7	8		16	17	18	19	20	21	22	23	24	25	26	27
renewal		1,0	1																					
tamping	# in service life	7,0					1				1		1				1			1			1	
grinding	# in service life	3,0									1		1								1			
half set of switches exchange	# in service life	1,0															1							
crossing nose exchange	# in service life	2,0																			1			
check rail exchange	# in service life	1,0																				1		
overlay welding	# in service life	2,0																	1					
lipping removal	# in service life	13,0			1		1		1		1		1		1		1		1		1		1	
small maintenenace	# in service life	28,0	0,5	0,5	0,5	0,5	0,5	0,5	0,5	0,5	0	1	1	1	1	1,5	1,5	1,5	1,5	1,5	1,5	1,5	1,5	1,5
ballast cleaning	# in service life	0,0																						
rail pad exchange	# in service life	0,0																						
screw hole rehabilitation	# in service life	0,0																						
set of slepper exchange	# in service life	0,0																						
single sleeper exchange	# in service life	0,0																						
inspection	# in service life	28,0	1	1	1	1	1	1	1	1		1	1	1	1	1	1	1	1	1	1	1	1	1

Fig. 4 Example of a Standard Element, depicting service life and maintenance demand

historic maintenance tasks, service lives, and renewal demands. Furthermore, track professionals with great experience were involved in the creation process to integrate their knowledge in the standardized descriptions. By this means the Standard Elements are able to describe a typical situation, here the type of turnout in the railway network is under consideration [16].

By multiplying the number of maintenance tasks within the working cycle with their specific costs, it is possible to calculate the life cycle costs (LCC) of the turnout types considered. Thereby, the life cycle costs can be calculated for a given parameter set and thus represent the average value for all turnouts with the same boundary conditions or parameters as shown in Fig. 4. The life cycle costs include investment (replacement and disposal), maintenance costs, and costs of operational hindrances, without considering the costs of development, testing, and homogenisation.

Discounting and summing up all the single payments to a reference year (in this case the year of the (re-)investment) allows the calculation of the net present value (NPV), assuming the definition of a discount rate. The annuity (average dynamic annual costs) is calculated by multiplying the NPV using the capitalizing factor (CF) (Fig. 5). The annuity can thus be used for defining net wide strategies regarding types of turnouts and their maintenance scheme. If these steps are carried out separately year by year, the so-called annuity monitoring [17] is obtained in order to identify the economic service life. This service life is reached as soon as the savings of the depreciation by increasing service life is smaller than the required additional maintenance.

According to this procedure, it is thus possible to determine the optimum point in time for reinvestment seen from an economic perspective. Due to the long service life and the distribution of the investment costs over these years, the influence of depreciation on the annual costs decreases with increasing asset age, following a function f = 1/n, where n represents the years of service life. Maintenance activities

"Südbahn", v <200km/h	standard turnout	R = 500 m	double track								
total load per day and track	profile	sleeper	crossing nose			substructure			fastenings		
55000 gross tons	60E1	concrete	Mn / rigid			good			indirect		
service life	**years**	**36,0**	**0**	**1**	**2**	**3**	**4**	**5**	**6**	**7**	**8**
renewal		1,0	1								
tamping	# in service life	8,0					1				1
grinding	# in service life	5,0							1		
half set of switches exchange	# in service life	1,0									
crossing nose exchange	# in service life	2,0									
check rail exchange	# in service life	1,0									
overlay welding	# in service life	2,0									
lipping removal	# in service life	17,0			1		1		1		1
small maintenenace	# in service life	36,0	0,5	0,5	0,5	0,5	0,5	0,5	0,5	0,5	0
ballast cleaning	# in service life	1,0									
rail pad exchange	# in service life	0,0									
screw hole rehabilitation	# in service life	0,0									
set of slepper exchange	# in service life	0,0									
single sleeper exchange	# in service life	0,0									
inspection	# in service life	36,0	1	1	1	1	1	1	1	1	

$$NPV = \sum_{t=1}^{n} \frac{R_t}{(1+i)^t} \qquad NPV = \frac{\sum € \text{ of year } 0}{(1+0.05)^0} + \frac{\sum € \text{ of year } 1}{(1+0.05)^1} + \frac{\sum € \text{ of year } 2}{(1+0.05)^2}$$

$$CF = \frac{(1+i)^n \cdot i}{(1+i)^n - 1} \qquad i = \text{discounting rate}$$

Fig. 5 Scheme of calculating the net present value for 3 years

which occur late in the life cycle have a major influence on the annual cost function. The reason is that the costs of maintenance at the end of the service life pushes up the annual costs more significantly than the decreasing depreciation can compensate.

The economic justification of an additional maintenance task thus depends on the cost of initial investment, the maintenance costs itself and the remaining life time. In the case of turnouts, the economic service life is very important, as technically the single components can also be exchanged separately. However, this would lead to an economically unacceptable high demand on maintenance. Furthermore, the experiences of renewals carried out can be used to depict the components mostly limit the economic service life.

7 Lifetime Limiting Components

As the life cycle considerations always depend on the condition of the components and the re-invest decision depends on the remaining service life plus the initial and maintenance investment costs, the life time limiting components must be identified. Three different methodologies were used for this identification.

7.1 Renewal Project List

Within the Austrian Federal Railways there is a list of executed renewals of turnouts available, covering a time span of several years. As those lists also include the inspec-

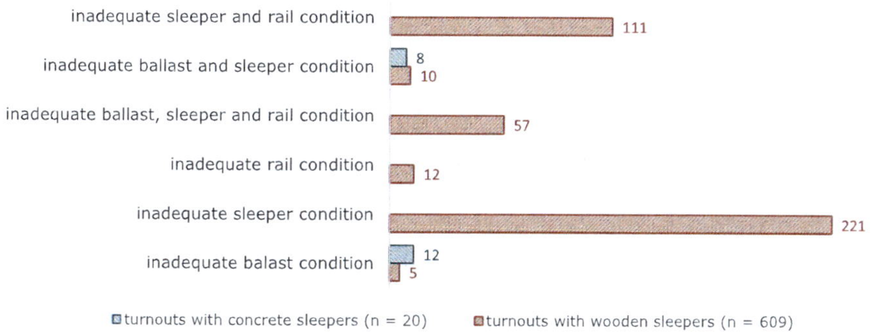

Fig. 6 Reinvestment reasons for different types of turnouts

tion data, one of those data sheets in this case from the year 2015, was used to identify the life-time limiting components. At this point, it has to be mentioned, that the information content of the entries within the renewal lists are widely spread. Some entries contain very detailed information, while a couple of entries have less or only general information. In most cases the reason for the reinvestment alone was mentioned without giving detailed information. Another fact, which has to be taken into account, is that turnouts are not only reinvested due to end of service life, but for example also due to major restructuring of railway stations.

Considering those mentioned facts, all of the turnouts within the renewal list of 2015 were subdivided according to their:

(1) material of sleepers
(2) different operation behaviour
(3) area of operation.

416 turnouts with wooden sleepers and 20 turnouts with concrete sleepers have been replaced in the course of reinvestment measures in 2015. It is of course obvious that the analysis of 20 turnouts with concrete sleepers is never as representative as the analysis of 416 turnouts with wooden sleepers. However, due to the limited reinvestment of only 20 turnouts with concrete sleepers in 2015, more turnouts could not be analysed. The subclasses have been analysed with the necessary information acquired and show the following reasons for the reinvestment (Fig. 6).

For turnouts with concrete sleepers the results show an unambiguous assignment for the reinvestment causes. 60 % of all turnouts with concrete sleepers were exchanged due to an inadequate ballast condition. The remaining 40 % were reinvested due to the ballast condition too, but combined with an inadequate sleeper condition. In the case of turnouts with wooden sleepers the reasons vary more widely. 53 % of turnouts with this sleeper material were replaced due solely to an inadequate sleeper condition. In 27 % the poor sleeper condition combined with a high rail wear was the reason for an exchange of the turnouts. Additionally, the condition of the ballast in combination with other components led in 17 % percent of all cases to a reinvestment, especially on tracks dealing with heavy traffic loads.

In summary, the ballast could be identified as the critical component for turnouts with concrete sleepers. In the case of turnouts with wooden sleepers, on the one hand, the sleepers themselves and on the other hand these components coupled with an inadequate ballast and rail condition was shown to be responsible for the end of the technical service life in most cases. Nevertheless, such project lists are greatly influenced by many boundary conditions and project specifications and this source reflects only the real state encountered. For this reason, it is necessary to verify the critical components on an economic basis. This requirement justifies the carrying out of further investigations based on other approaches. One of them, as mentioned before, is the application of Standard Elements.

7.2 *Analysis with Standard Elements*

Following the above-mentioned idea of Sect. 6, the methodology of annuity monitoring was subsequently used to identify the critical and thus lifetime limiting components of turnouts on an economic and network-wide basis. The following boundary conditions were chosen for both, for turnouts with wooden sleepers as well as for turnouts with concrete sleepers:

- Load class of 55,000 gross tons per day
- Radius of the diverging direction of 500 meters
- Rail profile of 60E1.

7.2.1 Turnouts with Concrete Sleepers

The Standard Work Cycle of the associated Standard Element shows a service life of 36 years for the mentioned parameter combination. Noted that increased maintenance has been required in recent years, specially in the last two. Using the knowledge from the reinvestment list analysis, an additional ballast cleaning task was intended to stretc.h the service life beyond 36 years. Experiences for this load class and the related boundary conditions have shown that the turnout has a service life of approximately 14 years after the maintenance task of ballast cleaning was carried out. As a result, the overall service life could have been extended up to 50 years.

After the extension of the service life the maintenance tasks were tightened and filled up likewise as they were done during the last 36 years, without changing all the other maintenance tasks. With this data base, the annual costs were calculated for this turnout under the given conditions. By using the dynamic annual cost monitoring method it was thus possible to identify the critical and life-limiting components of turnouts with concrete sleepers.

The diagram in Fig. 7 shows the annual costs function for a turnout with concrete sleepers. due to the increased maintenance in the last years a ballast cleaning was executed in the 34th year. The high costs of this maintenance task increases the

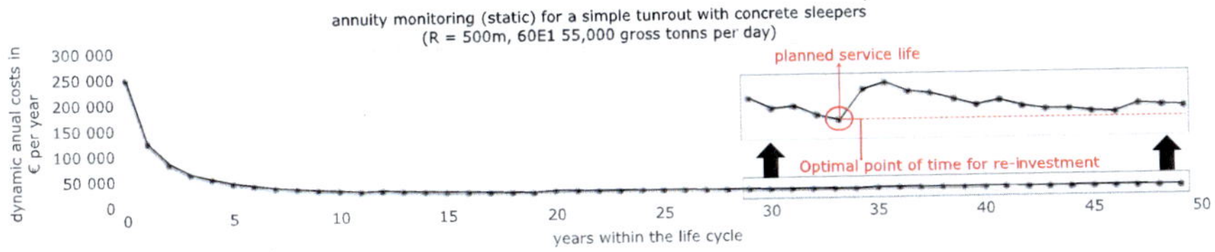

Fig. 7 Annuity monitoring of turnouts with concrete sleepers

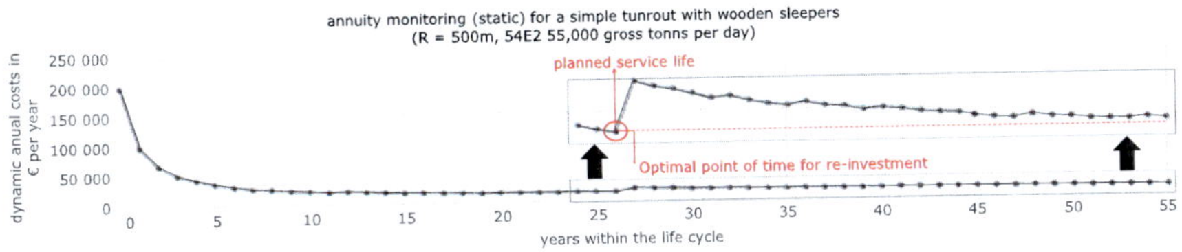

Fig. 8 Annuity monitoring of turnouts with wooden sleepers

annual costs to a higher level. Within the next 14 years, it is not possible to reduce the annual costs to a lower level as it was reached in the year 34 (depicted in the small figure).

As the experiences have shown that the ballast has a useful service life of 14 year after the cleaning process, an additional ballast cleaning would be necessary if the service life should be expanded one more time. Unfortunately, this action will increase the annual costs once again and they will not drop below the level of the year 34 at any time.

For this reason, it was possible to confirm the results on an economic basis. For turnouts with concrete sleepers the ballast represents the critical component.

7.2.2 Turnouts with Wooden Sleepers

For turnouts with wooden sleepers and equal boundary conditions, the working cycle of the Standard Elements illustrates an service life of 27 years. Referring to the reinvestment list, the entries of maintenance tasks include a full sleeper exchange in many times for turnouts with wooden sleepers. As a result, this maintenance task can stretch the service life significantly. Furthermore, experience has shown that turnouts with wooden sleepers have a useful additional service life of approximately 28 years after a full sleeper exchange. Due to this, the overall service life could be extended to 56 years.

Following the above executed procedure the annual costs were calculated. Similar to the ballast cleaning process at turnouts with concrete sleepers, the sleeper replacement for turnouts with wooden sleeper increases the annual cost drastically in the 27th year. Over the next 26 years, it is not possible to reduce the annual costs to a lower level than in those of the previous periods (small picture within the Fig. 8).

Due to the dynamic annual cost monitoring it is obvious that a poor sleeper condition at the end of the service life will lead to dramatically increasing costs. The sleepers thus represent the critical component in turnouts with wooden sleepers seen from an economic perspective, a general renewal should be facilitated at this point in time instead of a full sleeper exchange.

With the Standard Element analysis and the renewal list, it is now possible to depict the lifetime limiting components of turnouts. The above described approach illustrates the component causing a reinvestment is in case of turnouts with concrete sleepers the ballast condition. The service life of turnouts with wooden sleepers is badly affected by the rotting process of the sleepers. In both cases, it would be technically possible to prolong the service life by ballast cleaning respectively sleeper exchange. However, this would lead to higher annuities meaning the economic service life benefit would be cancelled.

7.3 *Questionnaires and Expert Interviews*

The Standard Element analysis which is based on statistical data with a network-wide mean was verified additionally by questionnaires and expert interviews. Within this investigation both, turnout experts and project managers responsible for turnout renewals were interviewed [18]. For this purpose, different questionnaires referring to the material property of the sleepers were prepared as shown in Fig. 9. To map the results not only on the experiences of the Austrian Federal Railways experts, experts and infrastructure managers across Europe were questioned. In most cases the interviewed person has filled in the questionnaires for both turnouts with wooden sleepers and those for turnouts with concrete sleepers. Nevertheless, in unusual situations the interviewed expert has answered questions only for one of the material properties as he/she was responsible for those turnout types. The questionnaires were supplemented by a personal interview in cases of unclear or ambiguous information. In sum a total amount of 30 questionnaires were evaluated, 16 for turnouts with wooden sleepers and the last 14 for turnouts with concrete sleepers [18]. Both, the surveys and the questionnaire, have shown similar results in the context of the life time limiting components as in the previous analyses, but they also have revealed a speciality too.

Previous studies declare the ballast condition for turnouts with concrete sleepers and the condition of the sleeper itself in case of turnouts with wooden sleepers as the critical component. During the expert interview, it turned out that in both cases the reason for the reinvestment is not only caused by the condition of those assets. The condition of several components combined lead to the need for a turnouts reinvestment, although this situation is mostly driven by the components mentioned above. In the case of turnouts with concrete sleepers, the sleepers themselves also have a great influence of the reinvestment decision (Fig. 10 and Table 1). For turnouts with wooden sleepers, the connection between the sleepers and the rails are also an important reason for the reinvestment. As the connection depends on the condition of

Institut für Eisenbahnwesen und Verkehrswirtschaft

BASIC ANALYSIS OF THE FINAL CAUSES FOR THE DECISION OF A TURNOUT RENEWAL

This questionnaire only refers to straight, single turnouts on the main tracks of the core network with wooden sleepers.

1. Main category:

a) Selection of the decision reasons

Which of these listed, technical categories are the final reasons for the decision of a turnout renewal?

Please enter the percentage, with which each of the chosen categories is responsible for the turnout renewal. Take care that the sum of the distributed percentages is 100.

RAIL ___%	WOODEN SLEEPERS ___%	FRICTIONAL CONNECTION OF THE RAILS AND THE SLEEPERS ___%
SWITCH RAIL ___%	FROG ___%	BALLAST ___%
SUBSTRUCTURE ___%	OTHERS: .... ___%	

b) Reason for the choice of main categories

Please fill in the chosen main categories from 1a) and check the reason for your selection.

Chosen main category:____________________

☐ *Inspection* ☐ *Experience* ☐ *Measurement:*__________ ☐ *Others*__________

2. Sub category: Further analysis of the chosen main categories

Please fill out the sub categories belonging to the chosen main categories.

WOODEN SLEEPERS:

a) If the wooden sleepers are the only cause for the turnout renewal:

Which maintenance tasks would be necessary instead of the turnout renewal?

☐ *Replacement of a single sleeper* ☐ *Screwlock renewal* ☐ *Replacement of multiple sleepers*

What was the reason for the turnout renewal instead of executing the maintenance tasks?

b) If the wooden sleepers are not the only cause for the turnout renewal:

Which criteria also affected the decision of the turnout renewal?

☐ *Rail* ☐ *Frog* ☐ *Substructure* → *Wooden sleepers*

☐ *Frictional connection (rails – sleepers)* ☐ *Switch rail* ☐ *Ballast*

Which maintenance tasks would be necessary instead of the turnout renewal?

Seite 1/8

Fig. 9 Excerpt of the questionnaire for turnouts with wooden sleepers [18]

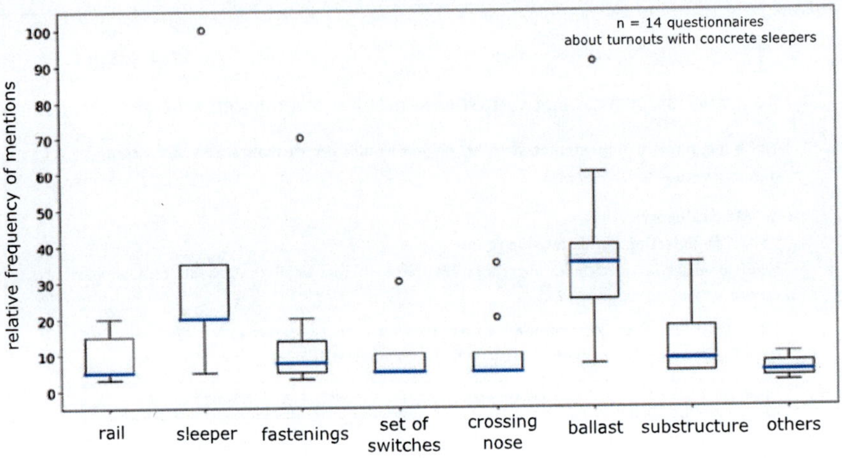

Fig. 10 Results of survey—turnout with concrete sleepers

Table 1 Results of survey—relative frequency of mentions for turnouts with concrete sleepers

Reason	Min.	1st Quantile	Median	Mean	3rd Quantile	Max.	Mentions
Rail	3.00	5.00	5.00	9.78	15.00	20.00	9 of 16
Sleeper	5.00	20.00	20.00	28.85	35.00	100.00	13 of 16
Fastenings	3.00	5.00	7.50	14.80	13.75	70.00	10 of 16
Set of switches	5.00	5.00	5.00	9.44	10.00	30.00	9 of 16
Crossing nose	5.00	5.00	5.00	10.50	10.00	35.00	10 of 16
Ballast	7.00	25.00	35.00	36.50	40.00	91.00	12 of 16
Substructure	5.00	5.00	8.50	12.67	17.50	35.00	12 of 16
Others	2.00	3.50	5.00	5.67	7.50	10.00	3 of 16

the sleepers, it can be said that the most important exchange factor for those turnout types are the sleepers themselves (Fig. 11 and Table 2).

In summary, the critical components are the ballast and the sleepers. It is clear that all individual components of a turnout must be seen as critical due to their influence of safety. It is therefore necessary to inspect the turnout continuously. The condition of the sleepers and the ballast also influences the life time of a turnout, since the maintenance tasks (ballast cleaning and sleepers exchange) lead to such a high annuity that it can no longer be economically justifiable once a specific point in the life time of a turnout has been reached. This knowledge alone permits an adequate prediction of maintenance demand and thus the prediction of the economic service life.

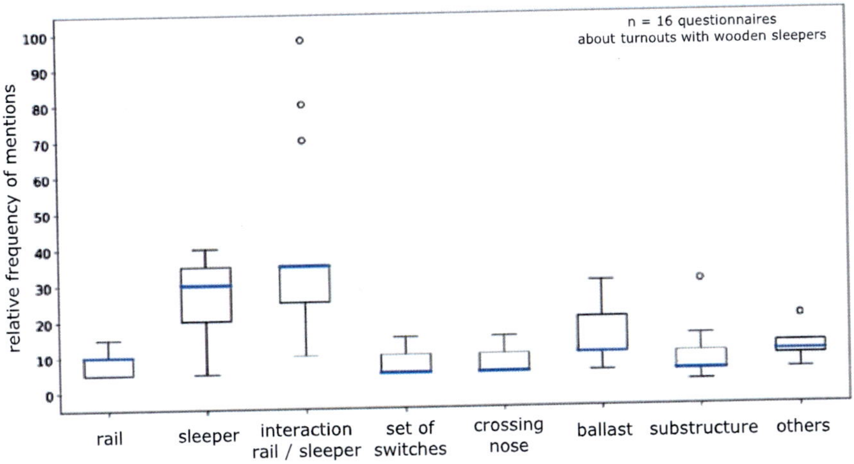

Fig. 11 Results of survey—turnout with wooden sleepers

Table 2 Results of survey—relative frequency of mentions for turnouts with wooden sleepers

Reason	Min.	1st Quantile	Median	Mean	3rd Quantile	Max.	Mentions
Rail	5.00	5.00	10.00	8.64	10.00	15.00	11 of 16
Sleeper	5.00	20.00	30.00	28.00	35.00	40.00	15 of 16
Interaction rail/sleeper	10.00	25.00	35.00	38.53	35.00	98.00	15 of 16
Set of switches	5.00	5.00	5.00	8.00	10.00	15.00	10 of 16
Crossing nose	5.00	5.00	5.00	8.00	10.00	15.00	10 of 16
Ballast	5.00	10.00	10.00	13.46	20.00	30.00	13 of 16
Substructure	2.00	5.00	5.00	8.50	10.00	30.00	12 of 16
Others	5.00	8.75	10.00	11.25	12.50	20.00	4 of 16

8 Summary

As turnouts are expensive and critical components in the railway infrastructure, maintaining these elements in a good condition is an essential task. The analysis of the inspection process has shown that turnouts are still inspected manually at fixedly defined time intervals and at least without any load-impact of the rail vehicle. Nevertheless, the possibility arises for an automatic and loaded inspection of turnouts. As the results have also shown that a combination of a significant number of inspection technologies is necessary to gather the actual condition of all assets automatically under load impact of the vehicle, the implementation is hardly possible at the point in time. Nevertheless, it is indispensable to guarantee a good condition of turnouts and thus a safe vehicle crossing. Many maintenance tasks thus need to be done during the life time of turnouts, but the economic justification for these tasks must be subject to a consideration of both the investment required together with the remaining life span. To know the remaining service life, the life time limiting components have to be identified. This was done with different analyses. The considered Standard Element analysis together with the interviews and questionnaires declared the condition of the ballast in case of turnouts with concrete sleepers, and the condition of the sleepers themselves in the case of turnouts with wooden sleepers as the major reasons for reinvestment decisions.

9 Future Work

The requirement for a combination of many inspection technologies in order to automate the inspection tasks for turnouts has been argued in this work. Furthermore, the critical components can be identified based on economic considerations only. The combined inspection data, however, is currently not fully available. Therefore it will be necessary to acquire and define usable data which allows deriving the condition of turnouts more precisely. As the data of the inspection technologies shown are not available for analysis, the track measurement car, which is normally used for the open track, will be analysed for its potential to collect information on turnouts. Following the idea to use this loaded measurements for the turnouts and to adapt the already used analysis methods of the open track, the data of the measurement car has been investigated if the condition of a turnout can be described and if it can be adapted for time series analyses [2].

References

1. Facts & Figures of the Austrian Federal Railway (2018) https://infrastruktur.oebb.at/en/company/facts-and-figures . Accessed 12 Dec 2018
2. Fellinger M, Wilfling PA, Marschnig S (2019) CoMPAcT—Data based condition monitoring and prediction analytics for turnouts. In: Intelligent quality assessment of railway switches and crossings. Springer Nature, United States
3. Deutschl E, Gasser C, Niel A, Werschonig J (2004) Defect detection on rail surfaces by a vision based System. In: Paper presented at the IEEE intelligent vehicles symposium, Parma, 14–17 June 2004
4. Tegelberg E, Le Flem J, Lases E (2017) We inspect: 100 switches, measuring in 6 hours. In: EURAILSCOUT inspection and analysis B.V. Online Fact Sheet of EURAILSCOUT, Amersfort
5. Auer F (2004) Berährungslos arbeitende Gleismessfahrzeuge und Auswertung der Messdaten. EIK—Eisenbahn Ingenieur Kompendium, Verband Deutscher Eisenbahn-Ingenieure e.V. (VDEI), Frankfurt am Main, pp 53–66
6. Cannito M, Ferrante L (2017) System operating and maintenance manuel—system E3 (VCUBE & RHIS). Manual of Practice of MER MEC S.p.A, Monopoli
7. Brightwell S (2016) Getting up to speed: Rail asset information hits the fast track. RailEngineer , Issue 143, Coalville, pp 66–67

8. Zarembski AM, Euston TL, Palese JW (2011) Development, implementation, and validation of an automated turnout inspection vehicle. In: Paper presented at the IHHA Conference, Calgary, 19–22 June 2011
9. Zarembski AM (2013) Improving inspection of turnouts. Railway track & Structures, New York, pp 24–27
10. DMA Srl Torino (2017) Turnout & crossing measurement system (TCMS). Product Brochure, Torino
11. Terra vermessungen (2017) Productvideo of SwissTrolleyII. http://www.terra.ch/de/bahnvermessung.html. Accessed 3 April 2017
12. Fritz P (2004) Efficient diagnostics for supporting the availability of turnouts and tracks. In: SIGNAL + DRAHT, Issue 96, Hamburg, pp 42–45
13. Papacek F, Pfleger H (2000) Wartungsarme Systeme zur Umstellung und Äæberwachung von Weichen. In: SIGNAL + DRAHT, Issue 92, Hamburg, pp 46–49
14. Stoll H, Bollrath B (2002) Weichendiagnosesystem SIDIS W. In: SIGNAL + DRAHT, Issue 94, Hamburg, pp 26–29
15. Wilfling PA (2017) Der Weg zur Smarten Weiche. Graz University of Technology. https://pure.tugraz.at/ws/portalfiles/portal/12105001/DerWegzurSmartenWeiche.pdf
16. Veit P (2016) The economic service life of track. In: Paper presented at the 11th edition of the World congress on railway research Milan, WCRR, 29 May–2 June 2016
17. Marschnig S, Veit P (2012) Life cycle management in der Realität. ZEVrail 136(9):335–341, Linz
18. Dittmer E (2018) Grundlagenanalyse der Ausbauentscheidung von Weichen. Graz University of Technology https://diglib.tugraz.at/grundlagenanalyse-der-ausbauentscheidung-von-weichen-2018

CoMPAcT-Data Based Condition Monitoring and Prediction Analytics for Turnouts

Michael Fellinger, Petra A. Wilfling, and Stefan Marschnig

Abstract Unloaded measurements of turnouts guarantee the adherence of safety critical geometric limits but do not allow behaviour forecasting. To push maintenance strategies for turnouts up to a preventive level, performance prediction is required. Therefore, it was necessary to identify data, being suitable to describe the condition of a turnout, based on loaded measurements. The measurement data from the standard track recording car EM250 of the Austrian Federal Railways are best suited for this purpose. Due to the limited length and the varying stiffness of a turnout, the measurement data must be positioned much more precisely. In this context, it was necessary to develop a methodology for a defined and comprehensible measurement data positioning. This method enables the positioning of all available measurement signals with a maximum deviation of one measurement point (25 cm) and ensures the synchronicity between each single measurement signal. Applying this positioning algorithm to the measurement data, it is possible to describe the behaviour of turnouts quite well. Different methods for describing the necessary tamping actions as well as the actual condition of the ballast are presented. These methodologies enable describing the present condition of turnout components and deliver statistical values, required for prediction algorithms.

Executive Summary

Turnouts are critical elements of the railway infrastructure. In order to reduce costs and to increase the service life, their behaviour over time must be understood. In addition to the manual inspection, data from the track measurement car are necessary

M. Fellinger (✉) · P. A. Wilfling · S. Marschnig
Graz University of Technology, Institute of Railway Engineering and Transport Economy, 8010 Graz, Austria
e-mail: michael.fellinger@tugraz.at

P. A. Wilfling
e-mail: petra.wilfling@outlook.com

S. Marschnig
e-mail: stefan.marschnig@tugraz.at

R. Galeazzi et al. (eds.), *Intelligent Quality Assessment of Railway Switches and Crossings*, Springer Series in Reliability Engineering,
https://doi.org/10.1007/978-3-030-62472-9_8

to analyse the behaviour during the passage of a train. Based on these measurements, the condition of individual components can be described. By applying the mentioned methodology over a certain period of time, condition development can be depicted and thus the next necessary maintenance task can be derived. Therefore, no new data sources or technologies are necessary. With smart evaluation techniques it is possible to understand the condition progress and to maintain critical components in an early stage. Thus, an economic optimisation of turnouts can be reached and this means a benefit for the whole railway system.

1 Introduction

Turnouts are important parts of the railway infrastructure. Their purpose is to give a rail vehicle the possibility to move from one track to another. Hence, each turnout consists of different parts such as the crossing nose, the stock and the switch rails as well as a switching machine (Fig. 1). In contrast to open track, moveable parts are necessary to allow the transition between two tracks and therefore, high static and dynamic forces occur while a rail vehicle is passing through [1]. Due to the higher load and the larger impacts, more frequent maintenance tasks are done in turnouts, compared with open track. Within the core railway network of the Austrian Federal Railways containing 9,782 track kilometres, 13,592 turnouts are situated [2]. This corresponds to an overall density of 1.39 turnouts per track-kilometre. Due to this high value as well as to the high frequency of necessary inspection and maintenance tasks, the infrastructure manager must give priority to these railway assets.

To convert the high investment costs into a correspondingly long service life and thus to ensure the economic efficiency of turnouts, life cycle considerations are necessary. For this purpose, all necessary activities throughout the entire service life must be known. This idea leads to the necessity of a data based behaviour prediction model. The data for these purpose can come from various sources. The most ordinary data origin to create a prediction model is the manual inspection. These measurements are determined unloaded and the overall condition of the turnout is assessed subjectively. The suitability as input for a prediction model has already been judged [4]. It is essential to move from a manual to an automated turnout inspection, which needs to take place under load. In this context, investigations are available, which assess the basic suitability of various inspection systems [5]. However, one problem

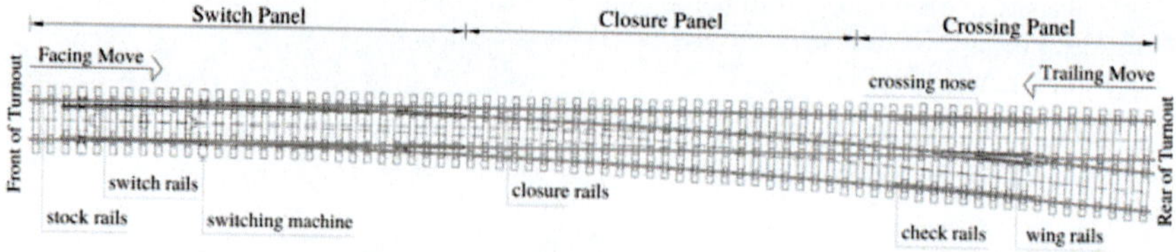

Fig. 1 Area specification and component nomenclature of turnouts

among others is, that the different measuring systems have to be combined to get a meaningful overview about the condition of the whole turnout.

The presented approach is thus to use track measurement data, recorded by a track measurement car (TMC) to assess the condition and to predict the behaviour of turnouts. The TMC offers the advantage that the measured data are determined under load and that all turnouts are measured between 4 and 6 times a year, depending on the track load. Therefore, these data should be best suited to describe the actual condition and to predict the behaviour of different turnout types by building up time-series. The individual sections of this paper therefore are:

1. Literature study on investigated methods for describing and predicting the turnout behaviour. Considerations on the open track as well as on turnouts are made.
2. Based on these findings, properties of measuring signals are discussed, being necessary to describe turnouts. This discussion leads up to the development of an algorithm to post-stationing the necessary measurement data.
3. By using these data, the possibility to rate the ballast condition is given and the effect of tamping can be analysed. By calculating the standard deviation of the longitudinal level, it is possible to identify areas with high rates of deterioration.
4. Finally, the presented methods are applied. The condition of four turnouts is described. Likewise, an attempt is made to generate functional knowledge about future developments from the actual condition as well as the past behaviour.

2 Literature Review

Compared with open track, turnouts are infrastructure components being maintenance intensive and thus have higher maintenance costs [7]. Similarly, different types of turnouts have different maintenance costs. Further consideration of turnouts thus has to be parameter specific [6]. As cost drivers for turnouts, the inspection on the one hand and the maintenance, e.g. tamping or welding, on the other hand, could be identified [7]. It is difficult to influence the cost of inspections, since these tasks should ensure compliance with safety-critical limits and the individual intervals are regulated within different standards. The possibility of influencing the costs is given only in relation to the maintenance activities. Using the annuity monitoring method [3] with some adaptations, it is possible to identify the critical components leading, in the event of a failure, to a renewal of turnouts. For turnouts with concrete sleepers, the critical component is the ballast. For turnouts on wooden sleepers, the sleeper condition or the condition of the fastenings should be considered critical [4]. Thus, it is clear that primarily these two components are considered.

In order to be able to optimise maintenance, knowledge about wear and tear of the currently considered components is necessary [3]. Related to descriptive models, these wear functions can only be derived from measurement data by forming time series. The possibilities of loaded measurements for the entire turnout as well as alternatives of the manual inspection were shown in [5]. However, it should also be noted that manual inspection data is not suitable for forecasting and therefore different data sources are required [8]. One way to generate new data is to inspect turnouts using video analysis [8, 9]. It is an innovative approach to automate the inspection of turnouts. However, loaded measurements of the track are required for a behaviour prediction, which probably cannot be measured using video systems. In [10], this idea was taken to describe turnouts under the usage of measurement data from the TMC. Also the problem of exact measurement data positioning was discussed within this study. In [11] the possibility for the geometrical description of turnouts by means of measuring car data in Sweden is discussed, whereby the main focus was placed on the evaluation of the longitudinal level signals.

The exact positioning of the measurement data is an important factor in order to derive useful information for turnouts. It is important to identify the turnout within the measurement signals [12] and to synchronise the individual measurement data for time series generation. In this context, synchronisation means ensuring that a specific measurement point at the position x in a first measuring signal corresponds to the same measurement point at the same position within a second signal. In [13, 14] the minimum of the longitudinal level is determined within a certain range and it is claimed that this corresponds to the position of the crossing nose. Based on this position, the measurement signal is subdivided into four different areas and subsequently analysed. Also other methodologies were used to position or synchronise measurement signals as only synchronised measurement data can be analysed within time series. In [15] the method of Dynamic Time Warping (DTW) was used and improved to synchronise measurement signals. However, this investigation refers to the open track, which is why the identification of turnouts within the measurement signal does not matter in this case. This task and the synchronisation of the signals tend to be a great challenge, which is why this topic is dedicated to a separate section within this paper. The measurement data positioning as well as the area identification should only be regarded as a means to an end. These two tasks only build the basis for the main work, the condition description of turnout components.

As shown in [16, 17], the longitudinal level correlates very strongly with the ballast condition. Above all, the wave range between 3 and 25 m is conspicuous by identifying faulty ballast areas. For turnouts, similar analyses have already been carried out with slightly different evaluation methods [13, 14]. Following this idea, it is also attempted within this paper to describe the condition of the ballast within turnouts by using the longitudinal level signal and some smart analyse methods.

3 Proposed Methodology

Since manual inspection data are not suitable for prediction, the possibility of describing the condition, based on measurement data for the open track, is discussed within this paper. The used data is explained in detail in Chap. 4. As shown in Chap. 2, precise positioned measurement signals are necessary. Since a variety of turnout components should be described, it is not sufficient to shift individual measurement signals based on specific characteristics. A smart and automated post positioning method is necessary positioning all signals of the measurement car in a traceable and accurate manner. The design and implementation of this algorithm is discussed in Chap. 4. The goal of this research is to develop a descriptive model for the deterioration behaviour of turnouts. Therefore, only the superstructure should be describable. This work is completed in Chap. 5 by an examination of how far the longitudinal level measurements are suitable for describing the ballast condition. In Chap. 6, the previously discussed and described methods are applied to four turnouts and the results are compared. Finally, in Chap. 7, results are discussed and an outlook on future work and further occupations is given.

4 Data Source, Collection and Cleaning

The goal is the description of turnouts, based on loaded measurement data. These data are measured by the track recording car EM 250 of the Austrian Federal Rail- ways [19]. The TMC is capable of measuring all relevant track parameters every 25 cm. In principle and is equipped with two inertial measuring units (IMU) and a profile and gauge measuring system [20]. The inertial measuring units are used to measure the longitudinal level, the alignment and all derived values. The gauge is measured with an optical system, equipped with several lasers and mirror systems. Due to the difficult optical conditions within turnouts, the gauge cannot be consistently measured over the entire length of the turnout. All measurement data are stored within a database and are available in time series since 2001. The main lines and thus all turnouts on these lines are measured by the EM 250 between three and four times a year. However, in most cases the through direction is measured. It is thus subsequently attempted to describe the behaviour of the entire turnout based on these measured data of this direction. The stored measurement data are already rough positioned, which serves as a link to the asset database. Within this database, the geometry of the considered turnouts, the year of installation and the type of superstructure is defined. As the turnouts within the measuring signals cannot be determined exactly and automatically within a given area, it is necessary to identify the turnouts within the measuring signals. It must be ensured that the necessary measurement data can be analysed in time series and that all signals from one mea- surement run are synchronous. Since this task has to be carried out for many turnouts, an important step within the data cleaning and treatment process is to develop a post positioning algorithm for the measuring data of the EM 250.

4.1 *Suitability for Turnout Description*

As mentioned, measurement data must fulfil the following requirements to enable the application for the condition monitoring of turnouts:

1. A turnout must be automatically identifiable within one measurement signal. Therefore, a unique point from a turnout must be found within each signal.
2. One measurement signal must be synchronous with each other measurement signal within one measurement ride.
3. Each measurement signal must be synchronous with the same measurement signal of past and future measurement runs.

Nowadays, it is not possible to identify turnouts within each measurement signals automatically. Hence it is necessary to search for characteristics within the measurements, which can be clearly assigned to a turnout. Since the gauge is measured by a laser, a closer consideration of this signal is expedient. To measure the track gauge (i.e. the distance between the two rails, 14 mm below the rail surface), the distance to the left and to the right rail is measured from the centre of the measuring frame [20].

Face to face with the crossing nose, a drop-out within the half gauge signal (Fig. 2-blue measurement signal between position 1060 and position 1080) is visible. This measurement failure corresponds to the check rail, since the laser measures a too small distance with the consequence that an error value is recorded. Opposite (location of the crossing nose—green measurement signal in Fig. 2) the laser measurement along the closure rail is recognisable and an error value in the area of the wing rail (position 1060–1080) is recorded. After the laser jumps from the wing rail back to the crossing nose, the half gauge is again measured correctly. Therefore, the position of the turnout can be clearly identified within these two measurement signals. This is possible since the distances between the beginning of the check rail and the theoretical crossing point can be deduced from the detail turnout drawing.

Thus, it is possible to identify a turnout within this signal. Based on the other two requirements, the synchronism within one and between different measurement runs

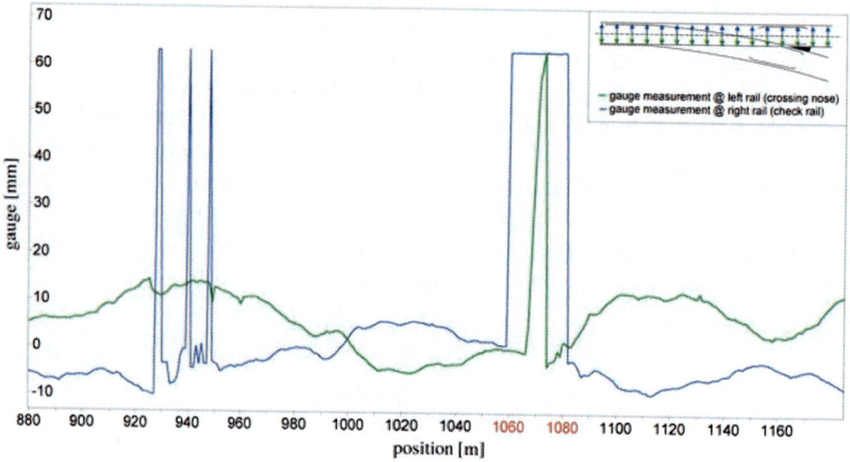

Fig. 2 Identification of a turnout within the gauge measurement signal

was evaluated. Both analyses show no synchronism and therefore, a post-positioning algorithm has to be developed for all measurement data.

4.2 Post-Positioning Algorithm

The already treated half gauge signals have been defined as reference signals. Within the following process, all measurement signals are related to them. This ensures that the identification of the different areas can also be transmitted to all other measurement signals. The main challenge is the mathematical handling of the different measurements. To assess the synchronicity, three different methods were analysed:

1. Cross-correlation function: This function rates the similarity of two data sets with different shifts in the x-direction. The cross-correlation function can be used to determine the magnitude of the shift between two signals. For this purpose, the maximum of the cross-correlation function is particularly interesting. Its position indicates how much one signal must be delayed in order to be synchronous with the other one. Within [12], this method was used successfully. Nevertheless, there is another method, which is also investigated.
2. Dynamic Time Warping: This methodology was developed for speech recognition and is among others used to extract individual letters from a spoken text. Following this idea, it should be possible to check whether some sections of a measurement signal can be found within others and thus whether they are synchronous or not. However, as this methodology stretches or compresses the signal, this method seems to be unsuitable for assessing synchronicity. Furthermore, it can be ruled out that no compressions or strains occur within the measuring signals.
3. Euclidean distance: The calculation of the Euclidean distance has proven to be an effective means of assessing whether synchronicity between two measurement signals prevails. In this case, a section with a length of 100 m is extracted from the master signal. In the test signal, a range of the same length is also extracted and always shifted by one measuring point. After each shift, the distance is calculated as the sum of the squares of the differences. If these values are calculated for each possible shift and displayed within a shift function, the position of the minimum can be used to determine whether the signals are synchronous or not. If the minimum of the shift function occurs at $x = 0$, there is synchronicity, otherwise not. However, even if there is no synchronicity, the necessary shift for synchronicity can be calculated.

Once the reference measurement signal has been determined and the chosen Euclidean Distance Method (EDM) turned out to be the one being able to describe the relationships between two measurement signals best, it is necessary to look for interdependencies between the individual measurement signals to judge their synchronicity. For this purpose, all relevant measuring signals were subdivided into measurement signal groups (altogether 3 different ones). Each measuring channel group stands for signals being recorded by the same measuring system.

In a first step, the synchronicity between the measuring signals of the first measuring channel group was checked. Within this group, the unfiltered half gauge signals (the reference signals) and the combined full gauge measurement are included. By calculating the Euclidean distance, it could be demonstrated that both, the filtered half gauge measurement as well as the full gauge measurement are synchronous with the reference measurement signal (unfiltered half gauge measurement).

In a further step, the synchronicity within the second measuring channel group was assessed. This cluster also contains optically measured signals; the rail foot distance, the rail inclination and an additional full gauge measurement, also for rail profile recognition. Based on this measurement, the synchronicity with the rail foot distance signal was evaluated by comparison. Of course, the gauge and the base gauge have different amplitudes, however, the characteristic is completely identical. It was found that these two signals are synchronous. To assess the rail inclination signals, the gauge was re-calculated from these signals in combination with the rail foot distance and the calculated one was compared with the measured one. This process also showed that the rail inclination measurements and the gauge signal in this measuring channel group and therefore all signals in this cluster are synchronous.

Since a gauge measurement signal is present both, in the first and second measuring channel group, the synchronicity can be validated, again using the calculation of the Euclidean distance. Deviations between two groups of up to 2.75 m occur. These shifts are not big, but too big to be disregarded. It was possible to show that these two measuring channel groups are not synchronous and it was possible to determine the necessary shift, so that synchronicity is given. Thus, the areas identified in the reference measurement signal can be applied to these measurement signals as well.

Still, the synchronicity between the measuring channel group 2 and 3 had to be judged. Within the third cluster, all measurement signals measured by the IMU (longitudinal level, alignment, twist) are included. By knowing that the characteristic of the half gauge measurement can be represented by wavelength filtering of the alignment signal, the necessary shift between these two measuring channel group can be identified. Here too, synchronicity was not given, so that the recommended shift needs to be calculated. These shifts range between −2.75 m and 8.25 m.

Finally, using the mathematical method described above, the synchronicity within the measuring channel group 3 could be verified. Except for a fixed offset of 1.5 m between the alignment and the longitudinal level measurements, all signals in all wavelength ranges within this cluster are synchronous. Thus the identified areas of the turnout can also be transferred to these measurement signals.

The developed methodology was validated for 15 turnouts. Using this algorithm, it is possible to position the measurement data for each turnout with a maximum deviation from one measurement point (25 cm) and to identify the areas within each measurement signal. Thus, applying this algorithm, data are ready for evaluation.

At this point, all possible signals from the standard track recording car EM250, are in repositioned form. Therefore, it seems time to consider which signals could be suitable for which purpose. In principle, the longitudinal level and the alignment measurements should be able to determine the necessity of tamping actions and give the possibility to describe the ballast condition. The modifications of these two signals (cant, twist) are currently not in use. The track gauge signal could be used to describe the condition of the sleepers and the fastenings. Finally, the signals of the Rail Surface Scan may be able to describe the geometric conditions at the rail surface and could therefore be able to describe the condition of insulation rail joints. However, since all measurements are precise positioned, further research could be done using all of this data and maybe new findings can be extracted out of them.

It should also be mentioned that the measurement direction while travelling through the turnout has an influence on the signal characteristics and amplitude. For this reason, only measurements with the same measuring direction are compared with each other. The same fact occurs by comparing measurements from the through and from the diverging direction. Therefore, only signals which were recorded while driving at the through direction are analysed and compared with each other.

5 Ballast Condition Monitoring

Based on the post-positioned and synchronised signals, an approach for describing the condition of turnouts is presented. As already mentioned at the beginning, the longitudinal level signal is used for the open track in order to assess the necessity of a tamping action [21] and to rate the condition of the ballast [18]. These ideas should be used for turnouts as well. Due to the limited length of a turnout (Fig. 1: from the front of the turnout FoT to the rear of the turnout RoT) and due to the variable stiffness, the methods for the open track cannot be fully adopted. Based on the logic developed in Chap. 4, it is possible to synchronise the measurement data. Thus, the circumstance of variable stiffness can be taken into account and each single area (Fig. 1: switch, closure and crossing panel) can be analysed individually. In addition to the three mentioned areas, an evaluation is carried out for the area in front of the turnout as well as for the area directly after the turnout, in order to be able to extract effects that do not occur directly in the turnout, but which definitely influence the behaviour of them. Due to the limited turnout length, evaluation methodologies for the open track with an influence length of some 100 m cannot be applied. Instead, consideration is given to the short length of a turnout and a moving standard deviation of the longitudinal level is calculated with an influence length of one meter. This calculation serves therefore the base for assessing the upcoming maintenance activities like tamping or ballast cleaning. Thus, the process of describing the ballast behaviour over time is composed of the following steps:

1. Calculation of the moving standard deviation of longitudinal level with an influence length of one meter to identify critical areas related to the ballast.
2. Calculation of the standard deviation of longitudinal level for each of the five mentioned areas to observe the deterioration within each individual area.

3. Using executed tampings to deduce the effect of this maintenance tasks. Also, statistical parameters for further analyses can be calculated parameter specific.

Thus, by means of these methods, the ballast behaviour can be described over time. Additionally, this method is well suited to generate statistical values being necessary for the later prediction process. The next step is to clarify the ballast condition. Attention should be drawn once again to the relationship between the wavelengths of the longitudinal level measurement and the ballast condition, which has already been described several times [13, 14, 16–18]. Following this idea, an averaged power density spectrum (PDS) for each turnout and each measurement run is calculated.

In order to exclude effects before and after the turnout, the area between the front and the rear of the turnout is extracted from the longitudinal level signal with a wavelength of 3–25 m. This part of the longitudinal level forms the base for calculating an average power density spectrum. For the calculation, a Fast Fourier Transformation (FFT) with a rectangular window function is used. Since the PDS is calculated referring to the different frequencies, small adaptations are necessary to calculate them referring to the different wavelength ranges. Subsequently, a statement about the actual condition of the ballast should be possible using this method.

In addition, all other signals, measured by the standard track recording car are available. If these are needed later on, they can be evaluated, since they are already available in a positioned and synchronous manner. For the time being, reference will be made to all measurement signals and evaluation methods already discussed.

6 Results and Discussion

The application of the presented methods is done and depicted on four different turnouts within the main railway network of the Austrian Federal Railways. These turnouts have an average load of roughly 50,000 gross tons per day. Their specifications can be found in table one. These four turnouts were selected to make initial investigations and to get a feeling of what the processed data are capable of (Table 1).

Table 1 Specification of the researched turnouts

Turnout	Turnout type	Sleeper material	Year of installation
1	EW-UIC54E-500-1:14	wooden	1986
2	EW-UIC60-500-1:12	concrete with USP[a]	2007
3	EW-UIC54E-500-1:14	wooden	2009
4	EW-UIC60-1200-1:24	concrete	2005

[a]Under Sleeper Pads

6.1 Critical Area Identification

Considering the calculated moving standard deviation over the entire length of the turnout as shown in the heat maps (Fig. 3), illustrating the time on the y-axis and the datasets on the x-axis. The aim should be to identify critical areas. The scaling of the standard deviation is adapted and thus the same for all analysed turnouts.

Referring to turnout 1 (Fig. 3—top left), two areas are noticeable—the front of the turnout (between −180 and −160 DB) and the crossing panel (between −20 and 20 DB). The critical area in front of the turnout is an insulated rail joint (IRJ). The standard deviation in this range is even higher than at the crossing panel. In principle, the high standard deviation could cause from the rail surface (i.e. from the geometric shape of the IRJ) or from the ballast. Since there is no information about the ballast condition, the geometry of the rail joint was described based on other measurements. Therefore, the data of a Rail Surface Scan were evaluated. These data are standard ised measured every 5 mm by an individual measurement system. The measurable wavelength range is between 30 and 1000 mm, which is why an anomaly in this measurement signal is most likely to be assigned to the rail surface. Since abnormalities could not be detected within this signal and the described areas for turnout 1 and the illustrated tamping actions improved the standard deviation, it can be concluded that the cause of the fault is located within the ballast bed.

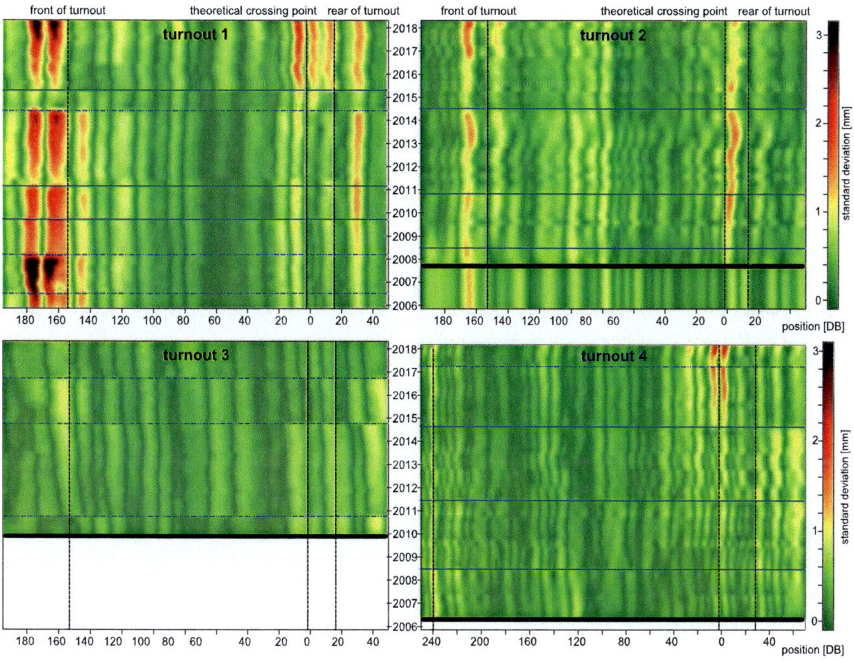

Fig. 3 Development of the moving standard deviation of the longitudinal level for different turnouts

It seems that due to the load impact, the ballast is presumably badly affected. The same phenomenon occurs, though not so significant, at the crossing nose. Again, it seems as if the constructional necessary gap causes a high ballast deterioration.

By looking at the analyses of turnout 2 (Fig. 3—top right), a similar picture emerges, although with a lower standard deviation level. Again, the area in front of the turnout where the IRJ are located (between −180 and −160 DB) seems to be critical. Again, the standard deviation in this area is higher than at the crossing panel. Likewise, the fast deterioration at the crossing nose as well as in the area of the welding joints within the turnout can be recognised again. The above made statement regarding the effect of tamping is also valid for this turnout, without having more accurate information about the current ballast condition.

Considering the heat map of turnout 3 (Fig. 3—bottom left), it should be noted that this turnout was built in late 2009. Up to this time there was open track only. This turnout shows slight noticeable areas at the front of the turnout and at the crossing panel, of course, with a lower severity than those previously considered.

However, it is interesting to note that there is also an area with a higher standard deviation at the end of the turnout, more precisely at the position of the last continuous sleeper (between 20 and 40 DB). At the other turnouts, this effect could only be anticipated. Again, the effect of a tamping can be shown very clearly, which is why the problems should be found again at the ballast.

Turnout 4 (Fig. 3— bottom right) is the only one analysed having a 1.200 m radius. Nevertheless, the heat map of this turnout shows similar results comparing with the previous ones. In this picture two critical areas—the area at the front of the turnout and at the crossing panel can be identified. Since there is no precise information about the position of the IRJ for this turnout, it must be assumed that the IRJ is the trigger for the higher standard deviation compared to the closure panel, in the region of the turnout's front. The previous analyses confirm this assumption. Otherwise, the turnout has a very homogeneous development of the standard deviation and it seems that the diverging radius does not affect the position of the critical areas.

In summary, regardless of the turnout's age and design, two areas can be identified which are critical, relating to the standard deviation of the longitudinal level:

1. The area in front of the turnout where the IRJs are located. Due to the geometric construction of the IRJ, a hit, which causes high dynamic forces and thus a very fast deterioration, occurs.
2. The entire crossing nose area. Within this zone, a constructive and geometrically necessary impact occurs, which leads to high dynamic forces and thus a high ballast deterioration.

Within a comparison of all four turnouts, turnout 1 has a significantly higher standard deviation than all the others. This is because turnout 1 will soon reaches the end of its service life and will be re-invested in 2020. Turnouts 2, 3, and 4 would be comparable in terms of their age. Since these turnouts are on the same main line, it can be assumed that, albeit roughly, all have experienced the same load. Within the comparison it can be seen that the standard deviation for turnouts on concrete

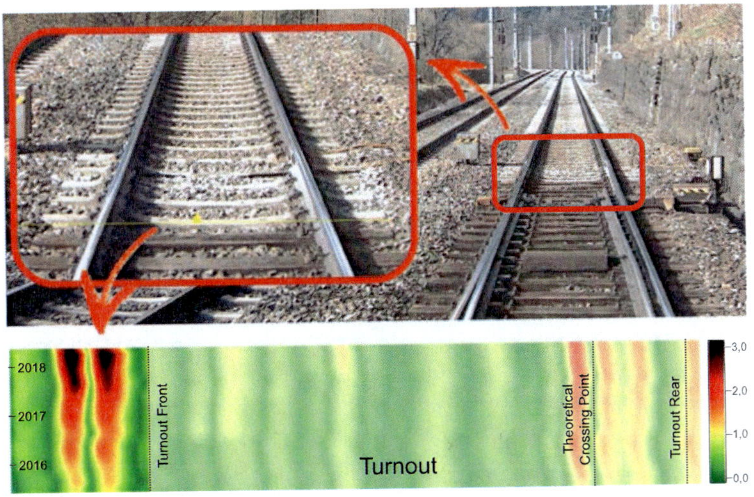

Fig. 4 Deterioration of longitudinal level at the turnout's front

sleepers with USP (turnout 2) and without USP (turnout 4) are almost identical. The standard deviation of turnouts with wooden sleepers (turnout 3) is lower.

Furthermore, the area at the front of the turnout seems to be very critical. At this position the insulated rail joints are situated. As shown in Fig. 4, it was also examined how this area looks like in reality. Within this part, at the picture rubbed ballast can be seen. This damage is noticeable by white spots in the ballast and white dust on the rail. The maintenance tasks already carried out in previous years at this section had only limited effects. This example also shows quite clearly, how critical a hit can be, for example relating to the condition of the ballast.

However, in order to be able to verify the development of the standard deviation exactly, the standard deviations for the different areas of a turnout were calculated in a next step. Since it has already been shown in the first evaluation that the area where the IRJs are situated and the crossing panel are critical, only the following analyses are presented and interpreted:

1. Standard deviation over the entire turnout between the front and the rear.
2. Standard deviation over a range of 10 m in front of the turnout.
3. Standard deviation from the second weld joint to the end of the turnout.

6.2 *Description of Tamping Actions*

Within Fig. 5, these evaluations are shown for all analysed turnouts. The top graph in each quadrant corresponds to the evaluation over the entire turnout, the one in the middle corresponds to the evaluation over the turnout's front and the lowest graph is to be understood as evaluation of the crossing panel. The solid vertical

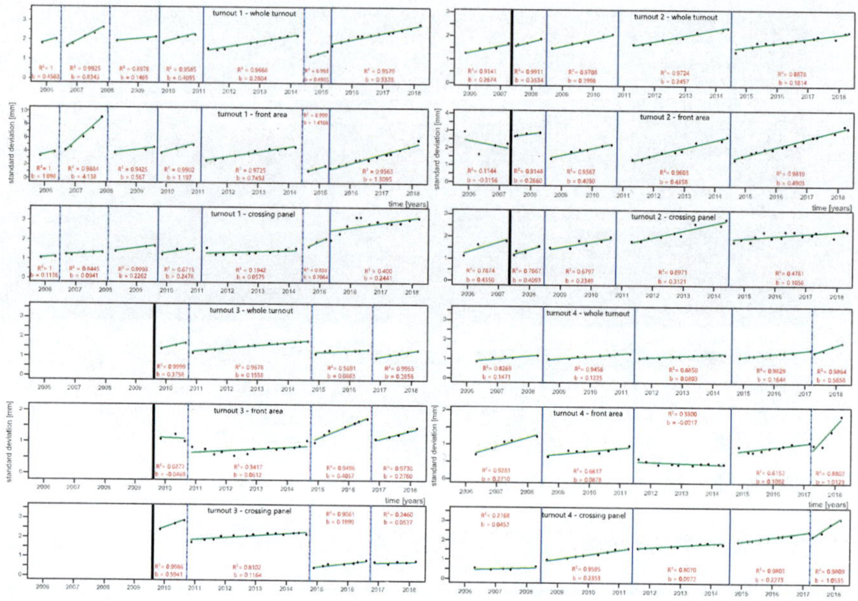

Fig. 5 Standard deviation of the longitudinal level for different turnouts and different areas

lines represent tamping actions, which were carried out and documented with date and used machine. The dash-dotted vertical lines are tamping actions, which were probably carried out, based on the longitudinal level development. Unfortunately, the corresponding documentation is missing. The thick vertical lines in black indicate a turnout re-investment. At the bottom of each graph, the coefficient of determination R^2 as well as the slope b of the deterioration function are shown in red.

By looking at the curves of the standard deviation of turnout 1 (Fig. 5—top left), the top graph shows the effect of tamping actions in relation to the longitudinal level. The tamping action in 2015 should be analysed more in detail, since the standard deviation has increased after this maintenance task. Beside this tamping action, all other maintenance tasks had an positive effect. Within the middle graph, a significantly higher scaling, compared to the one at the top, already shows the importance of monitoring this area. The standard deviation in the years 2007 and 2008 is higher by a factor of 4, compared to the entire turnout. The area in front of the turnout appears to be more critical and maintenance intensive. The analysis of the crossing panel generally shows that the tamping actions performed in this area had almost no effect. Also, by looking at the coefficient of determination R^2, it can be concluded that, compared to the previous evaluations, the bedding of the crossing nose is no longer ideal. The longitudinal level, especially in the last area, scatters very clearly and actually does not give a really coherent trend. This fact would also correlate with the age of the turnout and with the planned re-investment in 2020.

The analysis of turnout 2 (Fig. 5—top right) shows very uniform deterioration rates and the improvement of the standard deviation after a tamping action. Concentration on the curves after the re-investment, the coefficient of determination is always above 0.85 and the rate of deterioration range between 0.18 and 0.35. A very homogeneous deterioration of the standard deviation can be assumed. Even in front of the turnout, this fact is true, although the deterioration rates are even more homogeneous. However, the scaling of the standard deviation in this range is higher. At the crossing panel, a much higher variance can be seen. Especially toward the end, the spread increases significantly. All in all, once again it is confirmed that the crossing panel as well as the area in front of the turnout are critical.

With regard to turnout 3 (Fig. 5—bottom left), no abnormalities can be seen by analysing the entire turnout area. By looking at the area of the turnout's front, it can be seen that the deterioration of the longitudinal level after the tamping work in 2014 is significantly higher. This fact could be again caused by the bad geometry of the insulated rail joint in front of the turnout. Even after an additional tamping work in 2016, the deterioration rate cannot be reduced to the level before 2014.

The evaluations of turnout 4 (Fig. 5—bottom right) show, that the improvement after a tamping action is minimal, except for the tamping work in 2017. After this maintenance task, the deterioration rate increases extremely in comparison to the measurements before 2017. Even more clearly, this tamping action is recognisable within the evaluation of the front area, where the deterioration rate reaches 1.0 or more. This result indicates a bad insulated rail joint in front of the turnout.

In summary, it can be stated that the presented methodology is very well suited to describe tamping actions. Using this methodology, it is possible to deduce the need for the next tamping work, based on historical data. Considering the scaling of the standard deviation, the turnouts 2, 3 & 4 would be comparable to each other in relation to the total turnout area. With the shown method, the time for the next necessary tamping action, based on the data of this turnout, can be determined. However, this is only possible if at least one full cycle (re-investment to first tamping action) can be observed within the measurement data.

To enable the condition monitoring of new turnouts also, the parameter-specific statistical values can be determined from the evaluations, which in turn can be used for the description of new turnouts. However, the condition of the ballast should also be described, as this component is life-limiting. The idea is the description of the ballast condition by means of power density spectrum analyses. As there is currently no information about the actual ballast condition, only global analysis results can be presented and discussed.

6.3 Description of the Ballast Condition

For each of the four turnouts, a power density spectrum was calculated according to the methodology presented in chapter 5 (Fig. 6). Since there is no definite information about the ballast condition, the following assumptions were made:

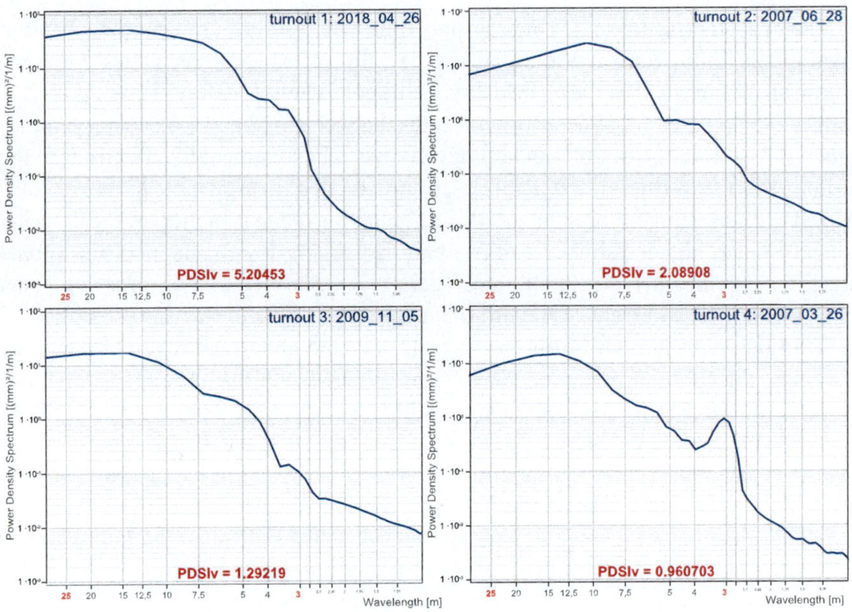

Fig. 6 Power Density Spectrum for different turnouts and different ballast conditions

1. Since turnout 1 will be re-invested in 2020, it can be assumed that the ballast shows signs of wear. No precise statement can be made related to how much the ballast has worn out. Although a clearly different characteristic within the wavelength consideration in comparison with a new ballast should be seen.
2. Since turnout 2 was renewed in 2007, it is primarily assumed that also the ballast was exchanged/cleaned. Even if it is not certain that this has actually happened, at least within a comparison between the year of the re-investment and the last measurement drive, a statement can be made about the condition of the ballast.
3. Turnout 3 was also installed within the measurement data record period (year 2009). Since there was no turnout at this area before, it is also assumed in the first place that the ballast was treated accordingly before the turnout installation.
4. Turnout 4 was installed in 2005. In turn, it is assumed that a new ballast in good condition was used at the re-investment.

The results from Fig. 6 are now interpreted in accordance with the made assumptions.In addition, in order to make the various power density spectra easier to interpret or compare, the integral under the function in the range between 3 and 25 m was calculated for each spectrum and named Power Density Spectrum Integral value—PDSIv. This value should provide information about the extent of change,especially in a direct comparison. It is shown in red in the lower range for each power density spectrum. Of course, it is possible that the ballast condition correlates better with a specific range in the power density spectrum.

The power density spectrum (PDS) of turnout 1 shows a rather constant progress in the wavelength range between 7.5 and 25 m and then drops steeply afterwards. The PDSIv is 5.2 mm^2. Since it is assumed that the ballast within the turnout is already worn, this value should rather represent the upper value range. In comparison with turnout 1, the PDS of the turnout 2 shows a maximum in the wavelength range of approximately 11 m. Again, from 7.5 m downwards, a very steep drop is recognizable. The evaluation of turnout 2 yielded a PDSIv of 2.08 mm^2which is significantly lower than that from turnout 1. The PDS of turnout 3 has an almost identical profile with respect to the characteristic as the one from turnout 1. This circumstance could be explained by the same sleeper material. Nevertheless, the evaluation of turnout 3 shows a PDSIv of around 1.3 mm^2, which in turn confirms the assumption of a good ballast condition. The analysis of turnout 4 show a very specific characteristic within the PDS, which is caused by the sleeper material. In the wavelength range between 3 and 4 m, a local maximum, not existing at the other turnouts can be seen. The PDSiv is 0.96 mm^2 and therefore the lowest level of all analysed turnouts. The value itself fits very well into the picture under the assumption of a good ballast condition.

Thus, there is a evidence that the PDS is able to classify the ballast condition. The individual results have to be contrasted and compared with other data sources, such as the ground penetration radar, but an approximate trend is already apparent. In a further step, the individual power density spectra were again calculated and this time, the spectra after the re-investment were compared with those of the most actual measurement runs. It should thus be shown that a deterioration within the longitudinal level in the spectra or by the PDSIv is recognizable and that the presumption of the connection with the ballast condition is justified.

Figure 7 shows the comparison of the power density spectra for all considered turnouts, where a re-investment was done between 2004 and 2018. In blue, the PDS after re-investment, and therefore the PSD from a supposed good ballast condition, is shown. In red those of the last and thus most recent measurement run is presented.

Within the direct comparison for turnout 2 (Fig. 7—top left), it can be seen very clearly that in the wavelength range between 25 and 12.5 m, the current PDS of the actual measurement drive is significantly higher than the one after re-investment, whereas in the range between 12.5 and 7.5 m, there are no significant differences. A discrepancy can also be recognised at smaller wavelengths in the range below 7.5 m. If the two PDSIv's are compared, an increase of the value from originally 2.09 to 3.25 mm^2 in a period of approximately 11 years (thus an increase of approximately 0.11 mm^2/year) can be shown. Within the evaluation, shown on the top right, the same facts for turnout 3 are depicted. By comparing, it can be seen, that the two curves have nearly not changed within a period of about 9 years. Both curves have an identical characteristic and similar amplitudes. Even when looking at the PSDIv a minimal change of around 0.07 mm^2 (which corresponds to a change of about 0.008 mm^2 per year) can be noticed. Finally, the same analysis was performed for turnout 4 (Fig. 7 - bottom). This analyses show a significantly changed characteristic.

Also, the difference within the amplitudes can be seen. When looking at the PDSIv, a change of around 0.17 mm^2 per year can be seen. By doing more analyses for more turnouts, it should be possible to find a range within the PDS, which is related to the

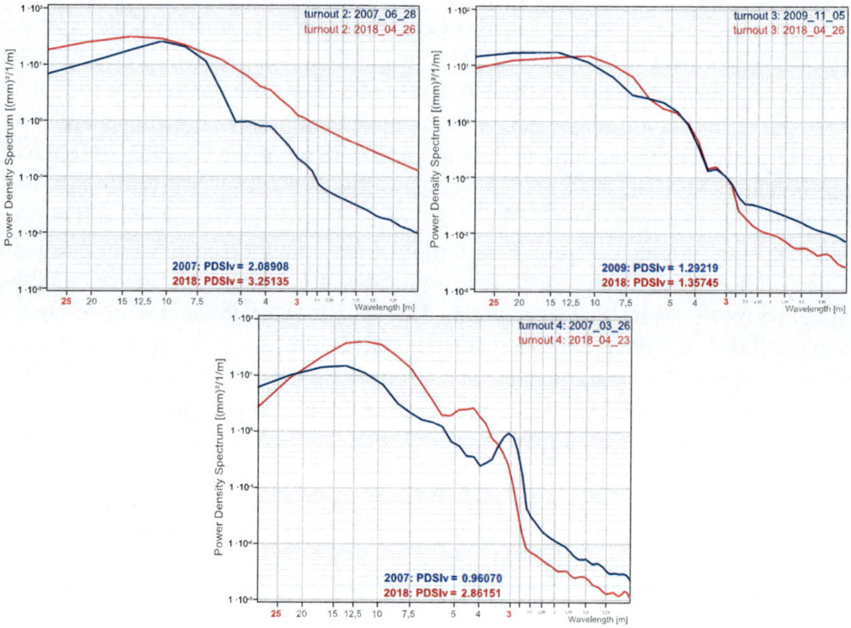

Fig. 7 Comparison of PDS for different turnouts and different measurement dates

actual ballast condition. Presumably, the degradation will be mainly in the higher wavelength ranges from greater than 10 m.

Of course, only a correlation of the power density spectra with the actual ballast condition, which is determined by ballast digging or ground penetrating radar measurements, provides information on how well these characteristics are suitable for describing the ballast condition or not. However, it seems to be a very good method to be able to make a statement about a component without a new measurement signal, which can be measured so far only with special measuring methods. The individual analyses have shown that the power density spectra have specific characteristics, depending on the construction of the turnouts. Within the comparison, a similarity in the order of magnitude between turnouts 2, 3 and 4 could be shown. Summarised, it seems that the power density spectra are capable to describe the ballast condition. Nevertheless, more detailed and in-depth investigations are still necessary.

7 Conclusion

The aim of this study is to show how a data-based condition monitoring for turnouts could be implemented, based on track recording car data. The track measurement data of the Austrian Federal Railways, available in time series since the year 2001, form the basis for this study. In order to derive information for turnouts out of this data, a complex post-positioning process is necessary taking into account the different relationships between the single measurement signals and thus enables an area identification within each signal. Only after applying this positioning process, it is possible to create reasonable evaluations for turnouts. Since the ballast and the sleepers could be identified as critical components for turnouts, an attempt was made to develop a methodology for describing the ballast condition and associated maintenance activities. For the description of necessary tamping activities, the standard deviation of the longitudinal level was used; calculated over a fix defined area within the turnout. For prior identification of critical areas, a moving standard deviation was calculated. With this method it was possible to identify the area in front of the turnout and the crossing panel as critical zones. Finally, power density spectrum calculations were used to check the condition of the ballast, based on the longitudinal level signal. Due to the lack of data concerning the actual ballast condition, further research activities are still necessary.

References

1. Zwanenburg WJ (2007) The Swiss experience on the wear of railway switches & crossings. In: 7th Swiss transport research conference
2. OeBB-Holding AG (2018) Bringing Austria together. OeBB compact 2017/18. Accessed 8 April 2019. https://konzern.oebb.at/file_source/corporate/presse-site/Downloads/Publikationen/
3. Veit P (2016) The economic service life of track. In: World congress on railway research (WCRR). https://www.sparkrail.org/Lists/Records/DispForm.aspx?ID=23608
4. Wilfling PA, Follinger M, Veit V(2019) Intelliswitch concluding book. Data sources and research model for turnouts, vol 1 Band. Springer Nature
5. Lakušić S (Hg.), Wilfling PA (2018) From manual inspection to permanent monitoring of turnouts. Zadar, Zagreb; University of Zagreb, Faculty of civil engineering
6. Nissen Arne (2009) Classification and cost analysis of switches and crossings for the Swedish railway: a case study. J Qual Mainten Eng 15(2):202–220
7. Nissen A (2009) LCC-analysis for switches and crossings: a case study from the Swedish railway network. Int J COMADEM 12(2):10–19
8. Molina LF; Resendiz E, Riley Edwards J, Hart JM, Barkan CPL, Ahuja N (2011) Condition monitoring of railway turnouts and other track components using machine vision. In: Proceedings of the transportation research board
9. Asplund M; Larsson D; Rantatalo M; Nissen A; Kumar U (2013) Inspection of railway turnouts using camera. In: World congress on railway research (WCRR)
10. Nissen A; Parida P; Kumar U (2010) Condition monitoring of railway switches and crossing by using data from track recording cars. Int J Cond Monitor Diagnostic Eng Manag 13 (July issue: ISSN 1363-7681)
11. khouy IA, Larsson-Kråik PO; Nissen A, Lundberg J, Kumar U (2014) Geometrical degradation of railway turnouts: a case study. Proc Instit Mech Eng [F] J Rail Rapid Transit 228(6)
12. Hovad E, Thyregod C, Andersen JF, Rodrigues AFDS; Ersbøll BK (2019) The effect of driving direction on spatially aligned track recording car measurements in turnouts. Int J Rail Transport
13. Minbashi Niloofar, Bagheri Morteza, Golroo Amir, ArastehKhouy Iman (2017) Geometrical degradation analysis of railway turnouts using power spectral density. J Transport Eng [A] Syst 143(8):4017035

14. Minbashi N, Bagheri M, Golroo A, Arasteh Khouy I (2016) Use of power spectral density for effective turnout geometry maintenance. Transport Res Record J Transport Res Board 2545 (1): 79–88
15. Xu P; Liu R; Sun Q; Jiang L (2014) Dynamic-time-warping-based measurement data alignment model for condition-based railroad track maintenance. IEEE Trans Intell Transport Syst:1–14
16. Landgraf M, Fabian H (2018) Fractal analysis as an innovative approach for evaluating the condition of railway tracks. Proc Instit Mech Eng [F] J Rail Rapid Transit 108:095440971879576
17. Landgraf M (2018) Smart data for sustainable Railway Asset Management. Railway track: assessment aggregation asset management. Graz: Verlag der Technischen Universität Graz (Monographic series TU Graz. Railway research)
18. Hansmann F (2018) The missing link between asset data and asset management (Monographic series TU Graz. Railway research)
19. Hanreich W, Mittermayr P; Presle G (2002) Track geometry measurement database and calculation of equivalent conicities of the OBB network. In: AREMA Washington DC 22–25 Sept 2002
20. Presle G (2000) The EM 250 high-speed track recording coach and the EM-SAT 120 track survey car. In: Rail engineering international, vol 29
21. Wen M, Li R, Salling KB (2016) Optimization of preventive condition-based tamping for railway tracks. Eur J Oper Res 252(2):455–465

Monitoring of Switches and Crossings/Tracks Using Smart Sensors

Elias Kassa, Jan Saramota, and Amund Skavhaug

Abstract Monitoring the condition of switch and crossings is a challenge for infrastructure managers (IMs). It requires specially trained switch inspectors to manually check the condition of the track. Hence, a fast and cheap system to continuously monitor the Switches and Crossings is a crucial matter for railway IMs. A methodology for continuous and real-time inspection of critical track infrastructures, such as switches and crossings, is developed using sensor technologies. This paper summarises the work done in developing a methodology to continuously monitor Tracks and Switches and Crossings, using inbuilt sensor technologies. The architecture of the sensor network developed contains communication platform to transfer data to a remote station in real time to follow up the status and condition of the track being monitored. A train-track interaction simulation tool developed to select suitable sensors architecture and communication, and to identify optimum sensor location along the track. The system was tested in a lab environment before the inbuilt technology was mounted in field demonstration. The system was installed in track infrastructure and the different stages and levels were tested in field. The monitoring system have been installed at a railway line in Germany to compare the vibration measurements with commercially available and expensive measurements as a validation exercise. The system has proved to have a potential for a smart failure and fault detection in rail infrastructure at network level.

Keywords Rail acceleration · Wireless vibration sensor · Switches and crossings · Condition monitoring

E. Kassa (✉)
Department of Civil and Environmental Engineering, Norwegian University of Science and Technology, 7491 Trondheim, Norway
e-mail: elias.kassa@ntnu.no

J. Saramota · A. Skavhaug
Department of Mechanical and Industrial Engineering, Norwegian University of Science and Technology, 7491 Trondheim, Norway

R. Galeazzi et al. (eds.), *Intelligent Quality Assessment of Railway Switches and Crossings*, Springer Series in Reliability Engineering,
https://doi.org/10.1007/978-3-030-62472-9_9

Executive summary
Monitoring of track geometric condition is performed by rail infrastructure managers (IMs) often using instrumented trains. Some IMs have a special train to inspect the track stiffness as a way of track condition monitoring. Continuous and more accurate monitoring of the track geometry and track stiffness by instrumented trains allowed the IMs to estimate the rate of track deterioration, to optimize track maintenance schedules and to replace only worn-out parts of the track. However, these vehicles are not fully applicable for condition monitoring of Switches and Crossings (S&Cs). Often the inspection at S&Cs is done manually by specially trained switch inspectors to check the condition of the track. This may lead to situations where just some parts of the rail track are inspected, longer track possession time, and decision is made based on many local measurements. Due to this fact, the whole S&C might often be replaced before the end of the life cycle. Therefore, there is a strong need for automated monitoring of S&C which requests a similar revolution as in assessment of geometric quality of plain line tracks.

This study aims to develop a methodology to continuously monitor tracks and S&Cs, using inbuilt sensor technologies. The key demand in this study is to create a system, which will be fast, cheap and provide continuous monitoring. Solution should be inbuilt and allow high integration into already existing infrastructure. System should provide real-time information about condition of the track to a remote maintenance centre. Communication system, data analysis and storage system should be part of the final solution. To fulfil all these demands, decision was made in this study to use as much of the already existing railway infrastructure as possible.

First, a train-track-substructure model was developed by coupling a Multi-Body-System dynamics tool Gensys with a Finite Element model in Abaqus. The analysis has been used to identify sensor location where acceleration signals vary less with respect to different variables and that satisfy the regulations by the railway administration. The rail acceleration is the main content of the track measurement, and it is used to evaluate the vibration characteristics of track structure or wheel-rail impact effect. Various position of the spatial rail will have different accelerations. The vibration analysis of the rail acceleration has been performed, considering track irregularity conditions, train speeds and contact patch location. Results of the rail acceleration under certain irregularity and different train speeds are presented to determine the feasible areas for sensor placement. Such results are used for continuously monitoring and inspection of the structural health of the rail and the supporting structures.

The second part of the study focuses on the development of the monitoring instrumentation. The methodology for continuous and real-time monitoring of S&Cs were developed. The system is based on battery powered sensor networks and a wireless communication system. Wireless sensors containing digital accelerometer and wireless module sense the vibration from the passing train for continuous monitoring and transmitting acceleration data to the network. Each sensor has own battery, that should power the device for many years. The important aspect of the sensor is ultra-low power consumption to achieve very long lifespan. Acceleration data are sent to the gateway and passed through GSM-R or WiFi to the dedicated server. The data are aggregated, analysed and stored.

Issues connected to outdoor environment were revealed during the first field test and were addressed during the second field demonstration. Due to much of the positive results observed, such as very cheap cost of the wireless sensor, reliability of data sampling, storage and transfer, this technology has a huge potential for the needs of IMs to monitor their track, and switch and crossing infrastructures.

1 Introduction

Monitoring of track condition is performed by rail infrastructure managers (IMs) using instrumented trains. The instrumented trains monitor the geometric quality of the track and some IMs have a special train to inspect the track stiffness as a way of track condition monitoring. Continuous and more accurate monitoring of the track geometry and track stiffness by instrumented trains allowed the IMs to estimate the rate of track deterioration, to optimize track maintenance schedules and to replace only worn-out parts of the track. However, these vehicles are not fully applicable for condition monitoring of Switches and Crossings (S&Cs).

There are a number of researches conducted in health monitoring of track infrastructures [1–4]. A survey of different railway monitoring applications using wireless techniques is conducted by Prabha and Chandana [5]. Victoria et al. [6] presented a review of the range of wireless sensor networks used for condition monitoring in railway sector. The review paper summarises list of sensor devises used for different fixed track infrastructures monitoring such as track, bridges, tunnels and for movable monitoring of vehicle components. A paper by Aboelela et al. [7] discussed an application of wireless network sensors for monitoring of ballast and subgrade material of a railroad system by using inclination sensors to predict track deflection. A railway track and bridge monitoring system was introduced by Bolle and Banoth [8] using wireless sensor networks based on ARM processor, which enable the collection and transmission of information through ZigBee and GPRS module. The presented system includes sensor nodes arrangement, collecting data, transmission method and emergency signal processing mechanism of the wireless sensor network. The system was designed to monitor a railway infrastructure for accident reduction and safety.

Existing methods and tools developed more than decades ago, to monitor geometric quality of a track give some insight to a real condition and quality of the track geometry. These methods and tools significantly help rail infrastructure managers to make informed maintenance decisions by understanding the real state risks of the track as well as its reduction in the operation performance. The structural health of the rail and the supporting structure therefore needs to be inspected and monitored continuously. Various position of the spatial rail will have different accelerations [9]. The rail acceleration is the main content of the railway dynamic test, and it is used to evaluate the vibration characteristics of track structure or wheel-rail impact effect. Remington [10] installed six acceleration sensors across the rail between two sleepers, and measured rail acceleration to predict wheel-rail noise.

This study is focusing on creating model infrastructure which will collect data from the sensors along the railway track and deliver these data to a dedicated server. Analysis and assessment of the data is not part of this study and will be treated in the next study. The main goals of this study are to develop methodology for continuous monitoring of switches and crossings; to develop system that is cheap, wireless, inbuilt, and allows real-time and continuous monitoring; to develop and install communication, data analysis and storage system; to test the system in lab environment; and lastly to demonstrate inbuilt technology at test sites.

2 System Architecture

Wireless sensor containing digital accelerometer, battery and wireless module sense the vibration from passing train. Acceleration data are sent to the gateway and passed through GSM-R or WiFi to the dedicated server. The data are aggregated, analysed and stored in the server. Afterwards, the status information with condition state is distributed to railway infrastructure and to the operating trains through GSM-R or programmable balise. Through this real-time communication, trains would get information of any defect on the track infrastructure long before approaching to the track. One advantage of this concept is utilization of many of the already existing infrastructure which leads to lower costs and better integration. The simplified infrastructure is shown schematically in Fig. 1.

2.1 *Wireless Sensor Network Hardware Platform*

The wireless sensors network are electronic devices that are attached to the rail or sleeper at different selected positions. The wireless sensor consists of electronic components to sampling and transmitting acceleration data to the network. Each sensor has own battery, that should power the device for many years. The important aspect of the sensor is ultra-low power consumption to achieve very long lifespan. The sensors must be cheap, small and easy to install. Significant precaution has been taken to the development of the casing. The sensor will have to operate in adverse weather environment for several years and it has to withstand very cold winters and very warm summers. Any damage or water ingress would lead to losing the sensor.

Fig. 1 Concept of monitoring system

2.1.1 Accelerometer

The digital accelerometer continuously monitor vibrations on the rail and once the values exceed the set vibration level indicating the incoming train, the sensor wakes-up the rest of the electronic. Woken microcontroller switches the accelerometer to high precision measurement mode to record the samples. Once the level of acceleration drops under a specified level after a train has passed, the sensor will set back to low-power measuring mode. The rest of the electronic will finish the tasks and fall asleep. Due to this, the accelerometer spends most of the time in low-power measuring mode waiting for the incoming train. It is crucial to choose accelerometer that has very low-power consumption especially in this mode.

The main concern was to select proper magnitude of gravity and sampling frequency. There are several working teams and many articles where the rail vibrations are sensed by highly sensitive instrumentation. Chong et al. [11] monitor rail vibration signals as a remote measurement system using device measuring the acceleration ±150 g and a sampling frequency up to 10 kHz. The magnitude in some cases come up to ±250 g and sampling sometimes reach dozens of kHz. The main goal of this task is however to investigate possibility to analyse condition of the rails with limited low-cost and low-power sensors. Chraim and Puttagunta [3] have been able to record usable signals using a ±2 g sensor attached to the rail web capable to sample on 1.8 kHz. For this study an accelerometer Analog Devices ADXL313 was selected.

2.1.2 Battery

To power up the wireless sensors for several years, special battery cell had to be used. Most of the classic batteries would self-discharge in few years even if they are not used. Therefore, the primary lithium-thionyl chloride batteries invented by Adam Heller were picked. These batteries are well suited to extremely low-current long-term applications where long life is a must. Output voltage necessary for proper function of all the components needs to be in range 3–3.3 V. Especially in cold winters the voltage of power cell can drop down under 3 V. Therefore, the Buck-Boost DC/DC convertor LTC3335 was chosen to deliver stable voltage on an output.

2.1.3 Wireless Module

Wireless communication is connection between two or more subjects realized without help of wires, cables or any other forms of electrical conductors. The carrier medium can be optical (light), radio (radio frequencies) or ultrasonic (sound). The communication distance can be between few meters and thousands of kilometres.

For this specific study, it is expected that there will be 192kB of data from each sensor for every train pass. This situation occurs when all 3-axes sample at 3,2 kHz for about 10 s. All these data need to be transmitted over hundred meters from the sensors placed on a track to the gateway. Alvi et al. [12] proposed an adaptive Time Division Multiple Access (TDMA) based Medium Access Control (MAC) protocol to carry out communication and to reduce the control overhead and to make the wireless sensor network energy efficient. The most suitable solution to transmit the data is to use some of the embedded low-power RF modules that perform well in given environment and can be relatively cheap, small and reliable. With so many standards and different protocols it is not easy to find the proper balanced solution especially for battery powered application. Comparing the pros and cons of each solution, the IQRF platform was chosen as the most appropriate. RF communication is performed in SRD-860 band on stated frequencies:

- RF Channel 1 52 (868.35 MHz with 20 kHz spacing)
- RF Channel 2 63 (869.45 MHz with 20 kHz spacing)

2.1.4 Printed Circuit Board (PCB)

The schematic circuit layout is shown in Fig. 2 and the printed circuit board (PCB) is shown in Fig. 3. The PCB is manufactured in two batches, the first ten boards as a prototype and second batch in bulk quantity for full scale tests.

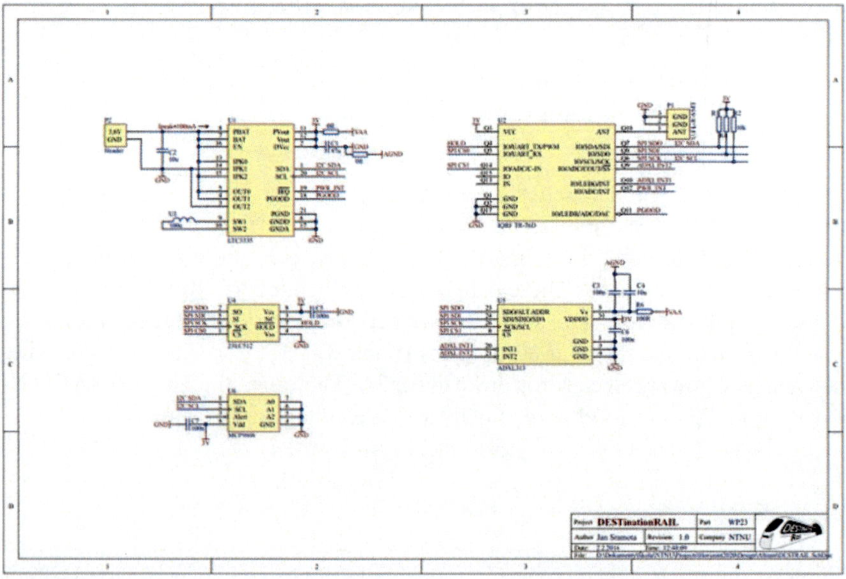

Fig. 2 Schematic circuit layout

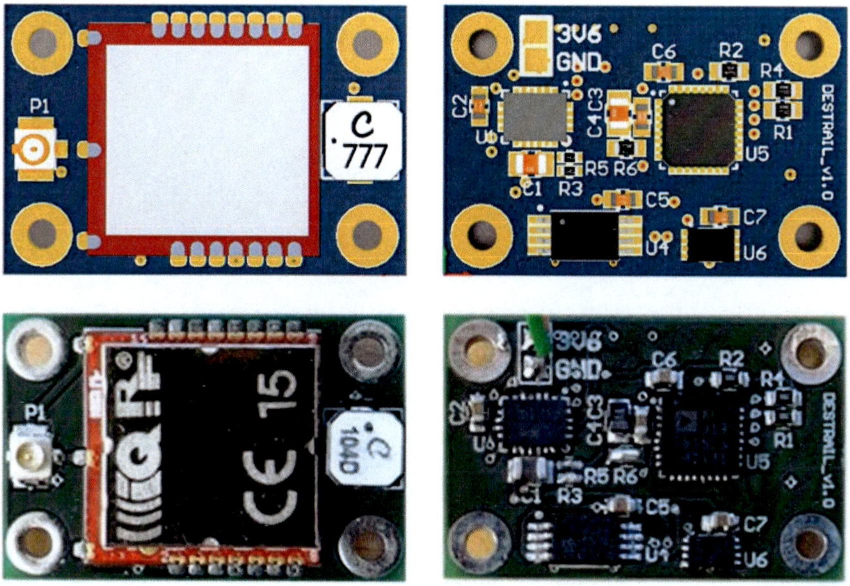

Fig. 3 Printed Circuit Board (PCB) simulation (top), manufactured board (bottom)

2.1.5 Casing

Design of wireless sensor casing is one of the most crucial and easily most underestimated part in many cases. However, faulty design of the wireless sensor casing can cause serious issues and can significantly affect performance of the system. Designers in similar projects often tend to focus on electronic (HW/SW) part even though one dysfunctional element in the chain will cause the entire system not working at all. Therefore, the design of the product and the choice of material has been done considering the required strength, external environment, weight of the casing, water tightness, and with due care to accommodate all the components in optimised space. Further, the recommendation of the numerical modelling work [13] and the suggested location of the wireless sensor on the rail is also considered [14]. Finally, the rail IMs regulation with regard to free clearance and structure gauge was considered based on the discussion with the Danish and Norwegian Rail Administrations.

There was concerns from railway infrastructure managers about the railhead position and the size of the sensor. Placing anything on railhead is generally considered as a bad practice. Therefore, it was decided to create as small casing as possible and place it on the rail web, rail foot and the sleeper, see Fig. 4. On the first design version, the battery was placed separately in a battery casing that will be buried into the ballast and connected to the sensor by cable. In the second design version, the acceleration sensor, battery and antenna were placed into one single casing to achieve compact design, see Fig. 4. Antenna was placed as far from conductive and ground plane as possible and into horizontal position. This should help to have signal less affected

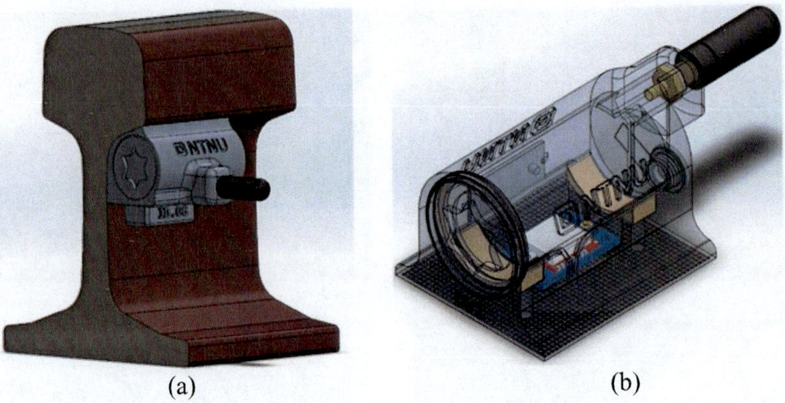

Fig. 4 Selected sensor casing 2nd version to be mounted at different parts of the track. **a** sensor placement on a UIC60 rail web, **b** sensor to be placed on sleeper

by conductive materials that are all around and should help with signal propagation to all direction around the transmitting position.

2.2 *Power Consumption of Wireless Sensor*

The wireless sensor is expected to have a very long operational lifetime to be cost-effective. Even though the power consumption of each electronic component used in wireless sensors design is known and the working conditions together with average self-discharge rate of the battery can be estimated, the real power-consumption will always vary on many of the parameters. The amount of data that needs to be transmitted, changes in temperature, and the level of optimization are some of the significant factors for such wide variation. As a brief guidance for the expected consumption, Table 1 shows the calculated expected data. However, the real revealing data will be obtained from a long-term monitoring.

3 System Evaluation and Validation

This section discusses the feasibility test and the whole system assessment from field test. Problems raised during the development of each parts of the system and suggested modifications are also presented.

Table 1 Power consumption of wireless sensor

The case of study		Results	
Trains per day	10	Battery capacity	7700 mAh
Data per train	192 kBytes	Consumption per year	1 746 mAh
RF datarate	2,5 kBs	**Lifetime**	**4,41 years**

			Sleep	Process	RX LP	TX STD	Total
IQRF TR-76D	time	[min]	1 404	1	21	14	
	current	[μA]	1	1 600	233	19 000	
	per day	[μAh]	23	27	83	4 496	4 628 μAh
	per year	[mAh]	9	10	30	1 641	1 689 mAh

			Sleep	Wait	Meassure	Total
ADXL313	time	[min]	1 338	100	1,667	
	current	[μA]	0,1	43	170	
	per day	[μAh]	2	72	5	79 μAh
	per year	[mAh]	1	26	2	29 mAh

			Sleep	Operating	Total
23LC512	time	[min]	1 440	0,016	
	current	[μA]	2	2 000	
	per day	[μAh]	48	1	49 μAh
	per year	[mAh]	18	0	18 mAh

			Sleep	Operating	Total
MCP9808	time	[min]	1 440	0,005	
	current	[μA]	0,1	200	
	per day	[μAh]	2	0	2 μAh
	per year	[mAh]	1	0	1 mAh

			Sleep	Operating	Total
LTC3335	time	[min]	1 438	2	
	current	[μA]	0,68	300	
	per day	[μAh]	16	10	26 μAh
	per year	[mAh]	6	3	9 mAh

3.1 *Feasibility Test of Concept*

The preliminary feasibility test of concept was performed during the first stage of the task. This was realized to evaluate the crucial components and their parameters. The main concerns were the electronics' capability to sense the incoming train, the accelerometer response on an overpassing train and the ability to transmit acquired data over the radio frequencies. Before the system was deployed at a field test at Tommerup test site in Denmark, it was fully tested for several days in a lab and outdoor environment.

As the first trial, the accelerometer's sampling frequency was set to 400 Hz while the magnitude of acceleration was set to $\pm$ 2 g (1024 values/LSB). It was equipped by four magnets with a thrust force 7 kg each and placed on the rail web with total aggregated force 28 kg, as shown in Fig. 5.

Measuring was triggered automatically by incoming train once the level of acceleration crossed over the specified threshold level. Data were automatically transmitted

Fig. 5 Passenger train passing over the test point (left), ADXL313 Evaluation Board attached on rail web (right)

over the whole structure into the C Sharp application running on a computer. The application recorded these data into the RAW file and displayed them in a chart.

3.2 *Antenna Test*

To be able to use the gateway outside of controlled environment e.g. faraday cage, all parts involved in signal propagation had to be measured in an antenna lab. This allowed to properly set the output constants and verify that not more than 500mW ERP will be radiated from the antenna. This is regulated by legislation to prevent RF interference between various systems, that might be dangerous especially in the railway system.

3.3 *Rail Coverage Test*

Rail coverage test was held on a track at Norwegian national railway infrastructure (Bane NOR). Sensors were deployed at 30, 60 and 90 m distance from the gateway that was attached on a pole of height 1.3 and 2.2 m, see Fig. 6. The main purpose of this test was to test the reliability of the communication between sensors and gateway. The sensor casing was also tested with regard to the usable distance and transmission properties.

The test was conducted with snow coverage of 10 cm layer, which didn't tend to signal damping. Further, passing trains during the active transmission didn't seem to interfere with the active communication. Reliable communication was achieved during the whole test with signal strength above −62 dBm (with gateway on 2.2 m pole).

Fig. 6 Antenna coverage field-test with sensors covered by snow

Fig. 7 System test in Munich—sensors mounted on rail and sleeper

3.4 Field Tests

A measurement campaign was carried out on one of the main railway lines around the Munich region. Frequency of trains at rush hours was around four to five trains in an hour for a single direction. The selected test site is a plain line track. Sensors were mounted at two locations to measure the acceleration parameters at the railway web and sleeper. Each of the test positions was equipped by the new developed sensors and a reference sensor which is known to give reliable data (see Fig. 7). The reference sensors are measuring single axis and were sampling on 25.6 kHz.

3.4.1 Sensor System Validation

During the one-day measurement test on the track, data were collected from all the sensors. About 14 train passages were recorded. Sensors were positioned at rail web (P1, P2), whilst two sensors were positioned at the sleeper (P3, P4). Reference sensors were also positioned at the rail web (inside the rail) and on the sleeper. The data from reference sensors is used for a comparison and validation of the sensors developed in this project. For data processing purpose, data from passing trains 4, 7 and 15 (time order) is processed. Records from other trains show comparable results.

Data records from train passage 4 using the reference sensor are shown in Fig. 8. Figure 9 shows the acceleration data recorded by the new sensors (DESTination-RAIL sensors). Vertical acceleration on the rail web over-leaped sensor g-range, therefore the output went into the saturation as shown in Fig. 9. Measurement of acceleration on the sleeper recorded signals within the range for all three axes.

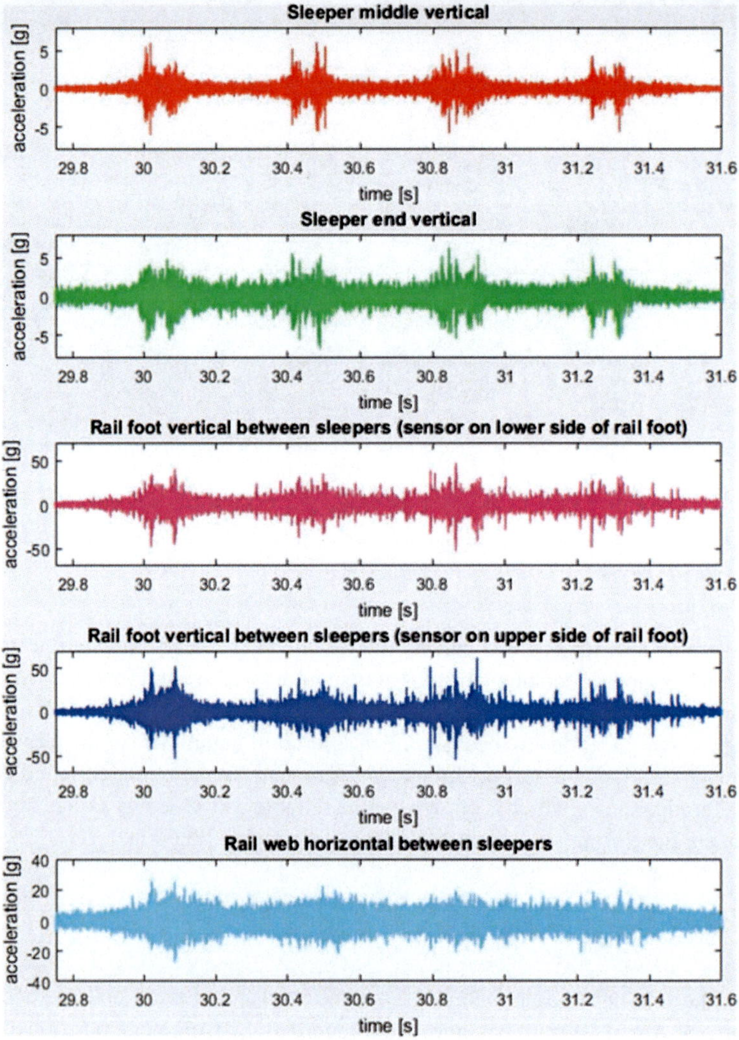

Fig. 8 Reference records from train passage 4

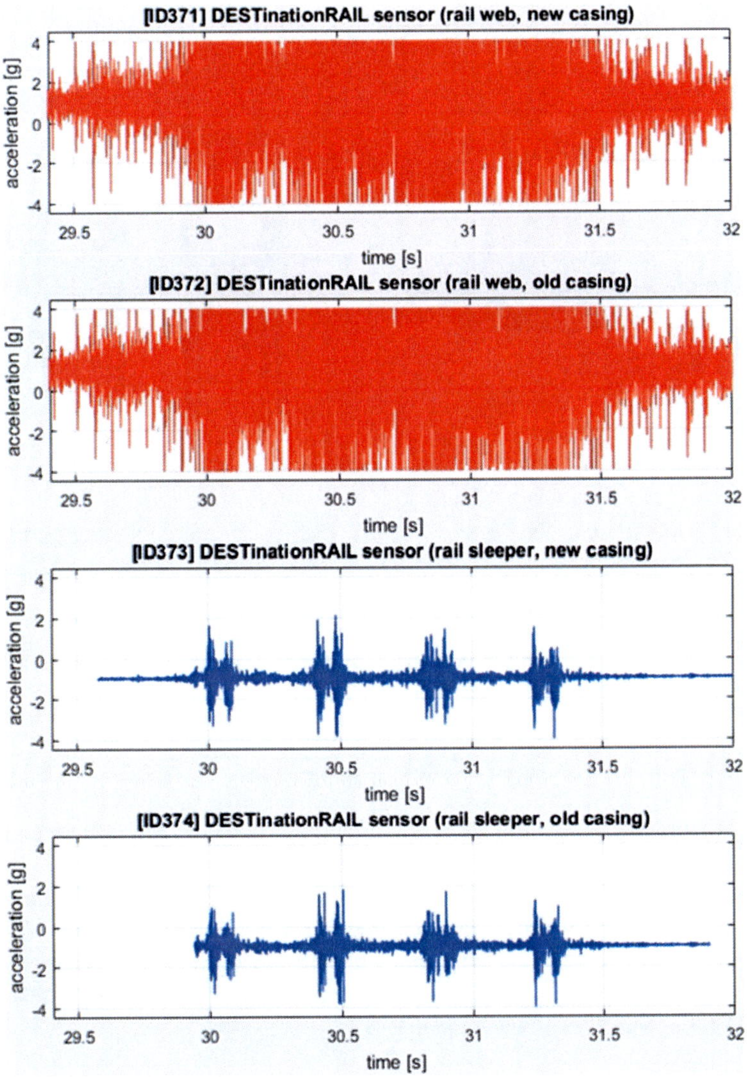

Fig. 9 DestinationRAIL records from train passage 4

The reference sensors were sampling on 25.6 kHz, so they contain 8 times more data than the new sensors that were sampling just on 3.2 kHz. For comparison purposes, the data from the two sensors were overlaid over the reference data (without any down-sampling), as shown in Fig. 10. Figure 11 shows the signal measured at the sleeper by the reference sensor and the new sensor.

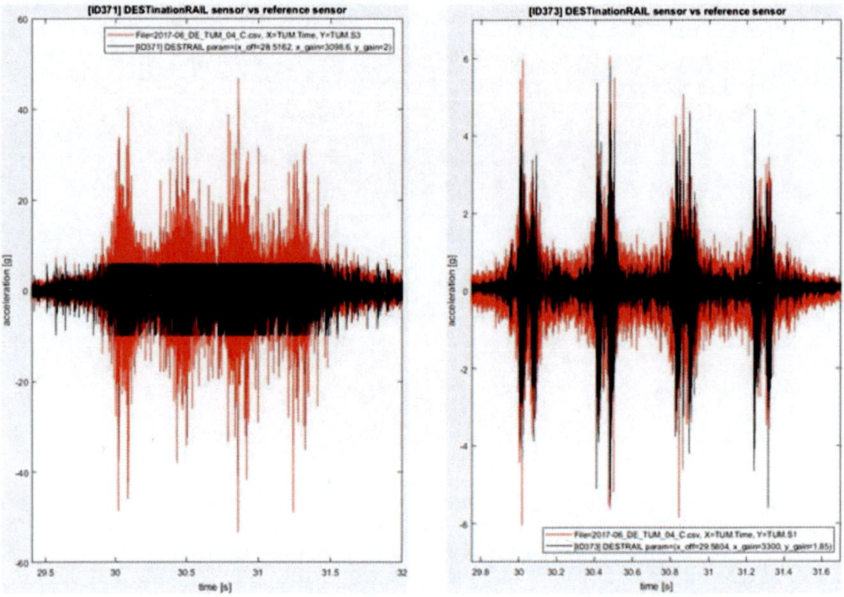

(a) (b)

Fig. 10 Comparison with reference sensor for train passage 4: **a** sensor at the rail web with new casing, and **b** sensor at the middle of sleeper with new casing

3.4.2 Measuring Abnormal Events

During the test day several train passes were measured. Some of the abnormal events from few trains pass with possible defects were also detected by the measured signal. Figure 12 shows the normal (healthy) train pass with distinct peaks for all axle passes. Figure 13 shows signals for two trains suspected with some form of fault. This was possible to notice during the test from the noise and vibration of these two specific train passes. These two specific trains (or one train passing twice) have either some defects on the wheels or in the suspensions.

4 Conclusions

This study introduces affordable low-cost technology for continuous monitoring of critical track infrastructure such as switches and crossings. Moreover, the system is able to be deployed to gather other telemetry from various sensors along the rail as well. The structure can be easily extended by another type of sensors allowing e.g. monitoring of landslides, train telemetry or any other low-amount type of data traffic.

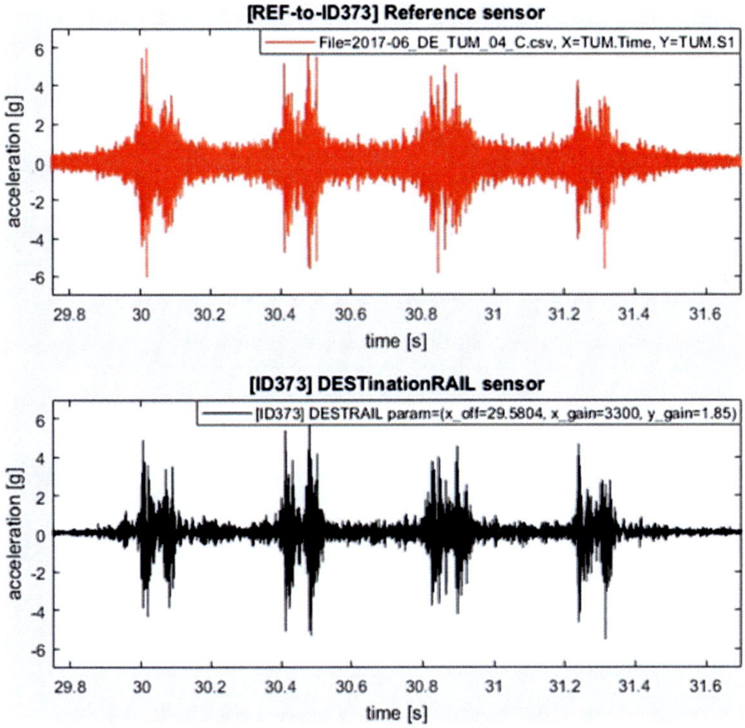

Fig. 11 Acceleration measured at the middle of the sleeper using: (top) reference sensor, (bottom) newly developed sensor

In general, wireless sensor network consists of sensors (wireless sensor), gateway and cloud (WCMS). System is very scalable and may be easily integrated into GSM-R network by switching the GSM modem to GSM-R version. To ensure, that the gateway is in accordance with the legislation, the whole gateway was tested in an antenna lab. During this the crucial parameters were set and validated against reference antenna in the controlled environment.

The current ± 4 g sensor can be used for measuring higher accelerations up to about ± 8 g. The sensors performed very well on the sleeper where 100% of the signals recorded were within the range. The biggest drawback was saturation of the sensor attached to the rail web. Either the sensor become overloaded and didn't record any usable data or the sensors recorded data that were saturated on the vertical axis and within the range for longitude and lateral axes. For future deployment of accelerometers on the rail, replacing the sensor with higher range is suggested. The other drawback was problem with desynchronization when the signals were compared against the records of the reference one. DESTinationRAIL sensor has less precise clock and therefore produced more or less than ideal 3200 samples per second. This should be subject for a future work.

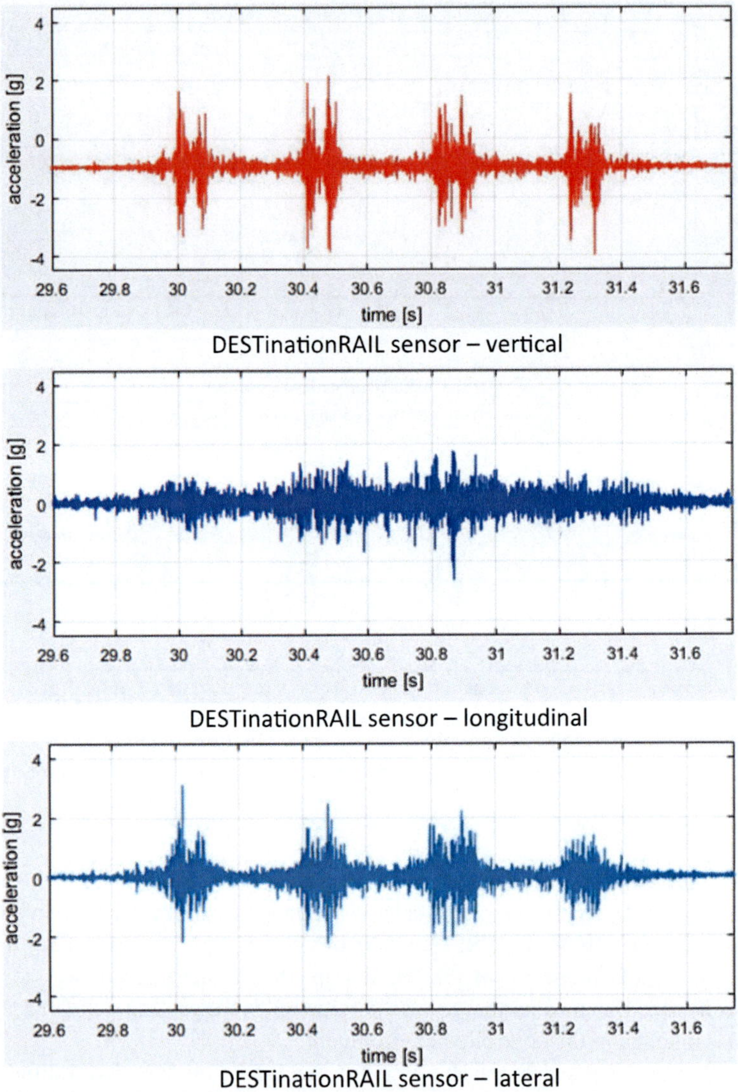

Fig. 12 Acceleration measured at the middle of the sleeper for a healthy train pass

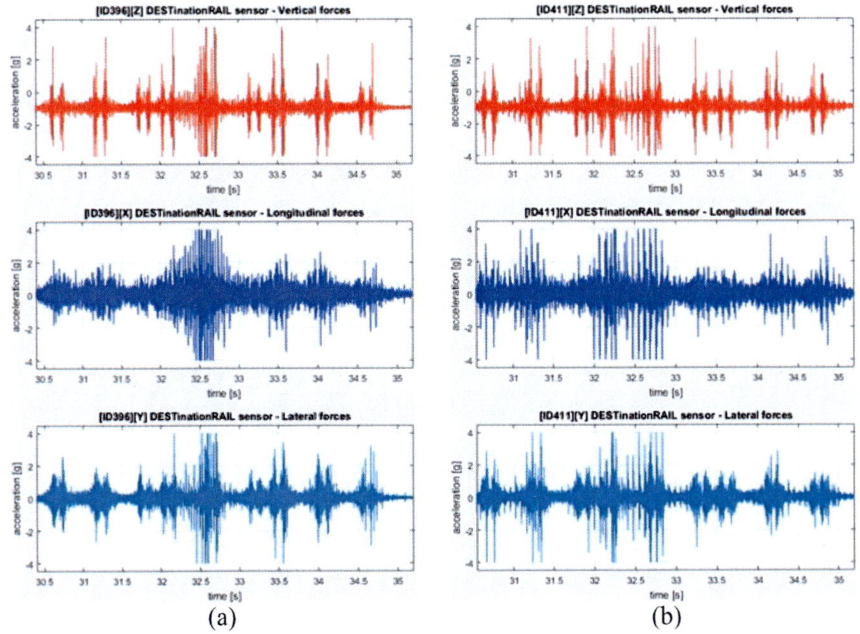

Fig. 13 Acceleration measured at the middle of the sleeper for: **a** faulty train 1, and **b** faulty train 2

Acknowledgements This work was part of the EC funded project DESTinationRAIL (No. SEP-210181866). Several project members have contributed in this study.

References

1. Barke D, Chiu WK (2005) Structural health monitoring in the railway industry: a review. Struct Health Monitor 4(1):81–93
2. Lee ML, Chiu WK (2007) Determination of railway vertical wheel impact magnitudes: Field trials. Struct Health Monitor 6(1):49–65
3. Chraim F, Puttagunta S (n.d.) Monitoring track health using rail vibration sensors. Academia
4. Tsunashima H, Naganuma Y, Matsumoto A, Mizuma T, Mori H (2010) Condition monitoring of railway track using in-service vehicle. J Mech Syst Transport Logis 3:154–165
5. Prabha R, Chandana DK (2015) Survey on different railway monitoring applications using wireless MAC protocols. IJLTEMAS IV(II):22–25. www.ijltemas.in
6. Hodge Victoria J, O'Keefe Simon, Weeks Michael, Moulds Anthony (2015) Wireless sensor networks for condition monitoring in the railway industry: a survey. IEEE Trans Intell Transp Syst 16(3):1088–1106
7. Aboelela E, Edberg W, Papakonstantinou C, Vokkarane, V (2006) Wireless sensor network based model for secure railway operations. In: IEEE International performance computing and communications conference (IEEE 2006), pp 623–628. https://doi.org/10.1109/.2006.1629461
8. Bolle V, Banoth SK (2016) Review on railway bridge and track condition monitoring system. Int Res J Eng Technol (IRJET) 03(08):1092–1095

9. Bracciali A, Cascini G (1997) Detection of corrugation and wheelflats of railway wheels using energy and cepstrum analysis of rail acceleration. Proc Instit Mech Eng [F] J Rail Rapid Transit 211(2):109–116
10. Remington PJ (1987) Wheel/rail rolling noise, ii: validation of the theory. J Acoust Soc Am 81(6):1824–1832
11. Chong L, Jiahong W, Zhixin Z, Junsheng, Tongqun R, Hongquan X (2016) Design and evaluation of a remote measurement system for the online monitoring of rail vibration signals. Proc IMechE [F] J Rail Rapid Transit 230(3):724–733
12. Alvi AN, Bouk SH, Ahmed SH, Yaqub MA, Javaid N, Kim D (2015) Enhanced TDMA based MAC protocol for adaptive data control in wireless sensor networks. J Commun Netw 17(3):247–255. https://doi.org/10.1109/jcn.2015.000046
13. Kassa E, Cai XP (2019) Modelling of the dynamic train-track-substructure interaction for track monitoring
14. Cai X, Kassa E (2016) Detection of acceleration sensitive areas of rail using a dynamic analysis. In: Proceedings of the 3rd international conference on railway technology: research, development and maintenance. https://doi.org/10.4203/ccp.110.275

Labelling the State of Railway Turnouts Based on Repair Records

Georgios Vassos, Emil Hovad, Pavol Duroska, Camilla Thyregod, André Filipe da Silva Rodrigues, and Line H. Clemmensen

Abstract Turnouts are the most expensive part to maintain on the railway track and therefore automated systems for detecting turnout defects are of great interest. Machine learning can improve predictive maintenance and is often used in automatic systems for precise prognosis. In this study, machine learning is used for identifying the condition of railway turnouts and potentially reducing costs by early automatic detection of defects. To train a machine learning algorithm, ordered, structured and categorized data (labelled data) are needed. A method is proposed to label the condition of turnouts in the Danish Railway based on a collection of repair records. This labelling of the turnouts is accomplished with unsupervised methods, namely a principal component analysis (PCA) followed by a cluster analysis. The labelling of the turnouts is investigated through comparisons of geometric measurements captured from the recording car. The difference in the physical properties illustrated by the geometric data indicates that the labelling is a good indicator of the relative condition of the turnout. When the data are labelled, supervised learning can be used to optimize the predictive power of machine learning algorithms (i.e. the algorithm learns from the labelled data) for classification of turnouts.

Executive Summary

Turnouts are the most expensive part to maintain on the railway track due to its labor intensiveness and therefore automated systems for detecting turnout defects are of great interest. In this study machine learning is used to identify the condi-

G. Vassos · E. Hovad (✉) · P. Duroska · C. Thyregod · L. H. Clemmensen
Department of Applied Mathematics and Computer Science, Technical University of Denmark, 2800 Kgs Lyngby, Denmark
e-mail: emilh@dtu.dk

C. Thyregod
e-mail: camt@dtu.dk

L. H. Clemmensen
e-mail: lkhc@dtu.dk

A. F. da Silva Rodrigues
Banedanmark, 2100 Copenhagen, Denmark
e-mail: AFSR@BANE.dk

R. Galeazzi et al. (eds.), *Intelligent Quality Assessment of Railway Switches and Crossings*, Springer Series in Reliability Engineering,
https://doi.org/10.1007/978-3-030-62472-9_10

tion of railway turnouts and potentially reduce costs by early automatic detection of defects. We take a more holistic approach than previous works and consider turnouts in their entirety rather than only single parts by using repair records (the stored repair information) as well as tamping and grinding gathered for several turnouts of the same geometric type. This study finds a suitable labelling of the turnouts from the collected multidimensional repair records, which gives information about their need for maintenance. The labelling of the turnouts is investigated with geometric measurements captured from the recording car and this labelling is shown to be a good indicator of the relative condition of the turnout. This is a first step towards using full scale machine learning algorithms (a system) through labelled turnout data. Machine learning systems can improve predictive maintenance and is often used in automatic systems for precise prognosis. To train a machine learning algorithm, ordered, structured and categorized data (labelled data) are needed and this is provided here.

1 Introduction

Maintenance and renewal of railway components consume a large part of the financial resources of the railway industry. Moreover, maintenance of turnouts takes a great portion of the budget al location in a railway business, implying that improvements in preventive maintenance of railway turnouts could be highly lucrative.

In the following, we attempt an overview of various techniques used to confront problems on different parts of the rail infrastructure. Some approaches model the behavior of turnouts based on simulating the forces in the wheel-rail contact in turnouts using a Finite Element (FE) model as in [1], or as in [2] where a 3-D FE model simulated a wheel rolling over a crossing rail and the model was validated with experiments. An example of simulating the ballast tamping with the Discrete Element Method (DEM) can be found in [3], where the ballast's vibration characteristic is related to the shape of the ballast particles.

Turnouts receive high mechanical loads and are therefore very prone to damages. Dynamic loads on turnouts were investigated through the use of field experimentation and failure statistics in [4]. The first part of the turnout is named switch blade region and consists of a movable part, called the point machine, which sets the direction of the train towards the straight or divergent direction and needs regular maintenance. Efficient fault detection and diagnosis via data mining techniques, using audio data from the point machine, is carried out in [5]. The switch blade region is further susceptible to problems stemming from wear degradation, which impacts the gauge alignment of the switch rail, or Rolling Contact Fatigue (RCF) on the crowns of the switch and stock rails [6]. The crossing panel is the last part of the turnout, where alignment is important for proper transition of the train wheel to the crossing nose. The alignment of the wheel that passes the crossing depends upon the direction of train. Surface damage due to wear and RCF, on the crossing nose, during the wheel transition to the crossing panel, was examined in [7] with the use of a dynamic finite element model. A three-dimensional vehicle-track interaction model was proposed in [8] to investigate the rail-pad stiffness in the crossing panel (crossing panel support structure) in order to decrease the ballast settlement from the ballast pressure. The crossing nose can also be severely damaged by the repeated passage of trains which can lead to the formation of surface or sub-surface cracks that may result in rail failure [9]. A wide range of general and more specific turnout problems are discussed in [10], while limits for maintenance and safety for turnouts in Denmark can be found in [11, 12]. Eventually, the need for tamping and grinding in the railway maintenance is explained in [15].

Preventive maintenance of railway turnouts is essential for managers in the railway industry to minimize repair costs, while at the same time ensuring the functionality and high quality of railway services. Great savings can be accomplished by applying predictive maintenance based systems through automatic condition based monitoring [13]. Data collected with the track recording car consists of a set of variables that are related to the geometry of the rails; standards for acceptable values of such variables, on a conventional track, are listed in [14]. In Denmark, the track recording car runs every three to six months, the exact date is called the campaign date, depending on the placement and usage (tonnage) of railway tracks, to monitor the rail tracks. A first step towards integrating condition monitoring of turnouts using track recording car data was performed in [16]. Condition assessment of conventional track with Statistical Process Control (SPC) was proposed in [17], where limits on the twist variables of the track recording are suggested. Track recording car data on railway turnouts was used in [18] for turnout ballast prognosis. Turnout degradation studied in [19] the rarity of reliable data, posing a potential limitation of modelling efforts.

In this study, we follow a more holistic approach and consider both the repair as well as tamping and grinding records, gathered from 166 turnouts of the same geometric type, throughout multiple campaigns. The goal of this study is to find a suitable labelling of the turnouts, using the collected multidimensional repair and tamping/grinding records, that can indicate the need for maintenance. To accomplish this we utilize unsupervised methods, in particular Principal Component Analysis (PCA) and Cluster Analysis, in order to group the turnouts as being degraded or not (high need for maintenance or low). Such machine learning methods are thoroughly explained in [20]. We first investigate the clusterability of the data and obtain indications of two clusters which we then identify with the k-means clustering algorithm and finally validate on the track recording car data. Our results illustrate differences in the quality of the two groups of turnouts, namely a separation between a "degraded cluster" and a "not degraded cluster". The treatment of the track recording car data in this study is based on [21] which unifies the data to a track recording car passing straight in a left-handed turnout data format. This greatly reduces the noise in the data, caused by temporal and spatial dependencies of the forces interacting at the train-turnout intersection when travelling trough a turnout.

2 Data

Data were collected from 166 turnouts with similar geometric design and components. The two data sets used in this analysis are (1) repair records and (2) track recording car measurements:

(1) The first data set contains SAP data (with manual repair information) and tamping/grinding data; we combine these two data sources in a single table to which we refer as repair records.
(2) The second data set contains track recording car measurements on the geometry from the turnouts.

2.1 Repair Records: Combined Repair and Tamping/Grinding Data

The first data set contains counts of the number of repairs (denoted `rps`) or tampings and grindings (denoted `TnG`) that took place at each turnout in different time windows, i.e. 125, 245 and 365 days, *after* a recording car measurement campaign, as illustrated in Fig. 1.

Table 1 contains the input data, before any clustering is done, captured from 166 different turnouts over 17 different campaigns dates. The unique combinations of 166 turnouts at the 17 campaign dates gives 1123 observations overall.

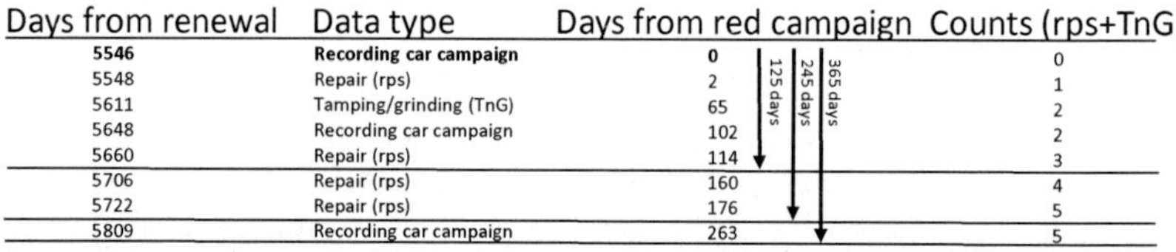

Fig. 1 An illustration of the three time windows for 125 days, 245 days and 365 days after the first campaign date shown. The repair records are counted in the respective time windows

Table 1 Repair records: Combined repair (`rps`) and tamping/grinding data (`TnG`) for 166 different turnouts (`Turnout`) with multiple campaigns totalling 1123 observations

Variable	Count data range	Median	Mean
`rps_125`	[0;14]	1	1.01
`rps_245`	[0;17]	2	2.01
`rps_365`	[0;18]	2	3.00
`TnG_125`	[0;3]	0	0.06
`TnG_245`	[0;3]	0	0.14
`TnG_365`	[0;3]	0	0.24

The repair on the turnout can be due to various reasons, e.g. fixing problems with the point machine, the crossing nose, loose bolts and fasteners etc. More detailed explanation of the turnout repairs, turnout maintenance limits and safety limits can be found in [10–12]. Tamping is performed to compact the ballast (the ballast is the foundation of the track) and to reorient the ballast stones to re-align the track, thus the ride becomes more pleasant for the passengers. Grinding is performed for smoothing the rail in presence of corrugation or other types of unevenness.

2.2 Recording Car Measurements

The recording car measures the features-outlined in Table 2. They describe the track geometry of 167 turnouts during multiple campaigns. `Turnout` is encodes the 166 different turnouts placed in Denmark, `km` is the relative position on the track which in our case is defined based on the Turnout, `Date` is the dates of the campaigns when the recording car measurements are captured inside the turnout, `Days` is the number of days past since the renewal date of the turnout, `Level_R_D1` and `Level_L_D1` are the level variables (vertical measurements) for the right and left rail, respectively, measured in short wavelength and similarly for the alignment measurements (horizontal measurements) `Alignment_R_D1`, `Alignment_L_D1`. `Gauge` represents gauge measurements, `Cant` is the level difference of the left and right rail and finally `Twist` is the measured twist at miscellaneous distances. A detailed description of the recording car data and its treatment can be found in [21], where the data format is unified to that of a track recording car travelling straight in a left-handed turnout. Further details about the track recording car geometric variables are elabo-

Table 2 Recording car information and variables (adapted from [21])

Variable	Description
`km`	From the first joint of switch blade panel
`Date`	The date of measuring campaign
`Days`	Days after renewal day of the turnout
`Level_R_D1`	Level data right short wavelength (3–25)
`Level_L_D1`	Level data left short wavelength (3–25)
`Alignment_R_D1`	Alignment data right short wavelength (3–25)
`Alignment_L_D1`	Alignment data left short wavelength (3–25)
`Gauge`	Gauge measurement between the left and right rail
`Cant`	Level difference of the left and right rail
`Twist.2m`	Change of cant (2 m)
`Twist.3m`	Change of cant (3 m)
`Twist.6m`	Change of cant (6 m)
`Twist.9m`	Change of cant (9 m)
`Twist.12m`	Change of cant (12 m)
`Twist.15m`	Change of cant (15 m)

rated in [14, 15]. The turnout geometry has a length of 43 m, while a campaign refers to the specific date when the track recording car is monitoring one or more turnouts.

Due to missing observations in the long range data, i.e. level data $D2$ and the alignment data $D2$, as well as the long wavelength (25–70) exceeding the length of a turnout (43 m), the long wavelength variables are discarded. The range of the spatial data km (relative position on the turnout) is selected from -10 m (before the first joint) to 53 m (10 m after the end of the crossing panel). This 10 m extension is to account for the possibility of anomalies in the recording car behavior when exiting the turnout, as well as for small misalignments that could be present in the spatial data [16, 21].

3 Methods

This section introduces the theory behind proposed method. First principal component analysis, which projects data into a low dimensional space is introduced, subsequently k-means clustering along with metrics for deciding the number of clusters are introduced. Finally, we describe local regression, which is used to illustrate the differences in the geometric measures for the obtained clusters.

3.1 Principal Component Analysis

Correlations between variables in the combined repair and tamping/grinding data as well as in the recording car measurements are investigated and Principal Component Analysis (PCA) is performed to reduce the dimensionality of the data.

PCA is a projection method which maps data points from an input space $\boldsymbol{X} \in \mathbb{R}^{N\times p}$ to a new lower dimensional space $\boldsymbol{Z} \in \mathbb{R}^{N\times d}$ with $d < p$. Furthermore PCA is an unsupervised/exploratory method since its objective is to maximize the variance.

Suppose $\boldsymbol{w}_1$ is the first principal component and $\boldsymbol{\Sigma}$ the covariance matrix of $\boldsymbol{X}$. If $\boldsymbol{X}$ is projected onto $\boldsymbol{w}_1$ then the spread between data points will be maximum [22]. The first principal component is the orthogonal eigenvector of $\boldsymbol{\Sigma}$ which corresponds to the largest eigenvalue, λ_1. Similarly the remaining principal components are the eigenvectors corresponding to the decreasing eigenvalues, $\lambda_1 > \lambda_2 > \cdots > \lambda_d$.

Usually only the leading k principal components that explain a sufficient proportion of the total variance are considered for further analysis. The latter proportion can be calculated as $(\lambda_1 + \cdots + \lambda_k)/(\lambda_1 + \cdots + \lambda_d)$ [22], where λ_i is the i-th eigenvalue, in decreasing order, of $\boldsymbol{\Sigma}$.

In the new input space, spanned by the k leading principal components of $\boldsymbol{X}$, each of the p features is a linear combination of these k principal components, $\alpha_1 \boldsymbol{w}_1 + \cdots + \alpha_k \boldsymbol{w}_k = (\alpha_1, \ldots, \alpha_k)$. Two measures useful for interpreting the results from PCA are described in [23]. The squared cosine, derived as $(\alpha_1^2, \ldots, \alpha_k^2)$, is used to indicate the quality of representation of some feature on the principal dimensions, while the contribution of each feature to some principal component is given as $\alpha_i / \sum_{j=1}^{k} \alpha_j$.

3.2 Clustering

Cluster analysis aims to divide the data points into groups, which are called clusters, such that those data points within some cluster are more similar than data points between different clusters [20].

3.2.1 K-Means Clustering

We choose to apply the k-means algorithm due to it's simplicity, low complexity and wide application. The goal of this algorithm is to split the N data points into K clusters such that the within cluster overall average distance between observations and the cluster center is minimized [20]. The squared Euclidean distance is used to calculate the distance between observations and cluster centers. K-means [24] assigns observations to cluster centers based on the distance. A cluster is represented by the center of all data points that belong to it and an observation will be assigned to the cluster with smallest distance to its cluster center. The k-means convergence is typically defined as when the clusters do not re-assign any observations between each other.

3.2.2 Clusterability and the Number of Clusters

It is worth noting that clustering methods will always partition the data, even if it is uniform. Therefore certain steps must be executed to ascertain the clustering tendency of the data before proceeding with applying the actual clustering. In addition after obtaining the results from clustering, validation must be performed to assess the quality of the produced clusters.

In this study the Hopkins statistic [25] is applied to assess the clusterability of the data and this is a statistical method for testing the hypothesis of uniformly distributed data.

Three internal measures are used for cluster validation and for finding the optimal number of clusters, namely the connectivity, Dunn and silhouette index. Connectivity reflects the extent to which data points are placed in the same cluster as their nearest neighbors. The Dunn index is given as the minimum between cluster (minimum separation) divided by the maximum within cluster (maximum diameter) pairwise distance. Compact and well separated clusters are expected to indicate large values of the Dunn index. The average silhouette method (silhouette index) [26] is used to derive the optimal number of clusters. Briefly, this method evaluates the average silhouette for a varying number of clusters. A good clustering will exhibit high average silhouette value, hence the optimal number of clusters is the one that maximizes the average silhouette. The silhouette of a data point is defined as:

$$s_i = \frac{b_i - a_i}{\max(a_i, b_i)}$$

where b_i is the minimum average distance of data point i to all data points inside those clusters to which data point i does not belong, while a_i is the average distance between data point i and the rest data points within the same cluster.

3.3 *Local Regression (Loess)*

A quadratic spline (smooth) approximation of each recording car variable is plotted as a function for the found clusters. Local regression (loess) is originally discussed in [27] as a non-parametric method for fitting regression models locally in some narrow range $(x - h_\lambda(x), x + h_\lambda(x))$ sometimes defined as a constant called span $h_\lambda(x) = \lambda$.

The kernel terminology used in [20] is applied to explain the local polynomial regression. Suppose N points (x_i, y_i) $\forall i \in \{1, ..., N\}$ are given and the goal is to estimate a point (x_0, y_0). To estimate y_0 based on x_0 using the observations x_i by weighing the distances to x_0 according to the Kernel:

$$K_\lambda(x_0, x_i) = D\left(\frac{|x_i - x_0|}{h_\lambda(x_0)}\right)$$

with $D(t)$ being the tricube weight function

$$D(t) = \left(1 - |t|^3\right)^3 \; if \; |t| \leq 1. \; Otherwise \; D(t) = 0$$

The y_0 value can be approximated by a any polynomial:

$$\min_{\alpha(x_0), \beta_j(x_0), j=1,\ldots,d)} \sum_{i=1}^{N} K_\lambda(x_0, x_i) \left[y_i - \alpha(x_0) - \sum_{j=1}^{d} \beta_j(x_0) x_i{}^j \right]$$

minimizing the sum of squares to estimate the coefficients α and β_j where we in this study use a quadratic $d = 2$. The solution is:

$$\hat{f}(x_0) = \hat{\alpha}(x_0) - \sum_{j=1}^{2} \hat{\beta}_j(x_0) x_i{}^j$$

The *local regression* method described above is used to illustrate the features, by cluster, of the railway turnouts measured by the recording car at different campaigns. A span of $h_\lambda(x_0) = \lambda = 0.1$ is applied for calculating an approximation of the the recording car variables using the statistical software R implementation [28] with

95% confidence band, similarly to the turnout application in [21]. Due to the large number of observations sampling is applied and for this purpose measurements from each cluster is sampled using 15,000 observations.

4 Results

First we describe the labelling/clustering into high and low maintenance need based on historical repair data (from the repair records) then we validate the labels through the track recording car data.

4.1 Labelling Through Historical Repair Records

The labelling method first considers the representation of the repair records through PCA and is followed by clustering which groups the repair records.

4.1.1 Principal Component Analysis of the Repair and Tamping/grinding Data

A correlation plot between variables of the combined repair and tamping/grinding data is presented in Fig. 2. It is apparent that repair counts are highly correlated and the same holds for tamping/grinding counts.

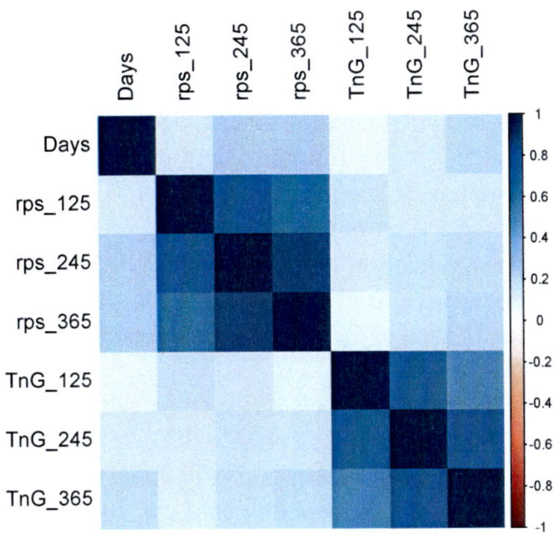

Fig. 2 The correlation of the days variable, the repair variables and the tamping/grinding variables

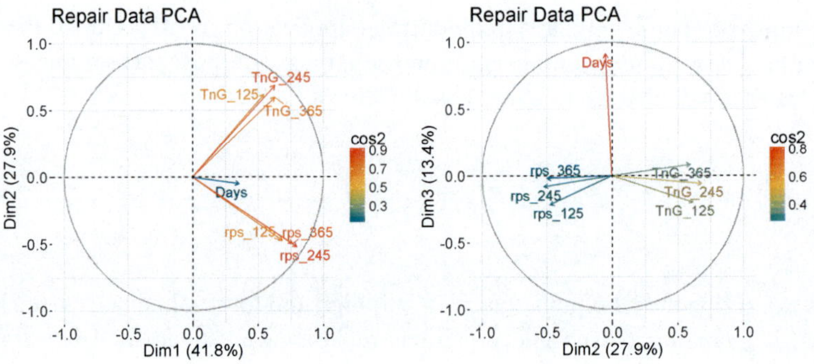

Fig. 3 Results from PCA on the repair and tamping/grinding data and its three principal components

PCA is performed on the combined repair and tamping/grinding data to reduce the dimensionality of the feature space. The projection of the initial variables to the plane spanned by the first three principal components can be found in Fig. 3. The first three principal components explain almost 85% of the total variance, while repair and tamping/grinding counts are mostly represented in the two first principal directions, as they are closer to the unit circle. The first dimension (Dim1) consist mostly of the repairs and the tampings/grindings variables. The second dimension (Dim2) separates the repair variables in the negative directions versus the tampings/grindings variables in the positive direction. The third dimension (Dim3) consist mostly of the Days variable (days after the turnout renewal date) with the repairs and the tampings/grindings variables going in the opposite direction.

4.1.2 Clustering of the Repair Records

The first three principal components of the repair records (the combined repair and tamping/grinding data) explains almost 85% of the data. Therefore, only the first three principal components of the repair records are used for cluster analysis. The aim is to obtain a (cluster) label for each pair of `Turnout` and campaign (`Date`).

Hopkin's statistics is calculated to 0.04 which falls in the rejection region (below 0.05), asserting non-uniformity and therefore clusters could be present in the data. Cluster validation with Connectivity, Dunn and Silhouette indexes are demonstrated for k-means clustering in Table 3. It is seen that all cluster validation measures suggest the existence of two clusters and thereby two clusters are applied for the following analysis.

Summary statistics are given in Table 4 of the two clusters where Cluster 1 has less observations namely $N = 177$ (N considers the unique combination of turnout and date), more repairs and tamping/grinding as compared to Cluster 2.

Figure 4 illustrates the results of this method with the two clusters. In general a clear separation can be seen of the two clusters in the first dimension (Dim1).

Table 3 Number of clusters (from 2 to 6) based on the three selected cluster measures for the k-means algorithm

Validation method	2 clusters	3 clusters	4 clusters	5 clusters	6 clusters
Connectivity	4.36×10	6.38×10	7.00×10	9.64×10	1.13×10^2
Dunn	1.13×10^{-2}	5.40×10^{-3}	5.30×10^{-3}	3.50×10^{-3}	4.40×10^{-3}
Silhouette	5.39×10^{-1}	5.16×10^{-1}	5.01×10^{-1}	4.13×10^{-1}	4.15×10^{-1}

Table 4 Summary statistics of the two clusters repair records. Cluster 1 has N = 177 observations recorded in 66 different turnouts (considering the unique combination of both turnout and date) and cluster 2 has N = 946 observations recorded in 165 different turnouts

Cluster 1 (N = 177)	Mean	St. Dev.	Min	Max
Days (age)	4,630	2,491	174	16,561
rps_125	2.147	2.081	0	14
rps_245	4.164	3.095	0	17
rps_365	5.684	3.957	0	18
TnG_125	0.395	0.623	0	3
TnG_245	0.780	0.650	0	3
TnG_365	1.006	0.687	0	3
Cluster 2 (N = 946)	*Mean*	*St. Dev.*	*Min*	*Max*
Days (age)	3,091	3,058	3	17,110
rps_125	0.798	0.840	0	4
rps_245	1.605	1.186	0	8
rps_365	2.471	1.566	0	10
TnG_125	0.000	0.000	0	0
TnG_245	0.022	0.147	0	1
TnG_365	0.093	0.298	0	2

Cluster 1 has more observations in the positive direction of the first dimension which suggest having older turnouts with more repairs and tampings/grindings occurring as compared to Cluster 2. This tendency is further shown in the second dimension (Dim2) where Cluster 2 is relatively more densely centred around zero again indicating that cluster 1 has more more repairs or tampings/grindings occurring. The third dimension (Dim3) is relatively similar for the two clusters although Cluster 2 has a small community of observations which are older. These observations from the small community come from two specific turnouts. The cluster frequency for the individual turnouts is discussed and illustrated in the appendix Sect. 5 together with a discussion of the small community observations.

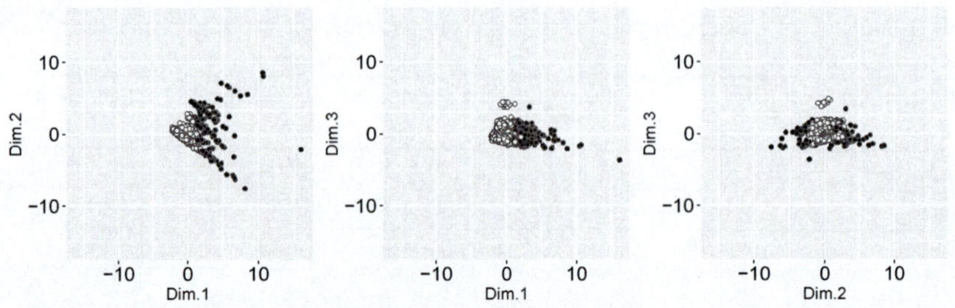

Fig. 4 k-means cluster result with 2 clusters. Cluster 1 observations are white circles (black edge) and Cluster 2 observations are full black circles

4.2 Validation of the Labelling Through the Track Recording Car Data

The proposed labels are validated through the track recording car data. First, the recording car data are investigated and as a result the track recording data are reduced in dimensions with PCA. Then the labelling is validated through plotting the two cluster groups on the recording car data (labelling of the recording car data).

4.2.1 Principal Component Analysis of the Recording Car Data

Correlations between the variables from the recording car data are shown in Fig. 5. It is revealed that the two `Level` variables are highly correlated. The six `Twist` variables also appear to be considerable positively correlated. The two alignments are slightly negatively correlated.

Due to the high correlations between the twist variables in the recording car, a PCA is performed on the six `Twist` variables. The two first principal components describe more than 85% of the total variance of all `Twist` and therefore these two principal components are used for further investigations.

4.2.2 Validating the Labelling Through the Recording Car Data

Summary statistics of the recording car for the two clusters are given in Table 5. The standard deviations are the largest in Clusters 1 for all the recording car variables which validates that this cluster has a poorer conditions and thus a higher need for maintenance.

The recording car variables are plotted against the relative position (km) on the turnout for the two clusters using Loess. The plots of individual features (variables) it is illustrated that a relative differences between the two clusters can be observed in Figs. 6 and 7.

More specifically Fig. 6a and b illustrate how the level measurements at short wavelength in Cluster 1 greatly oscillate at the crossing nose (at the gray vertical

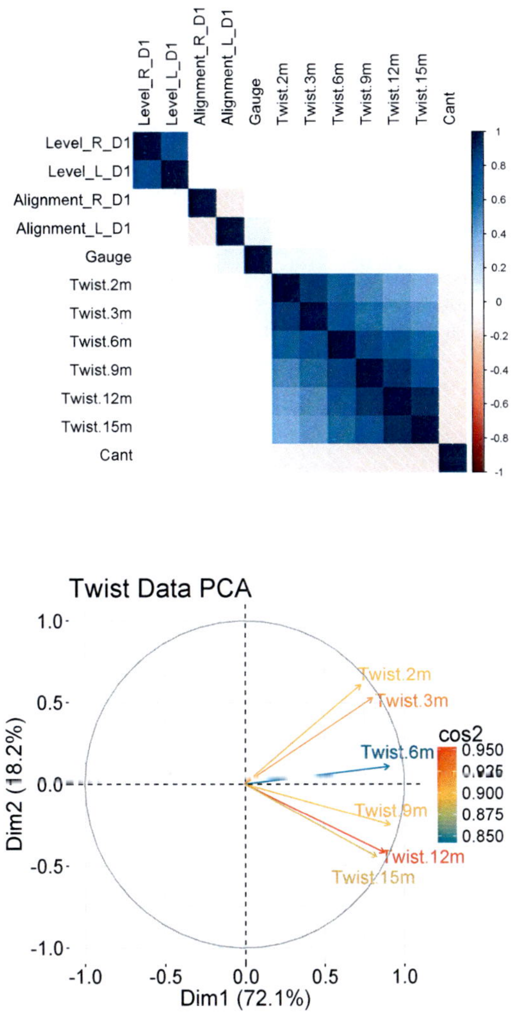

Fig. 5 The top figure shows the correlation of the recording car measurements. The bottom figure shows the results from the PCA analysis of the recording car data with respect to the twist variables

line) compared to those in Cluster 2. The alignment measurements, see Fig. 6c and d, oscillate substantially at the end of the turnout (last black vertical dotted line), and the behavior of the two clusters differs both at the beginning and the end of the turnout.

Smooth approximations of gauge and cant measurements against the relative position on the turnout are illustrated in Fig. 7a and b. Cluster 1 exhibits overall greater gauge values and furthermore has a sharp increase close to the end of the turnout. Cant measurements for Cluster 1 seem to have larger oscillations than those in Cluster 2. Finally, the principal components of the twist measurements are plotted in Fig. 7c and d. The essential information here is that both clusters exhibit some deviation around zero.

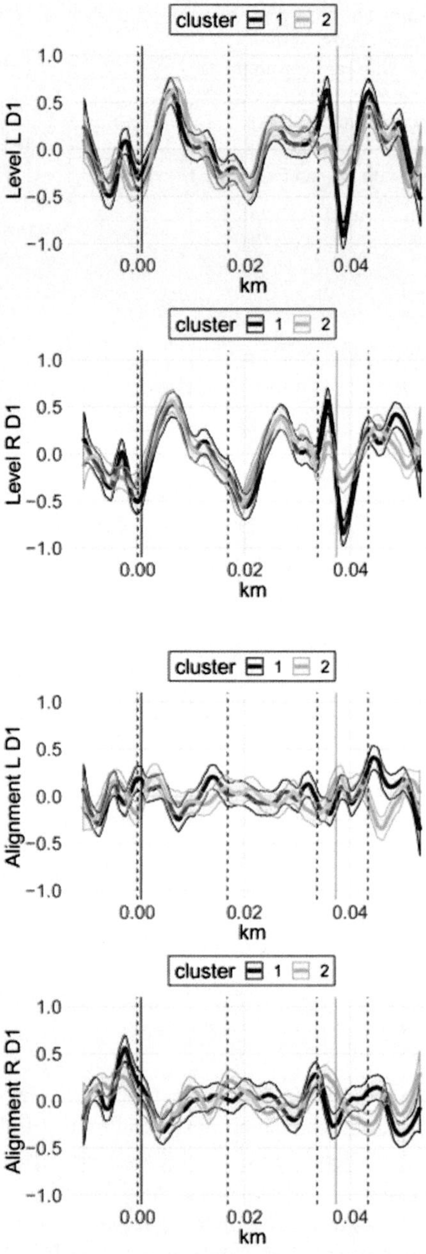

Fig. 6 Smooth approximations of `Level` and `Alignment` as a function of `km` and coloured by cluster. The turnout is divided into three regions by the black vertical dotted lines where the left region is the switch blade panel, the middle region is the closure panel and the right region is the crossing panel. The black vertical line on the left is the tip of the point machine and the gray vertical line on the right is the crossing nose position

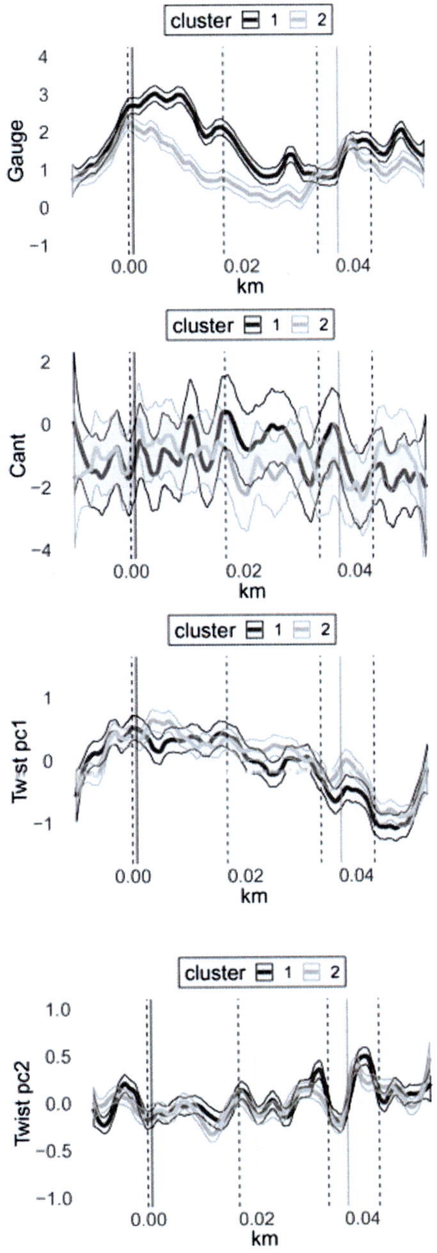

Fig. 7 Smooth approximations of Gauge, Cant and Twist as a function of km and coloured by cluster. The turnout is divided into three regions by the black vertical dotted lines where the left region is the switch blade panel, the middle region is the closure panel and the right region is the crossing panel. The black vertical line on the left is the tip of the point machine and the gray vertical line on the right is the crossing nose position

Table 5 Summary statistics of the recording car data for the two clusters

Cluster 1 (N = 44,452)	Mean	St. Dev.	Min	Max
Days (age)	4,625	2,483	174	16,561
Level_R_D1	0.014	1.612	−12.850	10.230
Level_L_D1	−0.003	1.690	−19.530	10.200
Alignment_R_D1	−0.008	1.572	−10.660	11.250
Alignment_L_D1	0.004	1.586	−37.150	11.520
Gauge	1.662	2.592	−7.540	44.690
Cant	−0.962	13.970	−92.420	50.550
Twist_pc1	−0.092	2.393	−13.550	16.289
Twist_pc2	0.020	1.131	−9.757	10.282
Cluster 2 (N = 237,903)	*Mean*	*St. Dev.*	*Min*	*Max*
Days (age)	3,090	3,056	3	17,110
Level_R_D1	−0.0003	1.476	−21.400	9.920
Level_L_D1	−0.008	1.543	−15.200	10.160
Alignment_R_D1	0.008	1.365	−12.850	15.620
Alignment_L_D1	0.006	1.521	−25.590	42.030
Gauge	1.002	2.480	−33.160	51.370
Cant	−1.260	13.724	−93.320	71.130
Twist_pc1	0.017	2.015	−15.805	17.249
Twist_pc2	−0.004	1.029	−7.723	8.552

5 Conclusions

Based on historic information of repairs and tamping/grindings it is possible to label a turnout's future need for maintenance. In particular the turnout data are split into two clusters; one with a higher maintenance activity than the other cluster. The essence of this study is that by counting the number of repairs and tampings/grindings for each turnout within certain time windows after the relative campaign dates, it is possible to obtain a new data set in which two clusters are present, while further investigation of the geometric track measures indicated that one of the clusters shows problematic behavior whereas the other shows non-problematic behavior. As no clear boundary exists between the two cluster borders and thus there is underlying continuous nature rather than discrete when moving from one cluster to the other, and this means that around the border of the two clusters it is hard to assign a specific cluster label correctly.

Cluster 1 shows higher counts of repairs and tamping/grindings up to a year after the recording car campaign, which indicates this cluster needs more maintenance as compared to cluster 2. This labelling can potentially be used to classify a turnout's future need for maintenance (one year ahead) based on current recording car data.

When plotting the recording car data for the two clusters, a clear differences are visible for most of the recording car variables. Cluster 1 generally shows larger peak values and larger fluctuations, which indicates a less pleasant ride for the passenger and potentially less safe as well.

Appendix: Clustering

Clustering Frequency for the Different Turnouts

The general cluster frequency for the turnouts can be seen in Fig. 8. Fewer turnouts are typically assigned to Cluster 1, but some turnout has a high frequency of Cluster 1. Typically a high frequency in Cluster 1 belongs to turnouts which are exposed to

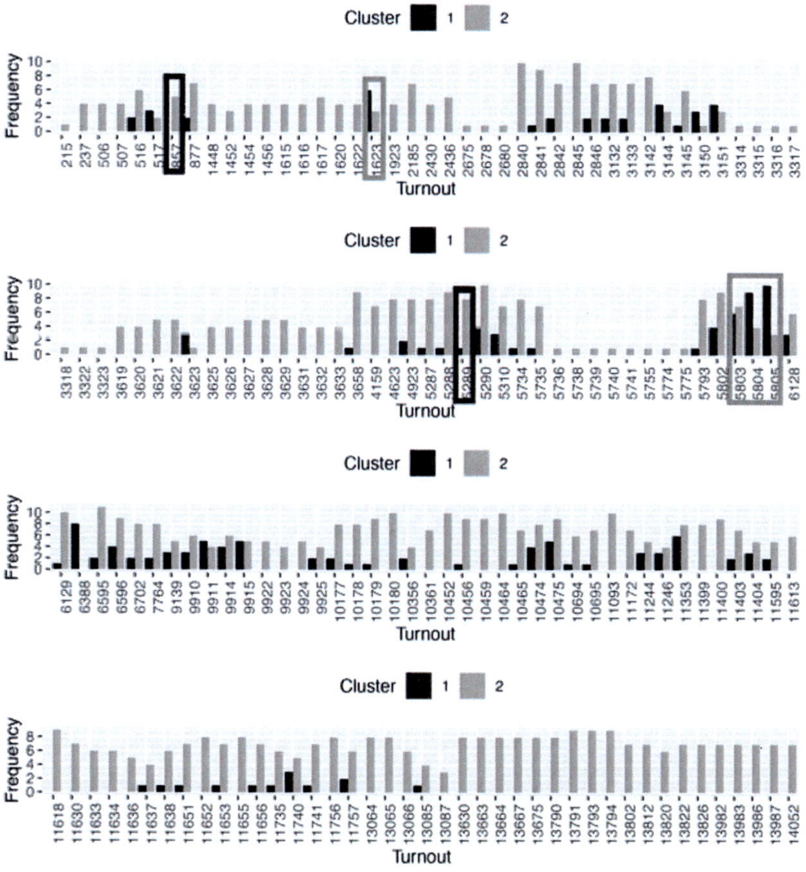

Fig. 8 Labelling frequency of all observations by turnout. The gray boxes shows turnouts with a high frequency of maintenance and the black boxes shows the two oldest turnouts having a low frequency of maintenance

a larger tonnage (used more) and thereby more frequent maintenance is needed for these turnouts. Specifically it can be seen that the turnouts with the numbers of 1623, 5803, 5804 and 5805 (shown with the gray boxes) are more frequently assigned to Cluster 1.

Cluster 2 small community which could be seen in Fig. 8, the small community was especially separated in the third dimension (Dim.3). This community consist of two older turnouts with the numbers of 857 and number 5289 (shown with the black boxes) having campaigns captured in the period of 15,145 to 17,110 days after the renewal day. These two turnouts only had a few repairs and tamping/grindings take place in the defined time windows.

References

1. Kassa E, Nielsen JCO (2009) Dynamic train-turnout interaction in an extended frequency range using a detailed model of track dynamics. J Sound Vibr 320(4–5):893–914
2. Ma Y, Mashal AA, Markine VL (2018) Modelling and experimental validation of dynamic impact in 1:9 railway crossing panel. Tribol Int 118:208–226
3. Wang X, Chi Y, Li W, Zhou T (2012) Study on the dem simulation of the granular railway ballast bed tamping. Adv Mater Res 524–527(4):3256
4. Cornish A, Kassa E, Smith R (2012) Field experimentation and analysis at switches and crossings in UK. In: 9th International conference on contact mechanics and wear of rail/wheel systems, CM 2012, pp 649–651
5. Lee J, Choi H, Park D, Chung Y, Kim HY, Yoon S (2016) Fault detection and diagnosis of railway point machines by sound analysis. Sensors 16(4):549
6. Nielsen JCO, Palsson BA, Torstensson PT (2016) Switch panel design based on simulation of accumulated rail damage in a railway turnout. Wear 366:241–248
7. Pletz M, Daves W, Ossberger H (2012) A wheel set/crossing model regarding impact, sliding and deformation-explicit finite element approach. Wear 294:446–456
8. Grossoni I, Bezin Y, Neves SGM (2016) Optimization of support stiffness at a railway crossing panel. In: Civil-comp Proceedings, 110 p
9. Dhar S, Zhang Y, Xu R, Danielsen HK, Juul Jensen D (2017) Synchrotron x-ray measurement of residual strain within the nose of a worn manganese steel railway crossing. In: IOP conference series: materials science and engineering, 219(1), 012016 p
10. Schmitt L, Létourneaux F, De Keyzer I, Crompton P (2016) CAPACITY4RAIL: towards an affordable, innovative and high-capacity European Railway System for 2030/2050. Mater Infrastruct 2(5B):105–114, Wiley Blackwell
11. Bo Nielsen NFN (2008) Lars Forsting, Banenorm bn1-14-2: Projektering, tilstand og eftersyn af sporskifter og sporskæringer, 64 p
12. Klaus Bergman NF (2008) Lars Forsting, Banenorm bn2-15-2: Projektering, etablering, tilstand og vedligeholdelse af sporskifter og sporskæringer, 31 p
13. Jensen PJ, Fongemie R (2016) Sporteknik: danske jernbanespors opbygning, konstruktion og komponenter. Banedanmark
14. Stanislav Jovanović, Dragan Božović, Mirjana Tomičić-Torlaković (2014) Railway infrastructure condition-monitoring and analysis as a basis for maintenance management. Hrvatski savez gradjevinskih inženjera, Cm 2012:347–358
15. CEN (2003) Railway applications—track–track geometry quality Part 1: Characterisation of track geometry. European Committee for Standardization (CEN); 2003 13848-1. https://standards.globalspec.com/std/13275308/en-13848-1

16. Nissen A, Parida A, Kumar U (2010) Condition monitoring of railway switches and crossing by using data from track recording cars. Int J Comadem 13(3):36–41
17. Bergquist B, Soderholm P (2015) Data analysis for condition-based railway infrastructure maintenance. Qual Reliab Eng Int 31(5):773–781
18. Seyed Asadzadeh M, Galeazzi R, Hovad E, Fink Andersen J, Thyregod A, Camilla FS (2018) Rodrigues, Ballast degradation modeling for turnouts based on track recording car data. In: Proceedings of the european conference of the PHM society, , Vol 4 no 1
19. Kumar U, Ahmadi A, Verma AK, Prabhakar V (2016) Current trends in reliability, availability, maintainability and safety, current trends in reliability, availability, maintainability and safety. Springer International Publishing
20. Friedman J, Hastie T, Tibshirani R (2017) The elements of statistical learning, 2nd edn. Springer series in Statistics, New York
21. Hovad E, Thyregod C, Fink Andersen J, Christian BL, Peter Spooner M, Rodrigues AFS, Kjær Ersbøll B (2019) The effect of driving direction on spatially aligned track recording car measurements in turnouts. Int J Rail Transport 1–15, Taylor and Francis Ltd
22. Alpaydin E (2009) Introduction to machine learning. MIT press
23. Kassambara A (2017) Practical guide to principal component methods in R: unsupervised machine learning, vol 1. STHDA
24. MacQueen J (1967) Some methods for classification and analysis of multivariate observations. In: Le Cam LM, Neyman J (eds) Proceedings of the fifth berkeley symposium on mathematical statistics and probability, vol 1, pp 281–297. University of California Press, Berkeley, CA
25. Hopkins B, Skellam JG (1954) A new method for determining the type of distribution of plant individuals. Ann Bot 18(2):213–227
26. Kaufman L, Rousseeuw PJ (1990) Finding groups in data: an introduction to cluster analysis. Wiley, New York
27. Cleveland WS, Devlin SJ (1988) Locally weighted regression: an approach to regression analysis by local fitting. J Am Stat Assoc 83(403):596–610
28. R Core Team (2017) R: a language and environment for statistical computing. R Foundation for Statistical Computing, Vienna, Austria

Automatic Detection of Rail Defects from Images

Emil Hovad, Helena Hansen, André Filipe da Silva Rodrigues, and Vedrana Andersen Dahl

Abstract In this study, images of the rails captured by a track recording car are investigated. The images are processed in order to asses and predict the rail quality by a simple method, which gives the possibility of fast and easy implementation. The first step of the method is detecting the rails from the images with an algorithm. The second step of the method is finding visually noticeable defects on the detected rail with automatized algorithms for defect detection. One of the algorithms for defect detection shows promising results and investigating the rail images could be a promising complementary method to the already used manual ultrasound measurements.

Executive Summary

The detection of defects on the rail surface is an important part of railway inspection, and early detection can prevent more severe defects occurring later (through predictive maintenance). Therefore, having systems which automatically detect the defects on the rail surface as e.g. microscopic cracks, squats and rolling contact fatigue is of great interest. An open-source computer vision system is developed for automatic detection of defects using images of conventional rail (straight track). The system detects the two rails and subsequently investigates the rail for defects. The system is validated with 40 images containing defects (a group of 28 images) and without defects (a group of 12 images). The system clearly gives different output for the two different groups of images. The fast implementation, automatic defect detection, promising results of detecting the defects and the ease of collecting images from the track recording car makes this approach attractive to apply as a complementary system to the standard approach with the manual ultrasound equipment. The system

E. Hovad (✉) · H. Hansen · V. A. Dahl
Department of Applied Mathematics and Computer Science, Technical University of Denmark, 2800 Kgs Lyngby, Denmark
e-mail: emilh@dtu.dk

V. A. Dahl
e-mail: vand@dtu.dk

A. F. da Silva Rodrigues
Banedanmark, 2100 Copenhagen, Denmark
e-mail: AFSR@BANE.dk

R. Galeazzi et al. (eds.), *Intelligent Quality Assessment of Railway Switches and Crossings*, Springer Series in Reliability Engineering,
https://doi.org/10.1007/978-3-030-62472-9_11

is implemented in the popular open-source programming language python 3 and thereby it is easily available for railway infrastructure managers.

1 Introduction

The railway track, which includes rail, sleeper, and fastener, is the most important infrastructure of the railway. In Denmark, over 3000 trains can be expected to run every 24 h on the rail network, which has a track length of 3476 km [1]. This railway network requires monitoring and maintenance in order to reduce faults that can cause problems with consequences extending to cost, disruption of transportation and safety [2]. The long-term impact of the train traffic and the natural environment on the railway will lead to a variety of defects and wear on the rail and its profile [3]. The mechanical properties such as indentation hardness of material composition used in rails and the contact friction conditions for this material composition is investigated in [4]. Deterioration of the rails can be caused by plastic deformation or wear and this deterioration is modelled in [5]. Different problems can occur with the rails such as corrugation and rolling contact fatigue (RCF) where microscopic cracks appear at the or close to the surface and these cracks can potentially grow and cause a rail break [6]. By detecting rail surface defects early more severe defects can be prevented and this may lead to a reduction of the maintenance cost and potentially lessen the chance of traffic disruptions and accidents. Therefore, having a system which automatically detects the defects on the rail surface (microscopic cracks) is of great interest.

Machine vision technology for detecting the condition of turnouts and its components is discussed in [7]. An example of a camera setup with a computer vision system for measuring the gauge of the track, which is discussed in [8]. Another example is a visual inspection system for detecting partially worn and completely missing fasteners which are proposed in [9] and the system ranks the fasteners into three levels of quality. A comparative study of the different techniques as deep learning with e.g. VGG16 network, faster R-CNN and standard computer vision techniques for investigating fasteners are performed in [10].

The possibility of computer vision based system monitoring the track gauge through drone imagery is discussed in [11]. In [16] a multi-stage system first using object detection for finding the cantilever joints and their fasteners and the last stage diagnose the fasteners' defects. In [12] an image acquisition device has been designed and the surface defects of the rails are optimized and the characteristics of the defects are obtained. In [17] an automatic railway visual detection system (RVDS) for surface defects and a region-of-interest detection region extraction algorithm by vertical projection and gray contrast algorithm is suggested. In [18] a deep convolutional neural network solution to the analysis of image data for the detection of rail surface defects is proposed. In this work the results of different network architectures characterized by different sizes and activation functions are investigated and the experimental results are promising. In [19] a vision-based method using a deep architecture of convolutional neural networks (CNNs) is developed for detecting concrete cracks.

Banedanmark uses a track recording car that captures images from the tracks, monitor the track geometry, measures eddy current and records sound from the wheel contact with the rails [2, 20, 21]. The standard approaches for investigating rail defects are eddy current for shallow defects (smaller than 3–4 mm) and ultrasound for the deeper part of the rail both collected with a track recording car. Due to the high speed the track recording car runs at and to ensure a precise diagnosis of the rail, additional manual ultrasound measurements are used to verify the data from the car. Ultrasound recordings of the track captured from a track recording car is in general a cost effective approach [22] and an additionally cost effective method could be analysis of the images.

In this study the images are processed by first detecting the rails by an automatic method (algorithm) and subsequently detecting the localization of the defects applying automatized methods as well (another algorithm). The automatized algorithms developed are easy to implement and the code is available in the appendix. The simple vertical geometry in the images are utilized to detect the rails by the low vertical standard deviation which is also shown in [12]. When the rails are detected the defects are found by the vertical pixel distribution. Due to vertically aligned rails in the images (anisotropic) with specific lighting for each individual image custom methods are developed for this application versus using standard computer vision methods as the Sobel operator [13], the Gabor transform [14] or filters as Top-hat transform [15] which showed less promising results. The method output preserves the same dimension as the original image and can thereby directly be compared to the original image and the severity of the defect can also be compared based on the quantitative output of the method at the defect position.

2 The Data

2.1 *The Rail Image*

The images of the rials are collected by the Eurailscout UST02 and preprocessing is performed prior to Banedanmark receiving the images. The preprossesing increases the visibility of the defects on the images. The image of the tracks, an example is shown in Fig. 1, is represented using gray-scale pixel intensities with pixel values being integer numbers from 0 to 255. The images have resolution of 1272×578 pixels, and the total number of rows and colums are denoted $N = 1272$ and $M = 578$. Indices i and j are used for rows and columns, such that $I_{i,j}$ denotes intensity of the pixel in the i-th row and j-th column.

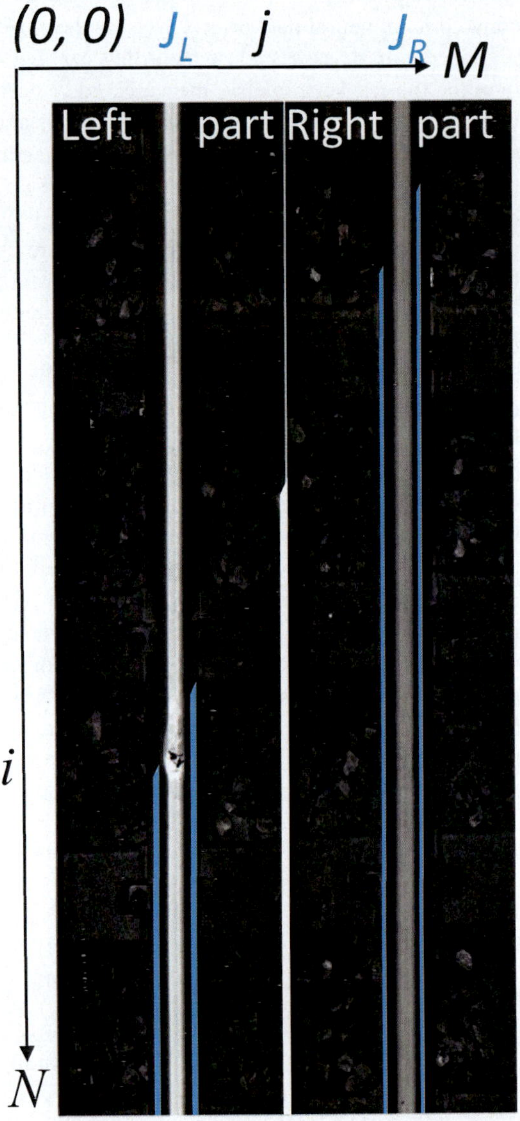

Fig. 1 Example of a rail image which depicts an area of approximately 2.503 m × 1.137 m and has the resolution of 1272 × 578 pixels. The image has a white vertical line separating the two rails and for illustration purposes, the white vertical line has been made thicker

Each image of the track is split in two parts by the white vertical line: a left part and a right part. The first part is the left side showing the ballast, the left side of the sleepers and the left rail (indicated in the Fig. 1 by J_L). The second part is the right side showing the right side of the ballast, the sleepers and the right rail (indicated by J_R). Each side of the image thus requires separate consideration for lighting conditions. The data only contains images of conventional track.

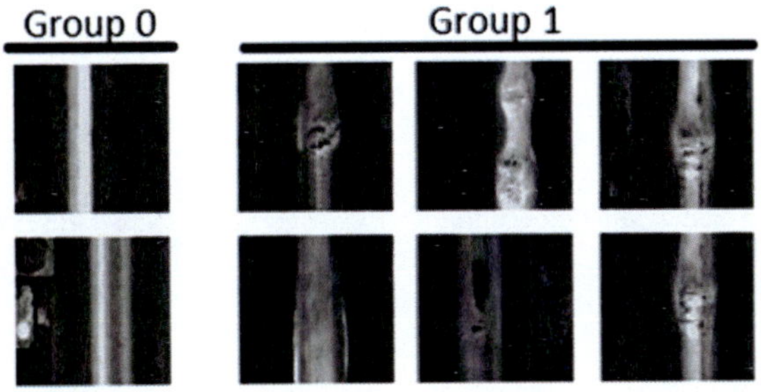

Fig. 2 (Left) Group 0 contains images of rail with no defects. (Right) Group 1 shows different examples of localized defects on the rails

2.2 *Image Groups 0 and 1*

The data set provided by Banedanmark contains two groups of images. The first group, denoted Group 0, features 12 rail images with no defects or visible surface deformation. The second group, denoted Group 1, features 28 images having severe defects such as cracks, squats and rolling contact fatigue that exhibit significant surface deformation. In group 1 there are 23 images associated with defects measured to be 10–20 mm deep from the rail surface which shows small sign of deformation in the rail top and 5 images with defects measured to be 20–30 mm deep, which represent a risk for rail break during service and are therefore removed. Defects deeper than 30 mm are removed in a very short notice due to high risk of rail breakage, so there are no pictures available. Severe defects as large surface cracks or large squats. All the images in Group 1 have clear visual signs of containing surface defects, because their goal is to calibrate the method to what is a clear indication of rail defects.

The defect can occur either on the left or right rail, but it is also possible for both rails to contain a defect. A selection of rails from Group 0 and Group 1 is shown in Fig. 2.

3 Method

3.1 *Detecting Rails in Images*

The rails can have variation in the position and therefore a method for finding the rails is described here. The rails in images are detected using a simple method, which allows a subsequent analysis focusing only on the rails. The first step of rail detection is computing a column-wise sum of absolute first-order differences

$$d_j = \sum_{i=1}^{N-1} \left| I_{i+1,j} - I_{i,j} \right|, \qquad j = 1, \ldots, M \tag{1}$$

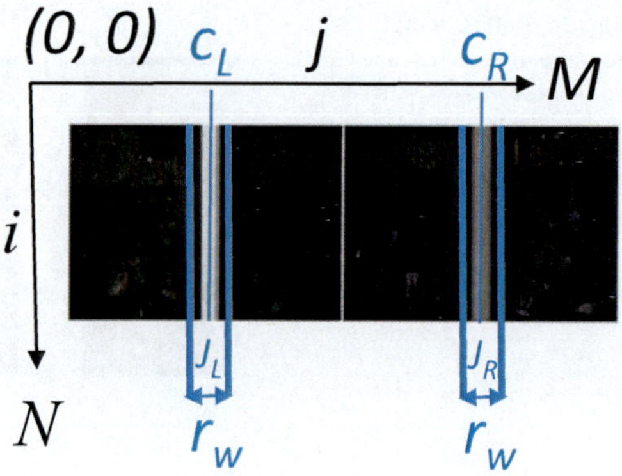

Fig. 3 An illustration of the notation used for detecting the rails. The center index of the left rail is denoted c_L, the center of the right rail c_R, the interval for the left rail is denoted J_L, the interval for the right rail is J_R and finally the rail width is denoted r_w

where $I_{i,j}$ is a pixel intensity (gray scale value) of a pixel in the column j and row i, where N is the total number of rows and M is a total number of columns.

The subsequent steps of rail detection involve processing d_j to identify the image columns containing rails, based on two assumptions. The first assumption is that image columns containing rails have rather uniform pixel values, while the image columns from the other parts of the image (containing ballast and the sleepers) show more variation in pixel values. In other words, the values of d_j are small for image columns containing rails.

The rail head is 72 mm wide and based on the image resolution $1.968 \times 1.968\,\text{mm}^2$ per pixel the rail width is 37 pixels. Thereby, the second assumption is that each rail occupies a 50 pixel wide area of the image which includes a margin of error of pixels on each side of the rail. This value is denoted $r_w = 50$, see Fig. 3 for the illustration of the notation.

Consequently, finding the left and the right rail requires identifying two 50-element wide intervals of d_j which contain relatively small values. Therefore, in the second step of rail detection d_j is convolved with the 1-dimensional kernel containing r_w ones. To preserve the image dimensions while convolving, d_j is padded on each side with $r_w/2$ elements having a very high value (in our case 10^{12}), since only low values are of interest in the next step. This gives

$$g_j = f * \hat{d}_j \tag{2}$$

where $\hat{d}_j$ is a padded version of d_j, while f is a convolution filter.

The third step of rail detection involves finding the two 50 pixel long intervals in g_j, denoted J_L and J_R for the left and the right rail.

These intervals are characterized by having small values of g_j. In a perfectly symmetric case, J_L and J_R would together account for the 100 smallest values of g_j equally divided between J_L and J_R. In a more realistic case, the 100 smallest values of g_j will be divided between the left and the right rail in a less balanced manner, but will still give a good indication of the rail position in the image. Therefore, we collect column indices corresponding to the 100 smallest values of g_j. To separate the left and the right rail, the k-means clustering algorithm is applied to the set of indices.

The k-means clustering algorithm is composed of the following steps (1) initialize the cluster centers (randomly or by prior knowledge), (2) iterate over all the observations and assign each to the closest cluster center, (3) calculate the new cluster centers based on assignment, (4) repeat steps (2) and (3) until convergence or for a pre-defined number of iterations. For the purpose of finding rail positions, the two cluster centers are initialized at round($\frac{m}{4}$) and round($\frac{3m}{4}$), corresponding to image columns where rails normally occur. Due to the convenient initialization and small number of columns in the images k-means clustering converges in a few iterations.

The two cluster centers, denoted c_L and c_R, correspond to the two image columns where centers of the rails are located. Subsequently finding the two rails intervals is based on the width of the rail r_w. Now the two rails columns intervals, namely J_L and J_R, can be found as

$$\begin{aligned} J_L &= \left[\mathrm{ceil}\left(c_L - \frac{r_w}{2}\right), \mathrm{ceil}\left(c_L + \frac{r_w}{2}\right)\right], \\ J_R &= \left[\mathrm{ceil}\left(c_R - \frac{r_w}{2}\right), \mathrm{ceil}\left(c_R + \frac{r_w}{2}\right)\right]. \end{aligned} \tag{3}$$

where J_L and J_R intervals are a subset of the column space having a total of M columns.

3.2 Detecting Defects on Rails

Having detected the rails in the images we move on detecting visually noticeable defects on the rails and quantifying the severity of the defects. Three defect detection methods are developed and tested. All three methods exploit the assumption that rails without defects have uniform distributions of pixel values, especially along the image columns.

The types of defects which are visually noticeable are cracks, squats and rolling contact fatigue. These defects are all detected in the same way by finding gray-scale threshold values based on the pixel value distribution.

3.2.1 Method 1: Column-Wise Standard Deviation

Method 1 consists of two steps. In the first step, for all the columns from the two rail intervals J_L and J_R the column-wise mean is calculated as

$$\mu_j = \frac{1}{n}\sum_{i=1}^{n} I_{i,j}, \qquad j \in J_L \cup J_R . \tag{4}$$

In the second step, column-wise standard deviation of pixel intensities is computed as

$$\sigma_j = \sqrt{\frac{1}{n}\sum_{i=1}^{n}\left(I_{i,j} - \mu_j\right)^2}, \qquad j \in J_L \cup J_R . \tag{5}$$

It can be shown that standard deviation may be computed slightly more efficiently using an expression

$$\sigma_j = \sqrt{\frac{1}{n}\sum_{i=1}^{n} I_{i,j}^2 - \mu_j^2}, \qquad j \in J_L \cup J_R . \tag{6}$$

The output of the Method 1 contains 50 values for the left, and 50 values for the right rail. The resulting values are positive numbers not larger than 255.

3.2.2 Method 2: Column-Wise Threshold

A method is constructed which exploits different brightness distributions of the pixels along the image columns containing rails. The method is given by

$$t_{i,j} = \max\left(|I_{i,j} - \mu_j| - \sigma_j, 0\right), \qquad j \in J_L \cup J_R \tag{7}$$

where σ_j is a previously computed column-wise standard deviation. The strength of this method is that the output is an image of the same size as the rails. Furthermore, the output will have values between 0 to 255, and can be shown as an image.

3.2.3 Method 3: Pixel-Wise Standard Deviation

This method computes standard deviation in a small neighbourhood around each pixel. The efficient implementation of local standard deviation utilizes an expression similar to Eq. 6. Since local averages may be computed using convolution, the pixel-wise standard deviation filter is given by

$$s_{i,j} = \sqrt{u * I_{i,j}^2 - \left(u * I_{i,j}\right)^2}, \qquad j \in J_L \cup J_R \tag{8}$$

where u is a local mean filter of a size w-by-h and $*$ indicates convolution. For the purpose of detecting rail defects it makes sense to use a filter which is elongated in the direction of a rail, i.e. filter width w can be much smaller than the filter height h.

While output of Method 3 contains pixel-wise values, due to the boundary effect of filtering the valid output area will be smaller than the rail area.

Table 1 The group statistics used for comparing the two groups of images using any of the three defect detection methods: maximal group value, the mean group value and the group standard deviation

	max	mean	std
Method 1	$\max(\max(\sigma_j)_k)$	$\text{mean}(\max(\sigma_j)_k)$	$\text{std}(\max(\sigma_j)_k)$
Method 2	$\max(\max(t_{i,j})_k)$	$\text{mean}(\max(t_{i,j})_k)$	$\text{std}(\max(t_{i,j})_k)$
Method 3	$\max(\max(s_{i,j})_k)$	$\text{mean}(\max(s_{i,j})_k)$	$\text{std}(\max(s_{i,j})_k)$

3.3 Comparing the Two Groups of Rail Images Using Three Defect Detection Methods

To compare Group 0 images (containing rails with no defects) and Group 1 images (containing rails with defects) a single value is needed per each image and for each of the three defect detection methods. Since defects may cover a relatively small area of the rail, it is of interest to establish whether defect occurs in any of the pixels containing rails, but the number of pixels affected by defect is of less importance. Therefore, a maximal response value is found for each image and each method also called $output_{max}$.

These per-image maximal values are denoted $\max(\sigma_j)_k = output_{max}$ for Method 1, $\max(t_{i,j})_k = output_{max}$ for Method 2 and $\max(s_{i,j})_k = output_{max}$ for Method 3. Here, k is used as an index to one image out of K images in a group, and it is assumed that while computing maximum one only uses the column indices for the detected rails in the image $j \in J_L \cup J_R$.

Having collected per-image maximal response ($output_{max}$), statistics can be computed for each of the two image groups. The statistics used include maximal value for the group, the mean of the group and the standard deviation of the group as described in Table 1.

4 Results

4.1 Detecting Rails in Images

Figure 4 exemplifies the performance of the method for detecting the rails. The rails are shown in the top image together with a definition of the image coordinate system.

The bottom plot shows the values of g_j. The bottom plot shows that it is possible to separate the rails from the background using the values g_j and the that two clusters of indices can be found corresponding to the left and the right. The rails are shown with the two horizontal lines. By visual inspection it is verified that the algorithm successfully detected all the rails in all the images.

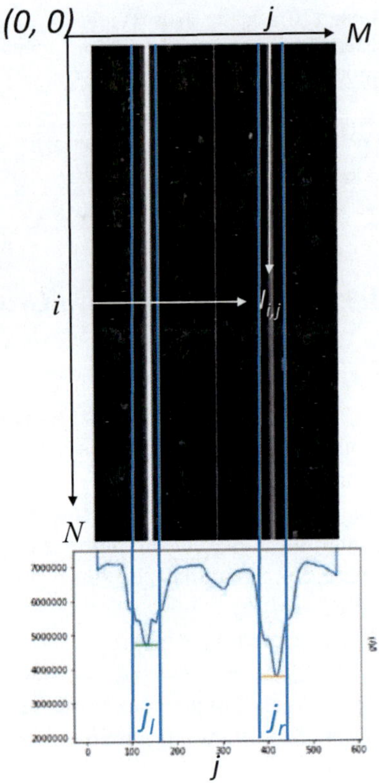

Fig. 4 Illustration of the rail detection algorithm

4.2 Detecting Defects on Rails

The result of the three methods per-image maximal response ($output_{max}$) are shown in Fig. 5 with box plots for the two groups. The round circles in the plots are outliers and the boxplots have confidence level (whiskers) chosen to the 25th and 75th percentiles. A clear distinction in the maximal values of the two groups images can be seen with Method 2: Column threshold (middle plot).

Table 2 shows a overview of the three methods examined with some varying sizes for the height and width (h, w) of the kernel for Method 3: local estimate of the standard deviation. It can be seen from the Table 2, that Group 0 and Group 1 have differences with respect to the average values and the standard deviations of the max output especially for the Method 2: column threshold method.

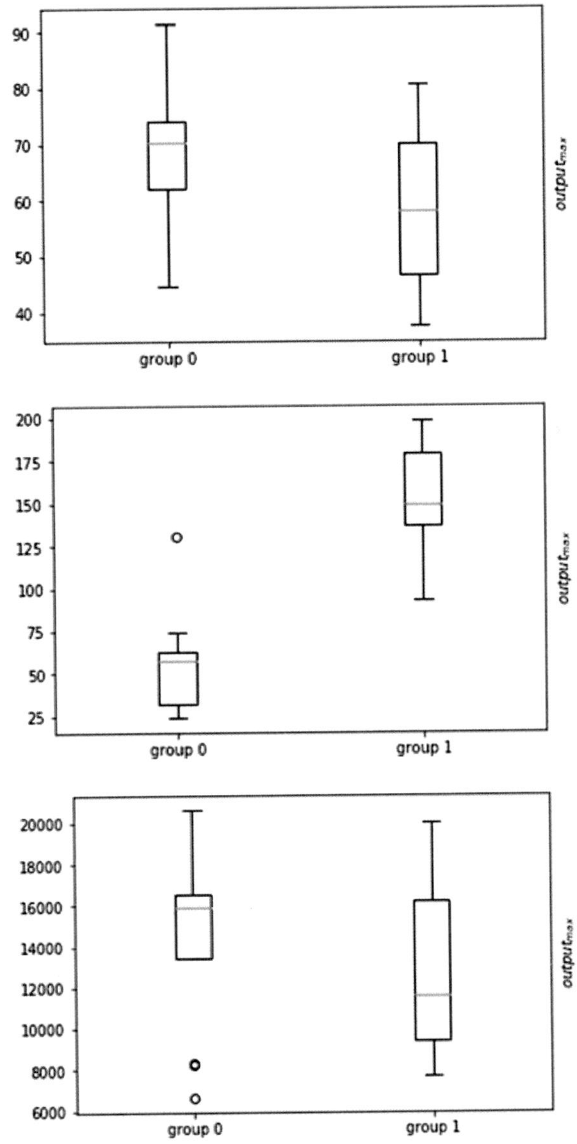

Fig. 5 Results for the two image groups and the maximal values ($output_{max}$) for each image computed with the three methods. (Top) Method 1: maximal column standard deviation. (Middle) Method 2: normalized column threshold. (Bottom) Method 3: local estimate of the standard deviation with mean filter $(h, w) = (100, 49)$

Table 2 Result of three methods for Group 0 and for Group 1

	max	mean	std
Group 0: Method 1:	9.14×10	6.79×10	1.30×10
Group 1: Method 1:	8.10×10	5.86×10	1.35×10
Group 0: Method 2:	1.30×10^2	5.57×10	2.79×10
Group 1: Method 2:	1.98×10^2	1.51×10^2	3.17×10
Group 0:			
Method 3 $(h, w) = (200, 49)$:	2.05×10^4	1.43×10^4	4.10×10^2
Method 3 $(h, w) = (100, 49)$:	2.06×10^4	1.45×10^4	4.11×10^2
Method 3 $(h, w) = (49, 49)$:	2.09×10^4	1.46×10^4	4.15×10^2
Method 3 $(h, w) = (20, 49)$:	2.13×10^4	1.48×10^4	4.20×10^2
Method 3 $(h, w) = (20, 20)$:	5.01×10^4	3.27×10^4	9.80×10^2
Method 3 $(h, w) = (100, 10)$:	6.03×10^4	4.31×10^4	1.31×10^4
Method 3 $(h, w) = (100, 5)$:	6.42×10^4	5.13×10^4	1.38×10^4
Group 1:			
Method 3 $(h, w) = (200, 49)$:	1.90×10^4	1.20×10^4	3.69×10^2
Method 3 $(h, w) = (100, 49)$:	2.00×10^4	1.27×10^4	3.82×10^2
Method 3 $(h, w) = (49, 49)$:	2.25×10^4	1.36×10^4	4.26×10^2
Method 3 $(h, w) = (20, 49)$:	2.60×10^4	1.50×10^4	4.94×10^2
Method 3 $(h, w) = (20, 20)$:	5.33×10^4	3.21×10^4	1.09×10^4
Method 3 $(h, w) = (100, 10)$:	5.92×10^4	3.57×10^4	1.29×10^4
Method 3 $(h, w) = (100, 5)$:	6.41×10^4	3.92×10^4	1.43×10^4

For Method 2 the higher standard deviation for group 1 is also expected due to the presence of the defects, but still reveals that the dispersion of pixel intensity is somewhat close to group 0. The advantages of Method 2 besides the best separation of the two groups is that it preserves the dimensions from the original image. This gives the possibility to directly visually compare the results to the original image as shown in Fig. 6. The high values on the right clearly shows the problematic areas on the original image on the left shown with the range from 0 non problematic to 93 the most deviating value and in this case also the most problematic on the shown image.

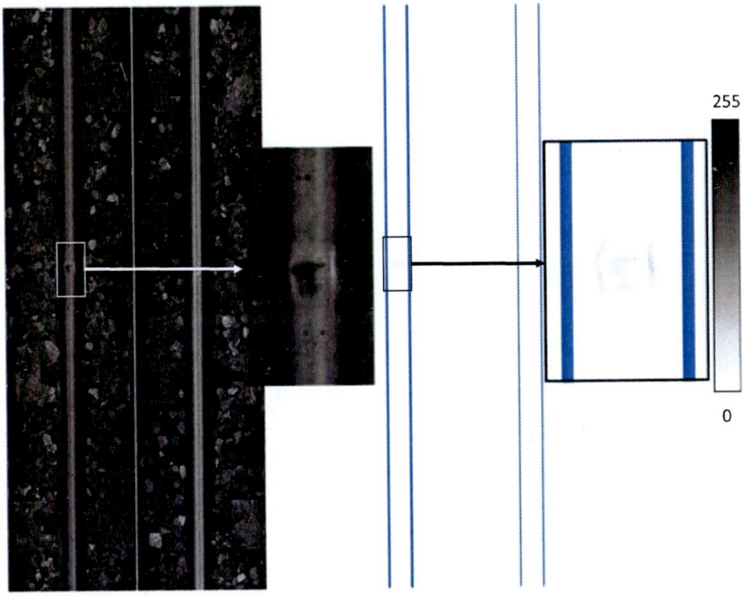

Fig. 6 (Left) Example of a rail image from group 1. (Right) Using Method 2 on the image shows the ability to detect the problematic area on the rail and the most problematic area is highlighted on the original images with a surrounding white box and the algorithm values for this area is surround by a black box. The blue lines are where the rails are detected. The range for the method 2 for this image example is from 0 to 93 and the gray scale can be seen at the right

5 Conclusions

The methods developed are simple due to the few computations needed to calculate averages and standard deviations where this easily can be computed on a newer laptop. The simple implementation and the promising results of detecting the defects and the ease of collecting image from the track recording car makes this approach attractive to apply as an complementary system to the standard approach with the manual ultrasound equipment. The implementation uses an open-source programming language Python 3 with the attached code placed in the Appendix.

In this study the images (conventional rail images) are processed by first detecting the rails by a automatic method and subsequently detecting the localization of the defects applying automatic methods as well. First, the rails are detected from the track images. It is possible to detect the rails in all the captured images of the conventional track. Image processing allows for detection of defects and deviating textures of the rails when comparing group 0 and group 1.

The best method found based on the separation of the two groups of images is Method 2: Normalized column threshold, which is based on how each pixel deviated from the mean and standard deviation row-vise. This method has many advantages besides the good results; it is simple, easy to implement, few operation are needed (mean and standard deviation) and it preserves the image dimension for a direct visual comparison to the original image. A maximal output threshold of e.g. 125 for this data set would give a good balance of recall versus precision. Now this maximum output threshold could be used for all images coming from the conventional track, although a larger dataset would help estimating a more robust threshold. Evaluation of images did not reveal any serious visual defects being missed, and the higher predictions did correspond with a larger maximal value for the deviating pixel in the image.

Two possible use cases of the approach: as a supplement to other defect-detection modalities, or as a pre-processing step in a more elaborate image-analysis based method (as deep learning methods) for detecting rail defects. The simplicity of this approach make it easy to directly apply as compared to neural networks tedious labelling process. This method could additionally be used for assisting in segmentation labelling of conventional rail images when finding the rails and the defects to create training data for deep learning algorithms.

Appendix: Python Code

```
# -*- coding: utf-8 -*-
"""
Code blocks
1) import data
2) define functions
3) create results for each group
4) create comparison results and show results
author: Emil Hovad
"""
#%% code block 1
############################################
####        start import packages ##########
import cv2
import numpy as np
import matplotlib.mlab as mlab
import matplotlib.pyplot as plt
from matplotlib import colors
import itertools
from PIL import Image
import glob
import os
from sklearn.cluster import KMeans
####         end import packages   ############################################

#%% code block 2
############################################
####         start functions   ################
#rail image detection function
def ImageSegmentation(im,railwidth):

    #Finding the stones versus rails based on low and high difference in
    #column pixels values
    # eq. 1 in the article
    SumDifferenceCol = np.sum(np.abs(im[1::1,:][0:-1,:]-im[2::1,:]),axis=0)

    #padding
    padding = np.ones(( np.int( railwidth/2)))*10e12
    padded_vector = []
    padded_vector.append(padding[:].tolist())
    padded_vector.append(SumDifferenceCol[:].tolist())
    padded_vector.append(padding[:-1].tolist())
    padded_vector = list(itertools.chain(*padded_vector))
    padded_vector = np.asarray(padded_vector)

    #Finding the best fit, the rails has a lower column standard deviation
    #as compared to the stones
    #make a convolution with the rail width
    convfilter = np.ones((railwidth))
```

```
    #Finding the lowest value for the two rails based on a convolution
    #In the article eq. 2
    convResult = np.convolve( padded_vector  , convfilter[:], 'valid')


    #list rail indices
    sortedlist = np.argsort(convResult)[:(railwidth*2)]

    # Initializing KMeans
    startpts=np.array([[int(np.round(im.shape[1]/4))],[int(np.round
    (im.shape[1]*3/4))]])
    kmeans = KMeans(n_clusters=2, init=startpts)
    # Fitting with inputs
    kmeans = kmeans.fit(sortedlist.reshape(-1, 1))
    # Predicting the clusters
    #labels = kmeans.predict(sortedlist.reshape(-1, 1))

    # Getting the cluster centers
    C = kmeans.cluster_centers_

    rail1start=np.int(C[0]-railwidth/2)
    rail1end  = np.int(C[0]+railwidth/2)

    rail2start=np.int(C[1]-railwidth/2)
    rail2end  = np.int(C[1]+railwidth/2)


    #output of eq 3
    #JL (the left rail)
    imgCut1=im[:,rail1start:rail1end]
    #JR (the right rail)
    imgCut2=im[:,rail2start:rail2end]

    return imgCut1, imgCut2, rail1start, rail1end, rail2start, rail2end,
    convResult

# Method 1)
# Column standard deviation
def method1std(imgCut1):
    std_values = np.std(imgCut1)
    return np.max(std_values)

# Method 2)
# Normalized column threshold
def method2threshold(imgCut1):

    col_mean1 = np.mean(imgCut1,axis=0)
    col_sd1 = np.std(imgCut1,axis=0)

    y = np.zeros(( imgCut1.shape[0] ,  imgCut1.shape[1] ))
    y_fault  = np.abs(np.abs(imgCut1-col_mean1)-col_sd1)
    larger_than= y_fault >  np.zeros(( imgCut1.shape[0] ,
    imgCut1.shape[1] ))
    y[larger_than] = y_fault[larger_than]
    y = np.asarray(y, dtype=np.uint8)
    return y

from scipy.signal import convolve2d

# Method 3)
# Local estimate of the standard deviation
def method3Filter(imgCut1):
# STD FILTER, SMOOTHING, THRESHOLDING
    #kernel = (25,25)
    filter_kernel = np.ones([100,5])/(100*5)

    I_double = imgCut1.astype(np.double)#/256

    E = convolve2d(I_double**2, filter_kernel, mode='valid')-convolve2d
    (I_double, filter_kernel, mode='valid')
    #E = np.sqrt(cv.boxFilter(I_double**2,-1,kernel)-cv.boxFilter
    (I_double,-1,kernel)**2)
    #mm=np.mean(E,axis=0)
    #S = cv.GaussianBlur(E,(29,29),50)
    #T = S>0.025
    return E

# Comparing the two groups
#This method finds maximal values for image in a group (need to be executed
for each group!)
#The values calculated is ymax,k, namely the maximum for each image
# (finding ymax for the k'th image out of the total K images)
def results(path_group, railwidth, method):


    image_list = []
    summary_list = []
    #loop over group 0
    countplot=0
```

```
if False==os.path.exists(path_group+'plots/'):
    #create plots directory for the image group
    os.mkdir(path_group+'plots/')

#insert your group 0 jpg images (we have jpg images) inside the folder
glob.glob('folder here')
for filename in glob.glob(path_group+'*.jpg'): #assuming gif
    im = cv2.imread(filename,0)
    image_list.append(im)

    # Segmentation
    imgCutL, imgCutR, rail1start, rail1end, rail2start, rail2end,
    convResult = ImageSegmentation(im,railwidth)


    if method == "method1":
        #get the results from the correct rail
        #std_column_value_L = method1std(imgCutL,imgCutL)
        #no difference but this is consistent with the code
        yL = method1std(imgCutL)# std_column_value
        yR = method1std(imgCutR)#std_column_value

    if method == "method2":
        # Filtering method
        yL = method2threshold(imgCutL)
        yR = method2threshold(imgCutR)

    if method == "method3":
        # Filtering method
        yL = method3Filter(imgCutL)
        yR = method3Filter(imgCutR)

    countplot+=1

    # finding ymax for the k'th image out of the total K images
    max_value         = max(np.max(yL),np.max(yR))
    railconcatonated = np.concatenate( (yL.flatten(), yR.flatten()) )
    mean_value =  np.mean( railconcatonated )
    std_value  =  np.std(  railconcatonated )
    #add values to the summary list
    summary_list.append([max_value, mean_value, std_value])

    #plot rail detection
    im_rgb = cv2.cvtColor(im,cv2.COLOR_GRAY2RGB)
    cv2.imwrite( path_group+"plots/"+str(countplot)+'normal.png',im_rgb)

    im_rgb[:,:,:] = 0

    #a direct comparison can only be made with method 2
    # preservs image dimensions
    if  method== 'method2':
        #draw faults with red
        im_rgb[0:im_rgb.shape[0],rail1start:rail1end,0] = yL
        im_rgb[0:im_rgb.shape[0],rail1start:rail1end,1] = yL
        im_rgb[0:im_rgb.shape[0],rail1start:rail1end,2] = yL

        im_rgb[0:im_rgb.shape[0],rail2start:rail2end,0] = yR
        im_rgb[0:im_rgb.shape[0],rail2start:rail2end,1] = yR
        im_rgb[0:im_rgb.shape[0],rail2start:rail2end,2] = yR

    im_rgb = cv2.bitwise_not(im_rgb)

    cv2.line(im_rgb,(rail1start,0),(rail1start,1271),(255,0,0),2)
    cv2.line(im_rgb,(rail1end,0),(rail1end,1271),(255,0,0)     ,2)
    cv2.line(im_rgb,(rail2start,0),(rail2start,1271),(255,0,0),2)
    cv2.line(im_rgb,(rail2end,0),(rail2end,1271),(255,0,0)     ,2)
    cv2.imwrite( path_group+"plots/"+str(countplot)+'edited.png',im_rgb)


    f = plt.figure()
    plt.plot(convResult)
    plt.ylim(ymax =  min(convResult)*2, ymin = min(convResult)/2)
    plt.plot([rail1start,rail1end],[min(convResult[rail1start:rail1end]),
    min(convResult[rail1start:rail1end])])
    plt.plot([rail2start,rail2end],[min(convResult[rail2start:rail2end]),
    min(convResult[rail2start:rail2end])])

    plt.text(1.02, 0.5, r"$g(j)$", {'color': 'k', 'fontsize': 10},
             horizontalalignment='left',
             verticalalignment='center',
             rotation=90,
             clip_on=False,
             transform=plt.gca().transAxes)

    plt.xlabel('j')
    plt.show()

    f.savefig(path_group+"plots/img_segmentation"+str(countplot)+'.png',
```

```
            bbox_inches='tight')

    return summary_list

####          end functions    #################
###############################################

#%% code block 3
###############################################
####          start create results    ##########
#create results for the two groups

#define railwidth (rw)
railwidth = 50

#method 1) Culomn standard deviation = 'method1'
#method 2) Normalized column threshold = 'method2'
#method 3) Local estimate of the standard deviation = 'method3'
#insert desidered method, namely the one of the strings
#'method1', = 'method2' or = 'method3'
method = 'method1'

#the path of the groups of images (good group)
path_group_0 = 'H:/BaneDk/WorkFiles/WorkFiles/Images/Examplesof0/'

#the path of the groups of images (defect group)
path_group_1 = 'H:/BaneDk/WorkFiles/WorkFiles/Images/Examplesof1/'

summary_list_0 = results(path_group_0, railwidth, method)
summary_list_1 = results(path_group_1, railwidth, method)


####          end create results     #############
#################################################

#%% code block 4
#################################################
####          start Show results   ###############

# use numpy
summary_0 = np.asarray(summary_list_0)
summary_1 = np.asarray(summary_list_1)

#The values calculated is explained in table 1, namely
#1) max(max(ymax,k)), 2) mean(max(ymax,k)), 3) std(max(ymax,k))
# calculate 1) max(max(ymax,k)), 2) mean(max(ymax,k)) 3) std(max(ymax,k))
# for the chosen method
# for group 0
max_0_of_max = np.max(summary_0[:,0])
mean_0_of_max = np.mean(summary_0[:,0])
std_0_of_max = np.std(summary_0[:,0])
# for group 1
max_1_of_max = np.max(summary_1[:,0])
mean_1_of_max = np.mean(summary_1[:,0])
std_1_of_max = np.std(summary_1[:,0])


#for inserting into the latex table in the article
print('group 0')
print(max_0_of_max, " & ", mean_0_of_max, " & ", std_0_of_max)

print('group 1')
print(max_1_of_max, " & ", mean_1_of_max, " & ", std_1_of_max)

#boxplots
maximum = [summary_0[:,0],
     summary_1[:,0]]
plt.boxplot(maximum)
plt.text(1.02, 0.5, r"$output_{max}$", {'color': 'k', 'fontsize': 10},
          horizontalalignment='left',
          verticalalignment='center',
          rotation=90,
          clip_on=False,
          transform=plt.gca().transAxes)
plt.xticks((1,2), ('group 0', 'group 1'), color='k', size=10)
plt.show()

####          end show results      ################
###################################################
```

References

1. https://www.bane.dk/da/Presse/Banen-i-tal
2. Stanislav Jovanović, Dragan Božović, Mirjana Tomičić-Torlaković (2014) Railway infrastructure condition-monitoring and analysis as a basis for maintenance management. Hrvatski savez gradjevinskih inženjera, Cm 2012:347–358
3. Olofsson U, Telliskivi T (2003) Wear, plastic deformation and friction of two rail steels-a full-scale test and a laboratory study. Wear 254(1–2):80–93
4. Bellemare SC, Dao M, Suresh S (2008) Effects of mechanical properties and surface friction on elasto-plastic sliding contact. Mech Mater 40(4–5):206–219. https://doi.org/10.1016/j.mechmat.2007.07.006
5. Brouzoulis J, Torstensson PT, Stock R, Ekh M (2011) Prediction of wear and plastic flow in rails-test rig results, model calibration and numerical prediction. Wear 271(1–2):92–99. https://doi.org/10.1016/j.wear.2010.10.021
6. Grassie S, Kalousek J (1993) Rail corrugation: characteristics, causes and treatments. Proc Inst Mech Eng Part F J Rail Rapid Transit 207(1):57–68
7. Molina L, Resendiz E, Edwards JR, Hart JM, Barkan CP, Ahuja N Condition monitoring of railway turnouts and other track components using machine vision, TRB 11-1442
8. Zheng S, Chai X, An X, Li L, (2012) Railway track gauge inspection method based on computer vision. In: 2012 IEEE international conference on mechatronics and automation. ICMA 2012(6284322):1292–1296. https://doi.org/10.1109/ICMA.2012.6284322
9. Feng H, Jiang Z, Xie F, Yang P, Shi J, Chen L (2014) Automatic fastener classification and defect detection in vision-based railway inspection systems. IEEE Trans Instrument Measur 63(4), 6630109: 877–888. https://doi.org/10.1109/TIM.2013.2283741
10. Wei X, Yang Z, Liu Y, Wei D, Jia L, Li Y (2019) Railway track fastener defect detection based on image processing and deep learning techniques: A comparative study. Eng Appl Artif Intell 80:66–81 Elsevier Ltd
11. Singh AK, Swarup A, Agarwal A, Singh D Vision based rail track extraction and monitoring through drone imagery, ICT Express
12. Min Y, Benyu X, Dang J, Yue B, Cheng T (2018) Real time detection system for rail surface defects based on machine vision. Springer International Publishing, Eurasip Journal on Image and Video Processing
13. Sobel I (1968) An isotropic 3 × 3 image gradient operator presentation at Stanford A.I. Project 1968
14. Gabor D (1946) Theory of communication. J Inst Electric Eng 93:429–457
15. Serra J (1982) Image analysis and mathematical morphology. Academic Press 610
16. Junwen J, Liu Z, Wang H, Nunez A, Han Z (2018) Automatic defect detection of fasteners on the catenary support device using deep convolutional neural network, Institute of Electrical and Electronics Engineers Inc. IEEE Trans Instr Measur 67:257–269
17. Zhang H, Jin X, Wu J, Wang QMY, He Z, Yang Y Automatic visual detection system of railway surface defects with curvature filter and improved Gaussian mixture model, Institute of Electrical and Electronics Engineers Inc., IEEE Trans Instrum Measur 67:1593–1608
18. Faghih-Roohi S, Hajizadeh S, Nunez A, Babuska R, De shutter B (2016) Deep convolutional neural networks for detection of rail surface defects. In: Proceedings of the International Joint Conference on Neural Networks. Institute of Electrical and Electronics Engineers Inc., pp 2584–2589
19. Young Cha J, Wooram O Büyüköztürk (2017) Deep learning-based crack damage detection using convolutional neural networks, Blackwell Publishing Inc. Computer-aided Civil and Infrastructure Engineering 32:361–378
20. Railway applications-track-track geometry quality—part 1: Characterisation of track geometry (n 13848-1 :2003+a1)
21. Jensen PJ, Fongemie R (2016) Sporteknik: danske jernbanespors opbygning, konstruktion og komponenter. Banedanmark

22. Kaewunruen S (2018) Discussion of "field test performance of noncontact ultrasonic rail inspection system" by stefano mariani, thompson nguyen, xuan zhu, and francesco lanza di scalea. J Transp Eng Part A-Syst 144(4):04017007. https://doi.org/10.1061/JTEPBS.0000026

Deep Learning for Automatic Railway Maintenance

Emil Hovad, Thomas Wix, Maxim Khomiakov, Georgios Vassos, André Filipe da Silva Rodrigues, Alejandro de Miguel Tejada, and Line H. Clemmensen

Abstract We propose a computer vision system which detects defects on rails using images. Banedanmark, the Danish railway manager, conducts regular inspections of their railway infrastructure to assess the safety and efficiency of operations. Indications of defects on rails are detected by a recording car which has a system based on an eddy current system for investigating the top part of the rail and an ultra-sound system for the lower part. The proposed computer vision system is a complementary and convenient method to identify and monitor defects on rails. The computer vision system is based on the deep learning framework called YOLO v3. The detected defects are represented by bounding boxes. The system can automatically detect surface defects as well as problematic isolation joints from images of the railway track, including complex geometric sections in turnouts. On an independent test set, it is shown that the computer vision system finds defects, with classification rates of 84% for surface defects, 71% for problematic isolation joints and 74% for non-problematic isolation joints, corresponding to a macro F1 score of 81.5%. This could potentially reduce maintenance cost through automatized fault detection from the developed computer vision system.

Executive Summary

Track monitoring is a fundamental activity which enables the railway infrastructure managers to plan maintenance operations efficiently and effectively. Traditionally,

E. Hovad (✉)
Department of Applied Mathematics and Computer Science, Technical University of Denmark, 2800 Kgs Lyngby, Denmark
e-mail: emilh@dtu.dk

T. Wix · M. Khomiakov · G. Vassos · L. H. Clemmensen
Department of Applied Mathematics and Computer Science, Technical University of Denmark, 2800 Kgs Lyngby, Denmark
e-mail: lkhc@dtu.dk

A. F. da Silva Rodrigues · A. de Miguel Tejada
Banedanmark, 2100 Copenhagen, Denmark
e-mail: AFSR@BANE.dk

R. Galeazzi et al. (eds.), *Intelligent Quality Assessment of Railway Switches and Crossings*, Springer Series in Reliability Engineering,
https://doi.org/10.1007/978-3-030-62472-9_12

track inspection is carried out by using human inspectors, installing sensors on the track sections under investigation, or through a recording car vehicle. Cost reductions, can be accomplished by automatic condition based monitoring on the captured data, as it reduces the need for manual analysis and has the potential to identify degradation at an earlier stage.

A computer vision system is proposed as a convenient method to automatically detect rail surface defects, as well as problematic isolation joints from images of the railway track, including complex geometric sections in turnouts. The computer vision system is based on the state-of-the-art deep learning framework called YOLO v3 which performs object detection. In this case, object detection means identifying the size and the placement of the defects with a surrounding bounding box. The system has been trained on labelled rail images with different defects such as surface rail defects on conventional rails, surface rail defects in turnouts and problematic isolation joints (supervised learning). After training, the algorithm can now detect the defects with a high precision degree and this could potentially reduce maintenance cost through automatized fault detection.

1 Introduction

According to [1], condition based maintenance of railway systems is increasingly important to infrastructure managers to enable a more cost-effective planning of the railway maintenance. Further, cost reductions can be accomplished by automatic condition based monitoring, combined with a suitable decision-support system [2]. Track monitoring is a fundamental activity which enables the railway infrastructure managers to plan and properly use the maintenance budget. Traditionally, track inspection is carried out by using human inspectors, installing sensors on the track sections under investigation, or through a recording car vehicle [3, 4].

Turnouts have a key role in any railway line to provide flexibility for the trains and therefore their behaviour has a major impact in the overall railway lines performance. Visual, instrumented and train borne inspection techniques are applied to the particular case of turnouts in [5].

Common damage mechanisms in railway turnouts are; plastic deformations of the wing rail at the crossing panel [6], wear due to high levels of friction between rail/wheel at the switch blade of the switch panel [7], fracture defects in the foot of the rail at the crossing panel, or railhead spalling due to crack propagation in the crossing nose are investigated in details in [8].

Rails constitute one of the main component of the railway infrastructure; their profiles are affected by wear over time resulting in a variety of defects [9]. The most common types of rail defects in the Danish rail network are single squats, multiple squats, squat in flash-butt weld and squats in aluminothermic welds. Squats are surface initiated defects due to rolling contact fatigue (RCF), and can be considered as one of the key issues in an efficient and robust infrastructure management [10–12].

Measurements on rails can be taken by non-contact ultrasonic inspection systems as discussed in [13].

An automatic system for detecting defects on rails could be based on images from a recording car, like the Eurailscout UFM120 used by Banedanmark. In addition to ultrasound recordings, a computer vision software provides further verification of defects indicated by the Eurailscout UST02 and it constitutes a useful tool for predictive maintenance activities. In [14] one can find an overview of existing computer vision technologies for inspection of rail tracks together with suggestions for turnout-specific inspection, through computer vision, and emphasizes that they can improve the efficiency and effectiveness of rail track inspection. Another computer vision system for measuring the track gauge (perpendicular distance between the inner faces of the rails) is proposed in [15], while in [3] drones were utilized for capturing images that can be used to quantify the track gauge as well.

One of the first studies to use convolution neural networks to perform Image classification of rail defects was [16]. Here images from a measurement car were used, with each image being assigned just one label. An overall accuracy of 92% is reported, while the accuracies of the squat-classes are 50–64% and the accuracy on the weld-class is 62%. Recently, in [17] classification of fastening system defects are performed with the deep learning architectures VGG-16 [18] and Faster R-CNN [19] where normal, missing, and defective fasteners are detected with high accuracy based on high resolution RGB images (3468 $\times$5472 pixels) with one class per image.

One of the current challenges in computer vision is multiple object detection which however has had great progress due to the advances in self-driving cars [20–22], with the first successful models like the YOLO [23] using a 7 by 7 grid of the image and a similar model called Single Shot detection (SSD) proposed in [24] accounting for several scales and aspect ratios. In [25] the object detection methods SSD and YOLO are used for detecting defective states of catenary support devices to localize the cantilever joints and their fasteners, it is likewise used to diagnose the fasteners. The images are high-resolution (6600 $\times$ 4400 pixels) and taken at night, but the promising results are encouraging us to utilize the YOLO framework.

This study focuses on condition monitoring of the rail surface and localisation of isolation joints based on images of the whole railway track, including complex geometric sections within turnouts such as the switch panel and the crossing panel. The condition monitoring performs object detection by predicting bounding boxes of surface defects, as well as good isolation joints and bad isolation joints. We decided to use the newer YOLO v3 architecture [26], an improvement of the original YOLO architecture, to effectively classify and localize the different types of defects that are of interest. Additionally, this network uses less computer resources at run-time compared to other state-of-the-art deep learning networks like Mask R-CCN [27], enabling for a more efficient method for edge computing deployment.

2 Data

This section describes the image data as well as the defects, non-problematic and problematic isolation joints.

2.1 Images from the Recording Car

Banedanmark is currently using the recording car Eurailscout UST02 from which images were collected with digital line scan cameras. These line scans are preprocessed into images of the rail, before Banedanmark receives them. An example of a rail image with a highlighted surface defect, can be seen in Fig. 1. The images are captured from a vertical view, mostly during daylight. The lighting is not uniform due to the different angles at which the sunlight rays are reflected on the rails. The images have very high contrast which greatly increases the visibility of the defects. The images consist of rail tracks including turnout sections, with dimensions of $578 \times 1273 \times 1$ pixels corresponding to the width and the height and one gray-scale channel, respectively. Each image covers approximately 2.507 m of track and consecutive images do not overlap. Both the right and left rails are included in a single image, while most of the area between the rails is removed to reduce the horizontal dimension to 578 pixels which corresponds to approximately 1.137 m in width.

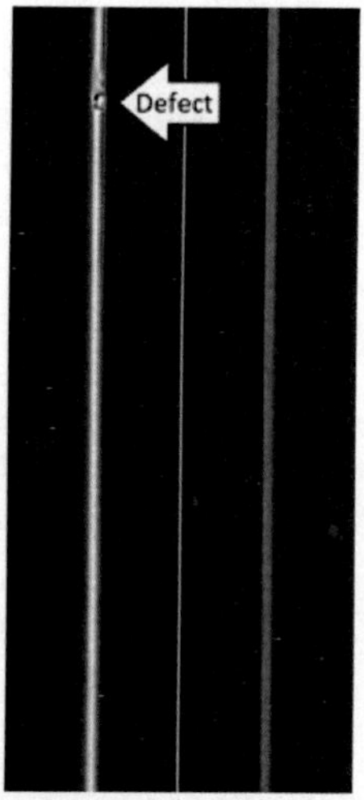

Fig. 1 Example of a rail image with a defect on the top part of the left rail

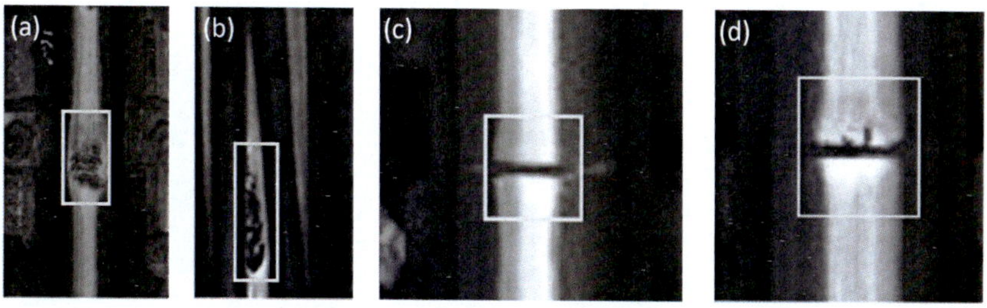

Fig. 2 Examples of the three classes **a** Class 1: A single squat, **b** Class 1: A crossing nose with a large defect, **c** Class 2: non-problematic isolation joint and **d** Class 3: A problematic isolation joint

Table 1 The number of images for different types of railways, and the number of objects/observations for the three different classes. Note that a single image may contain several object classes

Type	Conventional rail	Turnouts	Total
Images	1563	593	2155
Class	1. Defect	2. GoodIC	3. BadIC
Observations	2852	564	537

2.2 *Defects and Isolation Joints Classified on the Images*

Different defects are seen in the images as single squats, multiple squats, squats in flash-butt welds and squats in aluminothermic (aka thermite) welds, defects on the crossing nose and problematic isolation joints. Examples of the three different classes investigated are illustrated and surrounded by a bounding box in Fig. 2. The three classes illustrated are, Class 1: A rail surface defect shown in Fig. 2a and b, Class 2: A non-problematic isolation joint shown in Fig. 2c s(denoted GoodIC) and Class 3: A problematic isolation joint based on non straight cut in the rail shown in Fig. 2d (denoted BadIC).

The number of observations in each class are given in Table 1 together with the total number of images. Note that all the rail surface defects are treated and counted equally, each local defects is counted as a defect on the rail images.

3 Method

The YOLOv3 architecture for object detection uses supervised learning meaning that the model learns from examples also called a labelled data. The examples are used to learn the parameters in the network which can represent the data such that the three classes can be identified. This section describes the labelling process of the data and the data augmentation performed, as well as the performance metrics used

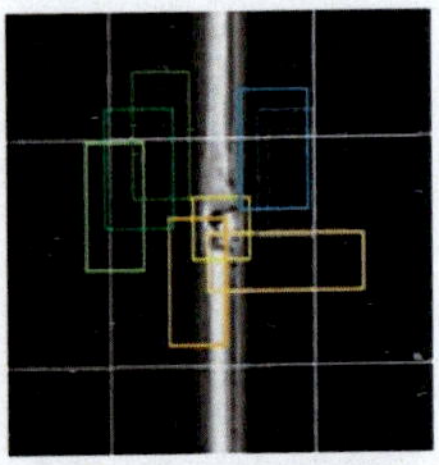

Fig. 3 A selected grid cell shown by the white lines which has several bounding box predictions, three bounding box predictions is predicted on the defect

for evaluating the final model. Overall, the object detection first splits the image into grid cells. Then in each grid cell, it identifies matching boxes around objects in the image and estimates a probability for each of the three classes shown in Fig. 3.

Section 7 describes the YOLO object detection model as well as the details of the data preparation, training the network together with the network architecture and the model parameters.

3.1 Labelling and Data Augmentation

The defects are tightly bounded by rectangles (boxes), one rectangle is used to bound a single defect. If several defects occur, each defect has an individual bounding box (multiple object detection). Examples of the four annotated images are illustrated in Fig. 2. The dataset was annotated in a collaboration between computer science experts and domain experts (the authors) with a label and a corresponding bounding box for each occurrence of the described classes.

The data are split randomly into three sets, 64% is reserved for the training, 16% for validation and 20% for the testing. We augment the training data with additional examples by mirroring the images vertically, horizontally and a combination of both, this allows for an introduction of additional examples for training, whilst also enabling the model to generalize better across different orientations of the defects. The number of images as well as labelled observations (objects) are summarized in Table 2 along with the distribution of images and objects into the railway types, the classes (defects, GoodIC and BadIC), and the training/validation/test sets.

3.2 Performance Metrics

To evaluate the object detection model, we use the confusion matrix as an overview of the classification performance, and the macro F1-score metric as the primary optimization objective. The general F1-score is given as

Table 2 The number of images are shown in the top table and the number of observations (objects) in each class is shown in the bottom table

Image types	Conventional rails	Turnouts		Total
Total	1563	593		2156
Training set	969	414		1383
Training set augmented	3876	1656		5532
Validation set	241	102		343
Test set	353	77		430
Objects in classes	1. Defects	2. GoodICs	3. BadICs	Total
Observations	2852	564	537	3953
Training set	1840	355	343	2538
Training set augmented	7360	1420	1372	10152
Validation set	496	96	89	681
Test set	516	113	105	734

$$\text{F1} = \frac{2(Precision \cdot Recall)}{(Precision + Recall)}, \text{ where} \tag{1}$$

$$\begin{aligned} Precision &= \frac{TP}{TP+FP} \\ Recall &= \frac{TP}{TP+FN} \end{aligned} \tag{2}$$

where true positives is denoted as TP, true negatives is denoted as TN, false positives is denoted as FP and false negatives is denoted as FN. In this study, we have a multi-class and multi-label case scenario. The macro F1-score is therefore given as

$$\text{Macro F1} = \frac{\sum_{i=1}^{C} \text{F1}_i}{C} \tag{3}$$

where C is the number of classes (in our case three) and F1_i is the F1-score for class i. That is, the macro F1-score is calculated as an average of the F1-scores for each class. Equal weight is given to all classes independent of the number of true labels in each class.

A true positive is defined as predicting the same class as the ground truth and additionally the bounding box of that prediction needs to have a minimum overlap area with the ground truth bounding box. This minimum overlap (threshold) is called the Intersection over Union, which is denoted as IOU_{thresh} (see Sect. 7 for further explanations).

4 Results

This section first contains a short summary of the training results, and subsequently the performance of the training, validation and tests are summarized for the individual classes along with an interpretation of these in terms of this application.

4.1 Training and General Performance Results

Two configurations of the network architecture and hyper parameters were tested.

The configuration with the highest input resolution showed the best performance and was therefore chosen for further investigation in this application. The training and validation results for the selected configuration are illustrated in Fig. 4. The F1-scores for both training (T) and validation (V) converge at around 7000 epochs (iterations of training with smaller batches of the data), and the training was thus stopped at this point. Further details of the network training and configurations can be found in Sect. 8.

The optimal network configuration gave macro F1-scores of 0.823, 0.815 and 0.829 with respect to the training, validation and test sets.

4.2 Performance Results for Individual Classes

In this subsection results from the performance of YOLO v3 on the training and validation data for only the selected configuration are summarized in Tables 3, 4 and 5 respectively. The confusion matrices show high accuracies 82–84% for the rail surface defects. Due to the visual similarities of GoodIC and BadIC there are some class prediction confusion of around 17–27%. However, the accuracies are still surprisingly high having a value in the range of 70% for the predictions of the two

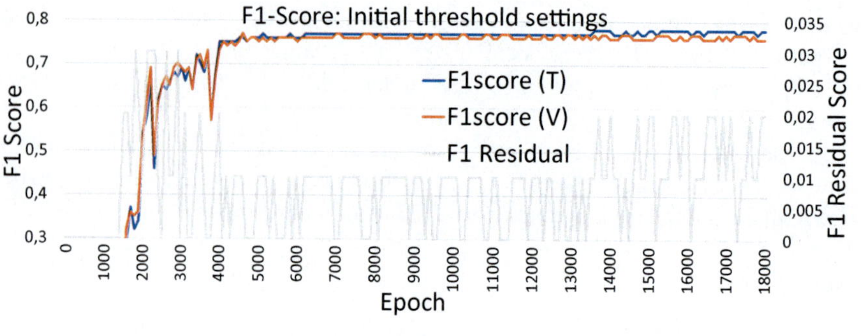

Fig. 4 F1-scores for the training (T) and validation (V) of YOLO v3, Configuration 2. These F1 scores (over all classes) are the output from Map in darknet [28]

Table 3 Confusion matrix for the Training case of the selected configuration. True observation on rows, and predictions in columns

Class		Predicted (training data)			
		Defect	GoodIC	BadIC	NoClass
True	Defect	0.82	0.00	0.00	0.17
	GoodIC	0.00	0.70	0.25	0.05
	BadIC	0.01	0.22	0.75	0.03
	NoClass	0.00	0.00	0.00	1.00

Table 4 Confusion matrix for the Validation case of the selected configuration. True observation on rows, and predictions in columns

Class		Predicted (validation data)			
		Defect	GoodIC	BadIC	NoClass
True	Defect	0.83	0.00	0.00	0.17
	GoodIC	0.01	0.70	0.23	0.06
	BadIC	0.01	0.27	0.69	0.02
	NoClass	0.00	0.00	0.00	1.00

Table 5 Confusion matrix for the Test case of the selected configuration. True observation on rows, and predictions in columns

Class		Predicted (test data)			
		Defect	GoodIC	BadIC	NoClass
True	Defect	0.84	0.00	0.01	0.15
	GoodIC	0.03	0.74	0.17	0.06
	BadIC	0.02	0.25	0.71	0.02
	NoClass	0.00	0.00	0.00	1.00

types of isolation joint (TP or TN). Due to a maximum of 6% of the isolation joints being missed (see NoClass (FN)) and less than 3–5% being misclassified as defects for the three performance metric cases.

Examples of correctly predicted classes are shown in Fig. 5 and examples of wrongly classified classes are shown in Fig. 6.

4.3 *Application Interpretation*

The number of true NoClass observations (possible bounding boxes without defects) is somewhat high, and thus the last rows in Tables 3, 4, and 5 appear almost without false positives. In fact, for the test set there is a total of 121 false positives. Out of the many possible discoveries this is a very low percentage ($\sim$ 0.00%), but if we

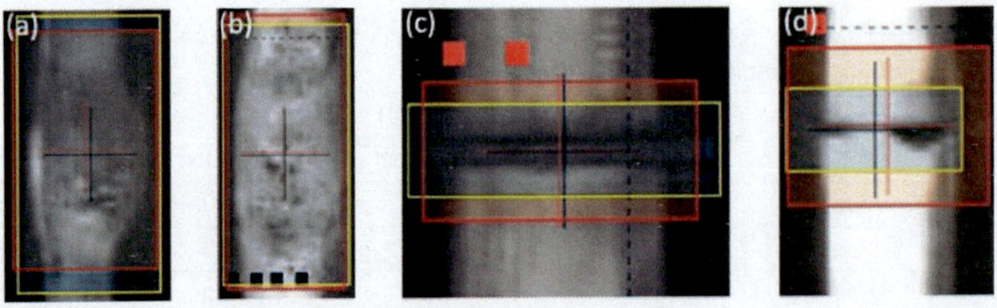

Fig. 5 Four examples of images with correct predictions where a yellow square is the ground truth and a red square is a prediction. **a** Correctly identified defect on a conventional rail (True positive of defect class), **b** Correctly identified defect on a crossing (True positive of Defect class), **c** Correctly identified good isolation joint (True positive of GoodIC class) and correctly identified faulty isolation joint (True positive of BadIC class)

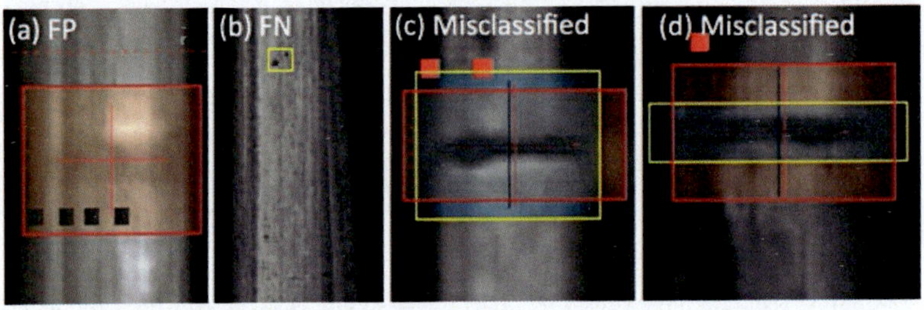

Fig. 6 Four examples of images with misclassifications where a yellow square is the ground truth and a red square is a prediction. **a** Incorrectly detected defect (False positive of Defect class), **b** Defect in a crossing which is not detected (False negative of Defect class), **c** Bad isolation joint incorrectly identified as good (False positive of GoodIC class) and **d** Good isolation joint which is incorrectly identified as faulty (False positive of BadIC class)

instead consider the amount of track investigated as 430 images $\cdot$ 2.51 m/image $\simeq$ 1078 meters of track investigated, this amounts to an average of around one false positive per 11 meters of imaged track. However, these false positives (FP) typically appear in the same part of the track as true positives, and thus we will investigate the number of false positives for tracks with no defects or isolation joints.

For a set of 2472 images without defects or isolation joints, the framework estimates 52 defects or isolation joints (false positives). Hence, on this data set consisting of 2472 images $\cdot$ 2.51 m/image $\simeq$ 6197 meters of track and with 52 false positives, there is an average of one false positive per 119 m.

If many defects occur in an image there is a tendency to miss some defects as is shown in Fig. 7a. If corrugation is present many false positives can occur as shown in Fig. 7b. In the case of Fig. 8a two predictions with a high overlap are shown, although their union covers the ground truth. In Fig. 8b, the ground truth could have been split up in two labels and only one defect was actually predicted.

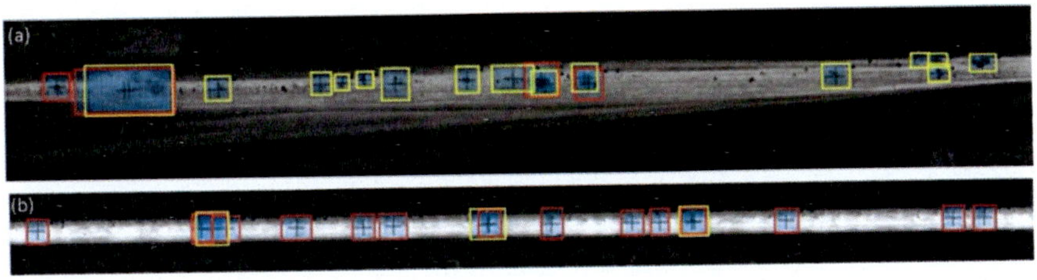

Fig. 7 Example of multiple defects which are shown with yellow squares (ground truth) and multiple predictions which are shown with red squares. **a** 3 predictions were made out of 15 labelled defects. **b** 15 predictions were made of 3 actual defects

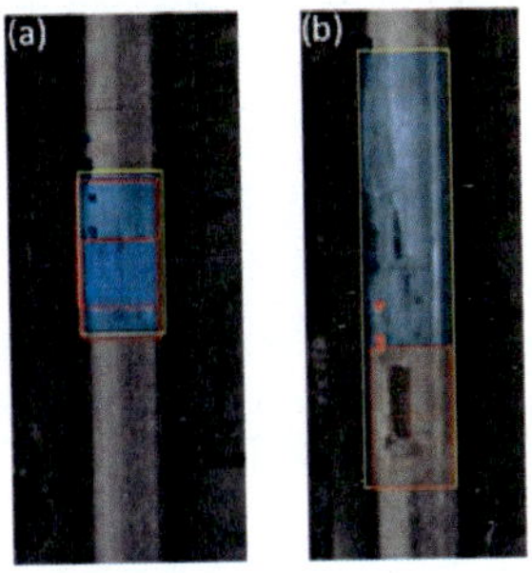

Fig. 8 Example of multiple defects which are shown with yellow squares (ground truth) and multiple predictions which are shown with red squares. **a** 3 predictions were made of 2 actual defects. **b** One prediction (red, small bounding box) was made of one actual defect (Yellow, large bounding box)

5 Discussion and Conclusion

A method for automatic detection of defects on rails is developed and promising results shown. In this study, the YOLO v3 framework is used to classify, localize and find the size of defects on the rails, localize isolation joints and furthermore classify the state of the isolation joints. The images are collected from all track infrastructure including turnouts and the results show the great versatility of the YOLO v3 framework. This versatility in the model can potentially reduce the need for visual inspections or find the location where visual inspection is needed earlier than planned in the entire railway infrastructure.

The recall rate for the defects is 84% which is promising, with respect to rarely missing surface defects. The recall rates for good and bad isolation joints are slightly lower (74% and 71%), due to difficulties in determining the state of the isolation joints to be either good or bad. For tracks which include defects, the false positives are on average one per 11 meters of track, and an error analysis suggests that improving these results require further labeling discussions related to the size of the bounding boxes as well as labeling of corrugation. On the other hand, the false positive rate is one per 119 meters of track for tracks with no defects. The results could potentially be complemented with other sources of investigation like the recording car's ultrasound system or visual inspections of the recorded images before field investigations are performed. Possible ways of improving the results are:

- Using progressive resizing (starting with a small resolution and then increasing the resolution during training) and end the training with higher resolution, larger than $w = h = 608$, although this requires more/a lot of GPU ram and takes more time.
- Having experts improving the labeling of the data as well as increasing the amount of training data.
- Using an image segmentation network (predictions made on individual pixels rather than in boxes). This might improve prediction results for especially corrugation, although labelling the data would be very time consuming due to a pixel-wise labelling process.
- Use an ensemble learner of the vision system results combined with predictions of defects from other data sources.

As a next step, the proposed model could be extended and used for quantifying the degradation (in terms of the different types of defects) of rails over time. Such a degradation model would require a collection of series of images of the same tracks over time.

6 Funding

Financial support for this study was provided by funding from "Innovation Fund Denmark" under grant number: 4109-0003B-INTELLISWITCH.

7 Appendix 1: You Only Look Once Object Detection

The following section explains how the YOLO network works by first describing the network architecture, and how the output is being computed across multiple prediction scales and anchor boxes, the non-max suppression, the loss function, and finally how the model learns using back-propagation.

7.1 YOLO Object Detecion Architecture

In order to fully understand the YOLOv3 neural network, it is useful to present a short review on the earlier versions of the YOLO architecture and methodology. In general and in the subsequent cases, object detection methods have consisted of three parts: Feature extraction, prediction of the presence of the object, and a

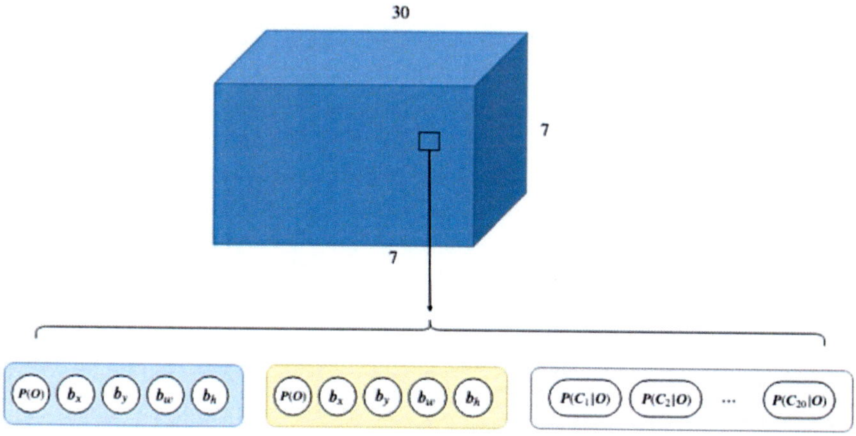

Fig. 9 Illustration of the final layer tensor in YOLOv1

localisation of the object in the image. The general idea is to divide the input image into $S \times S$ grid cells and for each cell predict B bounding boxes with C corresponding conditional classes. YOLOv1 [23] is based on a convolutional neural network, which acts as a feature extractor of the input image consisting of 24 convolutional layers, where the 24 convolutional layers have been pre-trained on the ImageNet-dataset for classification. These are followed by two fully connected layers and Illustrated above in Fig. 9, the final layer has dimensions $7 \times 7 \times 30$ which correspond to the SxS-grid which the input image was decomposed into, and a vector of length 30 at each grid cell consisting of $P(O)$, b_x, b_y, b_w, b_h encoding the probability that an object is present $P(O)$ also denoted p_c, a coordinate offset from the upper left corner of the grid cell b_x, b_y as well as the width and the height of the bounding box respectively b_w, b_h. In the paper the authors let $B = 2$ thus predicting two bounding boxes per grid cell (each 5 elements) following 20 elements of the conditional probabilities $P(C_i|O)$ for the occurrence of one of the 20 classes. Thus any given 7×7 grid cell can only become 'responsible' for one of the 20 classes, and for each of the 20 classes two bounding boxes are predicted. Hence, the final layer outputs an $S \times S \times (B * 5 + C)$ tensor predicting a total of 98 bounding boxes. The model was efficient but constrained by the coarseness of downsampling occuring in the convolutional layers (1×1 convolutions and pooling layers), thus having difficulties with detecting smaller objects (deformations such as problematic isolation joints).

The second iteration of the YOLO architecture, YOLO9000/YOLOv2 [29] attempts to alleviate some of the prior shortcomings by introducing a number of improvements. YOLOv2 improves on recall (in particular of smaller objects) by concatenating earlier layer feature maps (finer resolutions) forward to later layers (more coarse resolutions). In fact it introduces rescaling of input images every 10 epochs while training, as long as the image dimensions are a multiple of 32, the operations conform as expected. Inspired by the Region Proposal Network in networks such as Faster-R-CNN [19], the model also introduces anchor boxes for the localisation of the class object. Unlike, manually defined priors such as those in [19] the anchor boxes are computed automatically by k-means clustering on the training set to improve the priors for the training, which allow for easier training. The anchor boxes remove the fully connected layers from the network architecture presented in [23] and uses convolutional layers alone for the object detection. This makes the output tensor predict coordinates differently than given in YOLOv1. As we do not regress on the object coordinates directly with fully connected layers, this gives us a localised prediction per grid-cell in the feature map. Instead of predicting the centroid coordinate, in YOLOv2 we predict offsets to the anchor boxes.

7.1.1 YOLOv3

For our research we use YOLOv3. Similarly to before, we initially require a choice of feature extractor. We use the DarkNet-53 [30] architecture, which employs 53 convolutional layers. The feature extractor is pre-trained on the ImageNet-dataset [31]. The feature extractor architecture is shown in the left-most part of Fig. 10 with an example of an input image of size 416×416 pixels and 1 channel (gray scale).

The design of YOLOv3 as opposed to previous iterations introduces training across multiple scales of dimensions, i.e. we compute predictions across multiple scales in the network similarly to Feature Pyramid Networks [24]. Contrary to the multiple resolution training employed in YOLO V2, we do not change the dimensions of the input image batches between the epochs. The network is fully convolutional and can be described to consist of a number of repetitive operations involving 2D

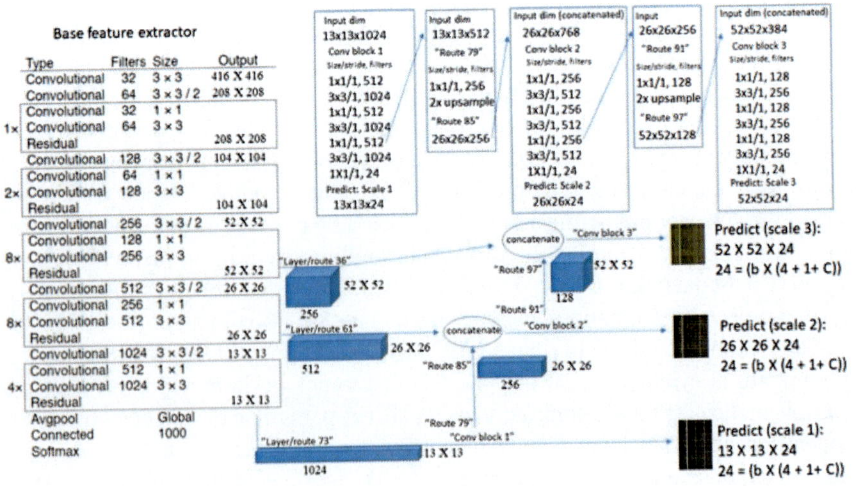

Fig. 10 The Yolov3 network architecture, adapted from [32]. The left part of the figure is the feature extractor shown in the table and the right part of the figure shows the prediction on the three scales

Convolutions with a 3×3 kernel and a 1×1 kernel, with batch normalization for each set of convolutions and a varying number of filters generally with a stride of 2 and a padding of 1. It employs residual (shortcut) layers similarly from ResNet-architectures [33], by adding prior layer activations to later layers, allowing for a greater depth in the network. Additionally, we apply what [26] refers to as 'routing'-layers, which resemble the longitudinal connections found in the Feature Pyramid Network architecture [34]. However, instead of performing addition from previous layer activations and passing it forward, we concatenate the feature maps from a previous layer activation to activations in a later layer of the network.

The full network consists of 106 layers (counting all layers including the YOLO layers). Using transfer learning, we want to reuse and thus fine-tune the DarkNet feature extractor, keeping parameter values of the pre-trained network while re-initializing the final layers prior to each prediction to conform with our classes. The layers (81, 93, 105) are in the final layers ("the YOLO layers 82, 94, 106) updated to $3 * (4 + 1 + C)$ for each of the C=3 classes and for each of the three anchors predicted for the scale, as shown in Fig. 10 on the right side.

YOLOv3 carries out predictions on three distinct scales (three prediction layers). The three prediction layers use feature pyramid networks by up-sampling the coarsest grid for a concatenation with the finer grid as suggested in [34] for a better detection on the finer scales. In [21] this was extended to five prediction layers, but we will use the three scales as in [26] for simplicity and speed.

7.1.2 Prediction Scales and Anchor Boxes

The output tensor of the neural network gives $S \times S \times [B \times (4 + 1 + C)]$ where $S \times S$ corresponds to the number of grid cells at each scale, computed at 3 scales with b_x, b_y, b_w, b_h and an confidence score p_c (objectiveness prediction) together with C class predictions. At each grid scale we have three anchor boxes (B). Each anchor box is defined by an x,y-coordinate, width and height respectively denoted as b_x, b_y, b_w, b_h. The output coordinates originates from $\mathbf{t} = (t_x, t_y, t_w, t_h)$ which is transformed by the anchor box "function" shown in Fig. 11 in the last YOLO layers (82, 94, 106).

The offset from the top left corner of the grid is defined as (c_x, c_y), where $\sigma(t) = \frac{e^t}{e^t+1}$ is the sigmoid function and p_w, p_h is the anchor box priors for width and height respectively. It has been suggested to use the k-means clustering algorithm to compute custom anchor sizes for the width (p_w) and height (p_h) when YOLO is applied on new data. However, in this research only nine default anchor boxes are utilized and no k-means customization is attempted for the priors.

A single prediction in one grid cell (at a certain prediction scale) for a single anchor, has four anchor box coordinates and an objectiveness score together with C class predictions conditional probabilities. Having $N \times N \times [B \times (4 + 1 + C)]$

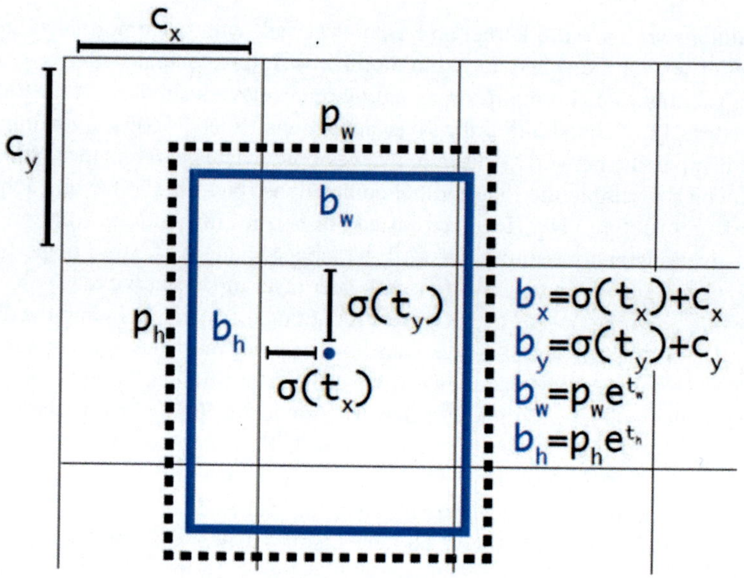

Fig. 11 Anchor box illustration. The figure is adapted from [29]

where $N \times N$ predictions is typically too many and thereby these predictions are suppressed by a threshold on the confidence score (p_c and also called objectiveness score), this threshold is denoted con_{thresh}.

7.1.3 Non-max Suppression on the Predictions

To suppress multiple bounding box detections for the same object in a gridcell, we use non-maximum suppression on the predicted bounding boxes. Given a number of predicted bounding boxes for an object in a gridcell, we iteratively remove boxes that overlap more than nms_{thresh} with the highest confidence score, measured in the Jaccard Distance metric, also known as Intersection over Union,

$$IoU = \frac{\#\ of\ Overlapping\ pixels}{\#\ of\ Union\ of\ pixels} \tag{4}$$

We repeat this procedure until only one bounding box is present. This procedure is visualized on one of the three scales in Fig. 12.

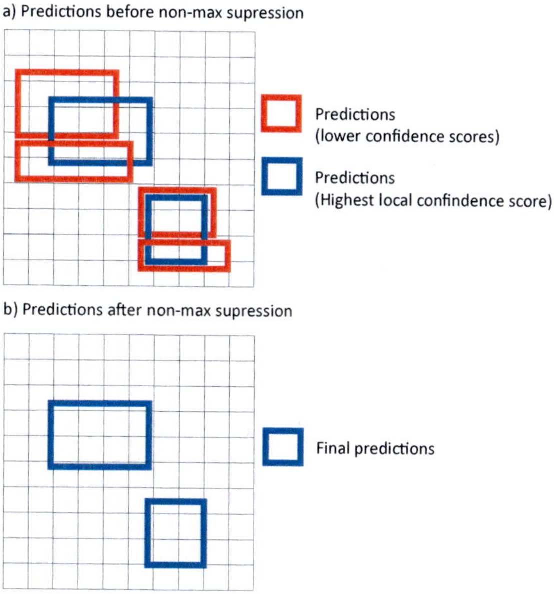

Fig. 12 Illustrations of Non-max supression. **a** Initial predictions before non-max supression and **b** Prediction after non-max supression

7.2 *The Loss Function*

Upon training the neural network, we need to have a measure of error from the ground truth to the predicted values, such that adjustments of the weights can be verified to go towards the minima of the error surface. The loss function is an innovative property of the YOLO architecture [23], allowing for a full end-to-end architecture for training an object detection model. Previously object detection was a multi-step procedure involving training of multiple models independent of one another (R-CNN). In YOLO v3, a number of modifications are made to the loss function, the first being the choice of using cross entropy (binary) for the class probability and objectiveness, whilst applying mean squared error loss to the spatial coordinates (b_x, b_y). The second is the fact that the loss is computed at all of the three different scales, meaning $S \in \{13, 26, 52\}$ for an example of an input image size of 416×416 pixels. If we define the cross entropy loss as given in Eq. 5

$$l(a_i, \hat{a}_i) = -a_i log(\hat{a}_i) + (1 - a_i) log(1 - \hat{a}_i) \tag{5}$$

Then we can write the full loss function as given in Equation 6.

$$L(\theta, X, y) = \lambda_{coord} \sum_{i=0}^{S^2} \sum_{j=0}^{B} \mathbb{1}_{ij}^{obj} \left(\left(\hat{b_x}_{(i,j)} - b_{x(i,j)} \right)^2 + \left(\hat{b_y}_{(i,j)} - b_{y(i,j)} \right)^2 \right.$$
$$\left. + \left(\sqrt{b_{w(i,j)}} - \sqrt{\hat{b_w}_{(i,j)}} \right)^2 + \left(\sqrt{\hat{b_h}_{(i,j)}} - \sqrt{b_{h(i,j)}} \right)^2 \right)$$
$$+ \sum_{i=1}^{S^2} \sum_{j=0}^{B} \mathbb{1}_{ij}^{obj} l(C_{i,j}, \hat{C_{i,j}})$$
$$+ \lambda_{noobj} \sum_{i=0}^{S^2} \sum_{j=0}^{B} \mathbb{1}_{ij}^{noobj} l(C_{i,j}, \hat{C_{i,j}})$$
$$+ \sum_{i=0}^{S^2} \mathbb{1}_{i}^{obj} \sum_{c \in classes} l\left(\left(p_i(c), \hat{p}_i(c)\right)\right) \tag{6}$$

$\lambda_{coord} = 5$ is chosen to increase the importance of the bounding box coordinates loss and decrease the loss importance for confidence predictions for boxes that do not have objects $\lambda_{noobj} = 0.5$ and $\mathbb{1}_{i}^{obj}$ denotes if the object appears in cell i (class) and $\mathbb{1}_{i,j}^{obj}$ for the coordinates loss, lastly $\mathbb{1}_{i,j}^{noobj}$ denotes that no object is present [26]. The loss is with respect to the bounding boxes, the squared error loss is used for the width and the height during training and if a ground truth box exists ($\hat{\boldsymbol{t}}_*$) the gradient used for back propagation is calculated based on the ground truth difference to $\boldsymbol{t}_*$ found by inverting the equations shown in Fig. 11.

7.3 Back-Propagation in Yolov3

For each iteration of a forward pass (passing an image through the network), the loss is computed as well as the derivatives w.r.t. the loss function parameters. This gives us a direction towards a lower loss on the error surface, used for updated the parameters with a learning rate stepsize α. We denote the network parameters as θ. Updating the parameters/weights is achieved by using Stochastic gradient descent (SGD) with momentum optimizer as initially suggested in [35] and given in Eq. 7,

$$V_t = pV_{t-1} + \alpha \frac{\partial L(\theta_{t-1}, X, y)}{\partial \theta_{t-1}}$$
$$\theta_t = \theta_{t-1} - V_t \tag{7}$$

where the momentum is denoted p (could be seen as a viscosity term), the learning rate is denoted α and the "learning" velocity as V_t where t is the current time step. The momentum parameter (p) acts as a "viscous damping term" which decreases the earlier velocity V_{t-1}. The learning rate decay γ reduces the learning rate α according to the learning rate scheduler implemented in [28].

7.4 Pre-processing

In order to use the recording car greyscale images in the YOLO v3 network the following pre-processing was performed. All images in the dataset were resized to have an input of 416 × 416 or 608 × 608, and all images in the dataset were mean centred and standardised. One input dimensions is used which was the greyscale channel.

8 Appendix 2: Results

This section gives detailed results from the network training.

8.1 Training and General Performance Results

Two configurations were tested, Configuration 1 with a lower image resolution and a smaller learning rate as compared to Configuration 2. All the training settings for the hyperparameters of the two configurations are shown in the upper part of Table 6.

Table 6 The training settings and performance metric results for the two tested configurations

Settings	Configuration 1	Configuration 2
Training parameters		
$w \times h$: Resolution	416 × 416	608 × 608
α: Learning rate	0.0010	0.0001
γ: Learning rate decay	0.0005	0.0005
p: Momentum	0.9	0.9
b: Batch size	64	64
sd: Subdivision	16	16
Initial threshold settings for F1-scores		
con_{thresh}: Confidence score threshold	0.25	0.25
nms_{thresh}: Non-max suppression	0.45	0.45
IOU_{thresh}: Intersection over union	0.50	0.50
Stopped at epoch	4700	7000
Selected settings for F1-scores		
con_{thresh}: Confidence score threshold	0.25	0.30
nms_{thresh}: Non-max suppression	0.45	0.45
IOU_{thresh}: Intersection over union	0.20	0.20
Results		
F1-score macro : Training	0.830	0.823
F1-score macro : Validation	0.810	0.815
F1-score macro : Test	0.790	0.829

The initial threshold settings were used for initial training and based on that it was decided at which epoch to stop training.

After stopping (at 4700 and 7000 epochs, respectively), additional threshold settings were investigated in order to improve the macro F1-scores. The tested settings were 0.2, 0.25, 0.3, 0.4 and 0.5 for both con_{thresh} (confidence score threshold) and IOU_{thresh} (intersection over union with the ground truth) giving a total of 50 additional settings for each configuration.

The threshold settings which gave the highest macro F1-scores for both training and validation were selected and these are shown in the bottom part of Table 6. Configuration 2 at the right of the table was preferred and selected for further investigation as it gave the best performance (highest macro F1-scores for training and validation). Note, the test macro F1-scores are listed for comparison, but these were not used for selection purposes in this work.

References

1. C for Rail C4R (2015) EU, Recommendations and guidelines for next generation monitoring and inspection. Towards an affordable, innovative and high-capacity European Railway System for 2030/2050, Tech. rep., International Union of Railways, UIC
2. Jovanović S, Božović D, Tomižić-Torlaković M (2014) Railway infrastructure condition-monitoring and analysis as a basis for maintenance management. Gradjevinar 66(4):347–358 https://doi.org/10.14256/JCE.959.2013
3. Singh AK, Swarup A, Agarwal A, Singh D (2019) Vision based rail track extraction and monitoring through drone imagery, ICT Express
4. C for Rail C4R (2016) EU, Monitoring-based deterioration prediction. Towards an affordable, innovative and high-capacity European Railway System for 2030/2050, Tech. rep., International Union of Railways, UIC
5. Rusu MF (2017) Automation of railway switch and crossing inspection. Ph.D. thesis, University of Birmingham
6. Cornish A (2014) Life-time monitoring of in service switches and crossings through field experimentation
7. Zerbst U, Lundén R, Edel K-O, Smith RA (2009) Introduction to the damage tolerance behaviour of railway rails-a review. Eng Fract Mech 76(17):2563–2601
8. C for Rail C4R (2015) EU, Operational failure modes of switches and crossings. Towards an affordable, innovative and high-capacity European Railway System for 2030/2050, Tech. rep., International Union of Railways, UIC
9. Olofsson U, Telliskivi T (2003) Wear, plastic deformation and friction of two rail steels-a full-scale test and a laboratory study. Wear 254(1–2):80–93
10. Molodova M, Li Z, Nunez A, Dollevoet R (2014) Automatic detection of squats in railway infrastructure. IEEE Trans Intell Transport Syst 15(5):6776541, 1980–1990. https://doi.org/10.1109/TITS.2014.2307955
11. Ishida M (2013) Rolling contact fatigue (rcf) defects of rails in Japanese railways and its mitigation strategies. Electron J Struct Eng 13(1):67–74
12. Freimanis A, Kaewunruen S, Ishida M (2018) Peridynamic modeling of rail squats. Sustain Solu Railwa Transport Eng:108–118. https://doi.org/10.1007/978-3-030-01911-2_10, https://doi.org/10.1007/978-3-030-01911-2
13. Kaewunruen S (2018) Discussion of "field test performance of noncontact ultrasonic rail inspection system" by stefano mariani, thompson nguyen, xuan zhu, and francesco lanza di scalea. J Transport Eng [A] Syst 144(4):04017007. https://doi.org/10.1061/JTEPBS.0000026

14. Molina LF, Resendiz E, Edwards JR, Hart JM, Barkan CP, Ahuja N (2011) Condition monitoring of railway turnouts and other track components using machine vision, TRB 11-1442
15. Zheng S, Chai X, An X, Li L (2012) Railway track gauge inspection method based on computer vision. In: 2012 IEEE International conference on mechatronics and automation (Icma 2012) 6284322, pp 1292–1296 https://doi.org/10.1109/ICMA.2012.6284322
16. Faghih-Roohi S, Hajizadeh S, Nunez A, Babuska R, De Schutter B (2016) Deep convolutional neural networks for detection of rail surface defects. In: International joint conference on neural networks (IJCNN) 2016, pp 2584–2589. https://doi.org/10.1109/IJCNN.2016.7727522
17. Wei X, Yang Z, Liu Y, Wei D, Jia L, Li Y (2019) Railway track fastener defect detection based on image processing and deep learning techniques: A comparative study. Eng Appl Artif Intell 80(2019):66–81. https://doi.org/10.1016/j.engappai.2019.01.008, http://www.sciencedirect.com/science/article/pii/S0952197619300089
18. Simonyan K, Zisserman A (2015) Very deep convolutional networks for large-scale image recognition. In: 3rd International conference on learning representations (Iclr 2015)—Conference Track Proceedings
19. Ren S, He K, Girshick R, Sun J (2016) Faster r-cnn: Towards real-time object detection with region proposal networks. IEEE Trans Pattern Anal Mach Intell 39. https://doi.org/10.1109/TPAMI.2016.2577031
20. Jensen MB, Nasrollahi K, Moeslund TB (2017) Evaluating state-of-the-art object detector on challenging traffic light data. In: IEEE Computer society conference on computer vision and pattern recognition workshops (2017) 8014856, pp 882–888. https://doi.org/10.1109/CVPRW.2017.122
21. Kim KJ, Kim PK, Chung YS, Choi DH (2018) Performance enhancement of yolov3 by adding prediction layers with spatial pyramid pooling for vehicle detection. In: 15th Ieee International conference on advanced video and signal based surveillance (avss) (2018), pp 1–6 https://doi.org/10.1109/AVSS.2018.8639438
22. Qu H, Yuan T, Sheng Z, Zhang Y (2018) A pedestrian detection method based on yolov3 model and image enhanced by retinex. In: 11th International congress on image and signal processing, biomedical engineering and informatics (cisp-bmei), pp 1–5 https://doi.org/10.1109/CISP-BMEI.2018.8633119
23. Redmon J, Farhadi A (2016) You only look once: Unified, real-time object detection. In: Ieee conference on computer vision and pattern recognition (cvpr) (2016), pp 779–788
24. Liu W, Anguelov D, Erhan D, Szegedy C, Reed S, Fu CY, Berg AC (2016) Ssd: Single shot multibox detector. Lecture Note Computer Science (including Subseries Lecture Notes in Artificial Intelligence and Lecture Notes in Bioinformatics) 9905:21–37. https://doi.org/10.1007/978-3-319-46448-0_2
25. Chen J, Liu Z, Wang H, Nunez A, Han Z (2018) Automatic defect detection of fasteners on the catenary support device using deep convolutional neural network. IEEE Trans Instrum Meas 67(2):257–269. https://doi.org/10.1109/TIM.2017.2775345
26. Redmon J, Farhadi A (2018) Yolov3: An incremental improvement, arXiv preprint arXiv:1804.02767
27. He K, Gkioxari G, Dollar P, Girshick R (2017) Mask r-cnn. In: Proceedings of the Ieee international conference on computer vision (2017) 8237584, pp 2980–2988. https://doi.org/10.1109/ICCV.2017.322
28. Alexey (2019) Darknet: Open source neural networks in c, https://github.com/AlexeyAB/darknet. Accessed on 2019 Nov 14
29. Redmon J, Farhadi A (2017) Yolo9000: Better, faster, stronger. In: Ieee Computer society conference on 2017 computer vision and pattern recognition workshops (cvprw), pp 6517–6525
30. Redmon J (2013–2016) Darknet: Open source neural networks in c, http://pjreddie.com/darknet/
31. Pre-trained weights for the darknet implementation of yolo v3, https://pjreddie.com/media/files/darknet53.conv.74

32. What's new in yolo v3 (2019) https://towardsdatascience.com/yolo-v3-object-detection-53fb7d3bfe6b. Accessed on 2019 Jan 19
33. He K, Zhang X, Ren S, Sun J (2016) Deep residual learning for image recognition. In: Proceedings of the Ieee computer society conference on computer vision and pattern recognition (2016), pp 770–778. https://doi.org/10.1109/CVPR.2016.90
34. Lin TY, Dollar P, Girshick R, He K, Hariharan B, Belongie S (2016) Feature pyramid networks for object detection, 10 p
35. Rumelhart D, Hinton G, Williams R (1986) Learning representations by back-propagating errors. Nature 323(6088):533–536. https://doi.org/10.1038/323533a0

The Effects of Renewal and Tamping on Ballast and Track Geometry in Turnouts

Seyed Mohammad Asadzadeh

Abstract The geometry of ballasted railway tracks is subject to gradual deterioration with the accumulated passing tonnage. Tamping is a corrective maintenance to restore track geometry to a good condition. A model of renewal/tamping effect in combination with track geometry degradation models is an integral part of track life-cycle cost analysis. However, the effect of renewal/tamping on track geometry in terms of the amount of improvement in different geometry parameters is not certain. Moreover, the effects of tamping on the degradation of ballast is of interest. Statistical models are needed to explain the uncertainty of these effects. This study investigates the effect of renewal/tamping in railway turnouts and puts forward a model for quantification of its uncertainty. The data from loaded geometry recordings are used. The study uses fractal analysis of track longitudinal level to distinguish between the effects of renewal/tamping on different wavelength contents of the geometry measurements and hence it quantifies the effects on ballast degradation. Furthermore, the influencing factors on renewal/tamping effect are specified and their contributions are quantified.

Executive Summary

Track geometry is regularly monitored by loaded measuring cars and maintenance is planned according to the revealed track condition. Ballast is an important component as it provides support for the superstructure. Therefore, evaluating the effects of tamping on geometry and ballast condition is necessary for life-cycle cost (LCC) analysis, especially when the decision should be made between either more tamping or complete renewal of the ballast. The research studies concerning geometry and ballast degradation in turnouts are limited. This study focuses on track geometry and ballast condition in turnouts, and on how they are affected by renewal and tamping. This study presents models to quantify the effects of tamping and renewal in turnouts. The effects of tamping/renewal are addressed from two different aspects:

S. M. Asadzadeh (✉)
Department of Electrical Engineering, Technical University of Denmark, 2800 Lyngby, Denmark
e-mail: semoasa@elektro.dtu.dk

R. Galeazzi et al. (eds.), *Intelligent Quality Assessment of Railway Switches and Crossings*, Springer Series in Reliability Engineering,
https://doi.org/10.1007/978-3-030-62472-9_13

(i) the short term recovery effect of tamping which is the amount of restoration in different track geometry parameters (longitudinal level, alignment, and twist), (ii) the longer term effect on geometry and ballast degradation rate. Specifically the rate of geometry degradation before and after both tamping and track renewal are compared and their difference is statistically explained. The results are based on maintenance and geometry recording data in 165 switches and crossings from the Danish railway network in time period 2012–2018.

Three geometric parameters in the turnout area are longitudinal level, alignment and twist (also referred to as Warp). The analysis and results are presented for these three parameters and the effects of tamping on these parameters are compared. The geometry measurements from one campaign to another for the same turnout may not be aligned and corrections should be made to have reliable data. This is especially important in turnouts with identifiable sections with different geometrical and load bearing characteristics. Tools based on mathematical methods are proposed to correct spatial misalignment of geometry data in turnouts. Ballast condition is evaluated by fractal analysis of the longitudinal level. The advantages are two-fold: (i) the method provides validated evaluation of the ballast fouled condition, (ii) the required data in most of railway networks are collected on a regular basis. Track geometry and ballast condition both degrade over time and the rate of degradation is an important parameter to make maintenance decisions. The degradation rates in this study are the amount of increase in standard deviation of track longitudinal level or increase in the degraded condition of the ballast per each passing tonnage.

The effects of both tamping and renewal on track geometry and ballast condition are reported in terms of simple regression models. These models relate the level of degradation after intervention to the level before intervention. The results for turnouts showed that on average 40–50% of the defects in the longitudinal level can be restored by tamping. A comparison performed for the effects of tamping in open track showed that more restoration (around 57–69%) is possible in open track. However, the uncertainty of tamping effect in turnouts is significant which means that there have been tampings that had failed to restore more than 20% of the degraded longitudinal geometry. The results may help infrastructure managers to find the root causes of the ineffective tampings. The results also showed that ballast condition is not majorly affected by tamping. However, the effectiveness of tamping in reducing ballast degradation rate is confirmed. This means that in turnouts when the degradation of ballast layer is happening at very high rate, tamping decreases the degradation rate. This may be because of the leveling effect of tamping in the longitudinal profile which helps reducing dynamic train/track interaction forces.

1 Introduction

Turnouts (also known as switches & crossings) are an essential part of railway infrastructure in best routing of trains and optimal track utilization. These systems are a vulnerable part of the infrastructure due to complex geometrical design, moving parts and rail discontinuity. Moreover their relatively significant share in total maintenance cost makes them critical components of the railway track.

In turnouts the general performance of system components is heavily dependent on the quality of support that the track substructure provides for the superstructure. Maintaining good quality of track geometry will ensure minimum dynamic forces between running vehicles and turnout components, and minimizes exposure to excessive failures of other system components. Tamping is a maintenance action to restore track geometry. Track renewal may also be considered as a remedy of geometric defects when long term effects are needed. The optimization of track geometry maintenance in a railway network often entails a decision between tamping and renewal. To help making the decision regarding renewal or tamping interventions, a common practice is to develop a life cycle cost (LCC) model for the track so that different decision making schemes can be simulated and their associated costs can be evaluated ([1]). Models of track degradation (for different components), tamping effects and renewal effects need to be embedded in the LCC model.

This study presents models to quantify the effects of tamping and renewal in turnouts. The effects of tamping/renewal are addressed from two different aspects: (i) the short term recovery effect of tamping which is the amount of restoration in different track geometry parameters (longitudinal level, alignment, and twist), (ii) the longer term effect on geometry and ballast degradation rate. Specifically the rate of geometry degradation before and after both tamping and track renewal are compared and their difference is statistically explained.

The results are based on maintenance and geometry recording data in 165 switches and crossings from the Danish railway network in time period 2012–2018. More than 3000 records of track geometry in these turnouts were available and after cleaning and alignment, these data were used to calculate the rate of geometry degradation, and to estimate the effects of tamping/renewal.

2 Data Cleaning and Alignment

The geometry measurements from one campaign to another for the same section of the track may not be aligned. Two types of data mis-alignment are shifted signal, and stretched or shrunk signal [2]. In order to correct positional errors of inspection data from a track geometry car, optimization methods are usually used to find the corrections which maximize some predefined "alignment criteria" [3, 4].

In the data available to this study, two types of positional errors need to be corrected (Fig. 1). First, the data from one campaign to another are not aligned i.e. the signal is shifted (Fig. 1a). It is noted that no stretched or shrunk signals have been observed in the data. This positional error in turnouts is well treated in Hovad et al. [4] where an optimization problem is solved to find the geometry parameter and also the amount of shift that maximize the cross correlation of geometry signals from consecutive

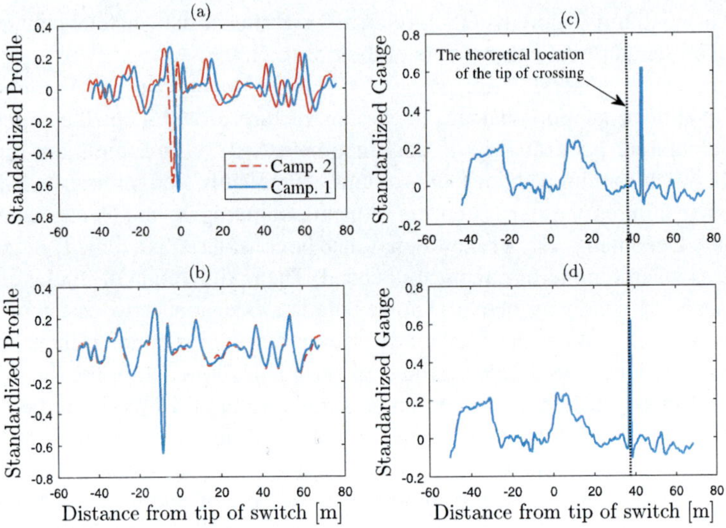

Fig. 1 Geometry data alignment; corrections for positional errors

measurement dates. In this study, the track longitudinal signals are used to calculate the shift that is needed to maximize cross correlation. An optimization procedure is employed to find the optimal shift in one signal to be maximally correlated with the signal from the previous campaign (measurement date) (Fig. 1b).

Second, the position of measurements in a specific campaign are not reliable. This correction is performed with respect to automated gauge measurements. Since the geometry measuring car uses a non-contact measuring system to measure gauge, the system is susceptible to outside interference and occasionally the equipment is confused by gaps in the rail. Because of this, the behaviour of loaded gauge measurements in the crossing area is unique with spikes at the tip of the crossing (Fig. 1c). This is probably because of the discontinuity of the rail in the crossing point. The distance between the nose of the crossing and the tip of the switch blade is a unique design specification of the studied turnouts, i.e. $X = 37.7\,m$. Accordingly, the signal shown in Fig. 1c should be shifted to the left so that the spike overlaps with the theoretical location of the crossing nose. Finally, all the previously aligned signals in Fig. 1b will also be shifted to the left, accordingly. It should be noted that the shifts found in each of the mentioned corrections are applied to all the geometry signals used in this study, i.e. longitudinal level, alignment, and twist.

3 Methodology

3.1 Geometric Quality Indicators

Three geometric parameters in the turnout area are longitudinal level, alignment and twist (also referred to as Warp). Twist is defined as the algebraic difference between two cross levels (Cant) taken at any two points within a chord length of 2 m. Twist is a crucial factor that is considered during derailment risk assessment and thus must not be ignored during the evaluation of track geometry [5].

The commonly used indicators for the quality of geometric parameters are the standard deviation (SD) and the mean-to-peak value which shows the maximum level of defects [6]. The quality of ballast is evaluated by fractal analysis of the track longitudinal level. Fractal analysis allows for analyzing wavelength content of the vertical track profile and distinguish between the track defects for short-, medium-, and long wavelengths. Standard deviation of track alignment is used to quantify geometric defects in the lateral direction. For twist, individual defects are represented by the amplitude from the zero-line to the peak value. The quality index is the mean number of defects for every meter of track length. An individual defect is recorded when the value of zero-to-peak exceeds a threshold value of 4 mm, which is an alarm limit in the European standard EN 13848-5 ([6]). The degradation and the recovery after tamping of different geometric parameters can be correlated. Andrade and Teixeira ([7]) reported that when used to trigger tamping actions the SD of the longitudinal level and alignment is a good predictor of maintenance needs for the rest of the track geometry parameters. Therefore, the information in one parameter may be used for estimation of tamping recovery in other geometric parameters. This aspect of correlation analysis will be addressed in more detail in the next sections.

3.2 The Applied Fractal Analysis Method

The divider method is the central part of fractal analysis [8–10]. It applies on the deviation pattern in the track longitudinal level. In this study, the divider method uses constant step length along the x-axis. Consider a signal that is measured using a constant step length λ. The length of the signal is approximated by connecting the consecutive points on the signal corresponding to the starting and ending points of the steps, forming a polygonal chain (Fig. 2). As the step length λ decreases towards the minimum possible length (which is 25cm in the case of track recording car data), more accurate approximation of the pattern emerges.

Figure 3 is the fractal plot that results from the fractal dimentioning illustrated in Fig. 2. This plot in which both the x- and y-axis are presented in logarithmic form, is known as a Richardson plot and is typical of profile data in turnouts in that it often exhibits bi-fractal pattern with two distinguishable linear parts. The relationship between the step-length values and the corresponding measured length is different

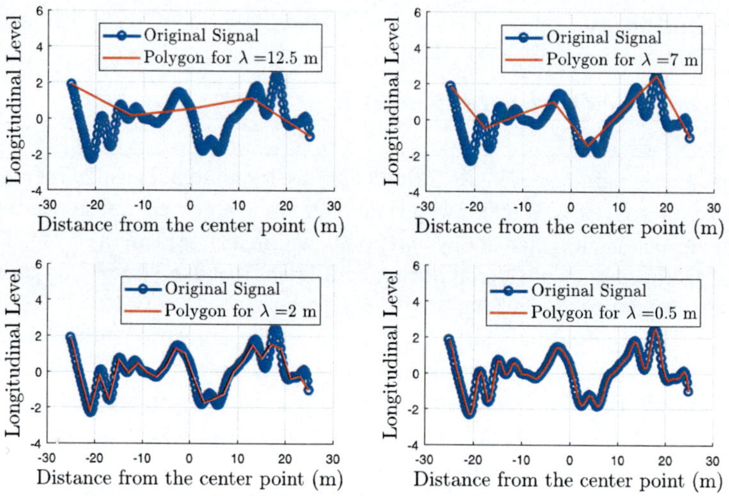

Fig. 2 Fractal dimensioning for an example geometry signal, subplots show the polygonal chains for different divider length labelfig

for these parts. Specifically, the fractal value is determined from the rate of change in the length of polygonal chain to corresponding step-length λ, based on a linear regression model with logarithmic variables:

$$log10(L(\lambda)) = \alpha + \beta log10(\lambda) \tag{1}$$

in which $L(\lambda)$ is the length of polygonal chain when the step size is λ, and α and β are the intersect and the slope of the regression line fitted to polygonal chain's length graph, respectively. As the polygonal chain's length is always decreasing with the step length, β is a non-positive value. Usually the degree of roughness in each of the fractal dimensions is calculated as $D_{Rj} = 1 - m_j$; $j = 1, 2$ in which m_j is the slope of the fitted line in fractal dimension region j.

The application of fractal analysis with the algorithm described so far to railway track geometry data in turnouts entails two orders of roughness. The small-scale roughness $D_{R2} = 1 - m_2$, as hypothesized by Hyslip [9], may be affected by the condition of ballast. Field measurements conducted by Landgraf et al. [11] in Austria showed that three orders of roughness are identified from the open track data and they were shown to be correlated with the condition of the sleepers, the ballast health condition, and the condition of sub-ballast and the soil.

In the analysis performed in turnouts, the fractal dimensioning results in two degrees of roughness. It is observed that the second fractal dimension $D_{R2} = 1 - m_2$ corresponds to the wavelength between 2 and 20 m. Accordingly, as an index for ballast quality (Q_{bl}), we refer to the slope of the fitted regression line in the second fractal region (usually $log10(\lambda) < 1.08m$) (see also Fig. 3). It is noted that more steep slope is hypothesized to be related to more degraded ballast.

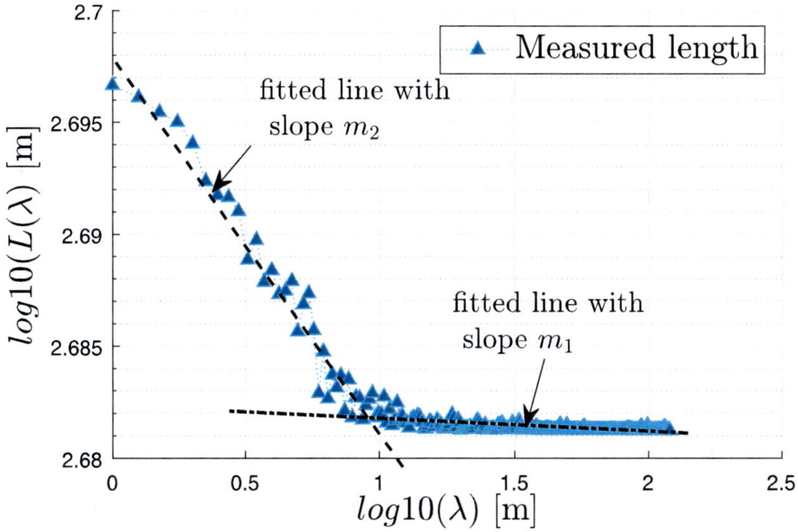

Fig. 3 Fractal plot for an example geometry signal

The procedure described above calculates Q_{bl} for the origin of the horizontal axis in Fig. 2, i.e. the center point of the geometry profile signal. In turnouts, this center point is 22 m after the tip of switch blade, and the length of geometry signal for which fractal analysis is performed is 75 m.

3.3 *Track Geometry Degradation*

Between every two consecutive interventions (either tamping or renewal), track geometry degrades proportionally with the accumulated tonnage. The rate in which the track geometry degrades is an important parameter for LCC modelling. Different mathematical models have been proposed to describe geometry degradation including linear, quadratic, and exponential models. A linear model reads

$$Q_{LL}(t) = \alpha + \beta t + \epsilon \tag{2}$$

in which $Q_{LL}(t)$ is the geometric quality index for longitudinal level and t is accumulated tonnage, ϵ is the error term and normally distributed random variable with mean zero and variance σ^2. The regression coefficient β represent the linear degradation rate. The model in Eq. (2) may also be represented in a quadratic or exponential form as in Eqs. (3) or (4), respectively.

$$Q_{LL}(t) = \alpha + \beta t + \gamma t^2 + \epsilon \tag{3}$$

$$Q_{LL}(t) = e^{\alpha+\beta t+\epsilon} \tag{4}$$

The Wiener process model for $Q_{LL}(t)$ is presented as

$$Q_{LL}(t) = Q_{LL}(0) + \mathcal{N}(\mu t, \sigma\sqrt{t}) \tag{5}$$

in which $Q_{LL}(0)$ is the initial degraded quality, μ is the drift coefficient, and σ is the diffusion coefficient. In the Wiener process, a Brownian motion model represents the stochastic dynamics of the degradation process. In model (5), the drift coefficient μ represents the degradation rate.

Audley and Andrews ([12]) showed that between linear and exponential models, the linear is preferred as in the geometry degradation process, there is no curvature to be captured by a non-linear model. In the study by Soleimanmeigouni et al. ([13]) the degradation rates from a linear model and a Wiener process are estimated to be very close. This result also suggests that a linear model is an acceptable model to calculate the degradation rate of track geometric parameters.

3.4 Renewal/Tamping Effect

The approaches to model the effects of renewal/tamping can be deterministic or probabilistic. The amount of recovery after tamping in any of the geometric parameters is the response variable. In a deterministic approach, the recovery variable ($\Delta\, Q_{LL}$) is described by a linear model with track quality before tamping ($Q_{LL}^{[b]}$), degradation rate before tamping ($\beta^{[b]}$), the line speed (v), and maintenance history ($\mathcal{M}$), as its predictors:

$$\Delta Q_{LL} = \mathcal{G}\left(Q_{LL}^{[b]}, \beta^{[b]}, v, \mathcal{M}\right) \tag{6}$$

Maintenance history is usually quantified by the number of past tampings since the last track renewal. In a probabilistic approach, the quantity of interest is the probability density function of the recovery variable (ΔQ_{LL}), its parameters are described as a function of track parameters:

$$P(\Delta Q_{LL} \le d) = F(d, \theta) \tag{7}$$

$$\theta = \mathcal{H}\left(Q_{LL}^{[b]}, \beta^{[b]}, v, \mathcal{M}\right) \tag{8}$$

The commonly used distribution functions to approximate F are lognormal, Weibull, and Gamma distributions. Furthermore, a linear form for function $\mathcal{H}$ is used to estimate θ ([12–14]). Figure 4 illustrates some of the variables involved in the mod-

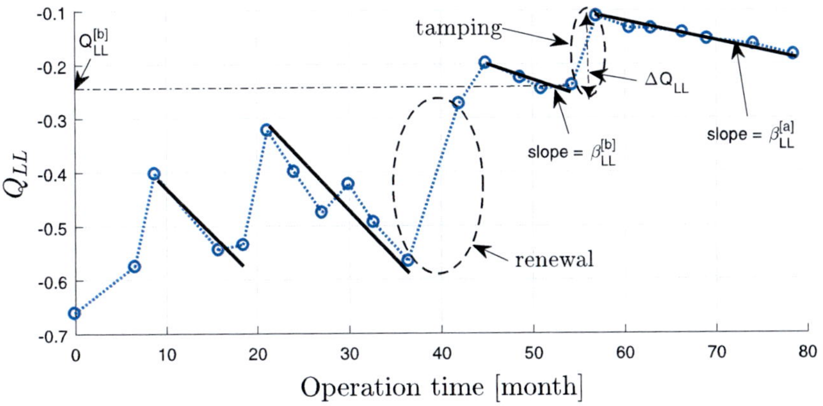

Fig. 4 Track longitudinal geometry degradation and the effects of renewal/tamping

elling of track geometry degradation and renewal/tamping effect. It is noted that in this figure, the horizontal axis represents the track operating time but the calculation of geometry degradation rate $\beta_{LL}^{[b]}$ and the ballast degradation rate $\beta_{bl}^{[b]}$ is based on accumulated tonnage i.e. the change in the standard deviation or the fractal values for every tonnage of the passing traffic.

4 Results

In this section, the effects of renewal/tamping are presented from two different aspects. First, the recovery effects that renewal/tamping have on different geometric parameters and on the condition of ballast. Second, the effects of renewal/tamping on degradation rate of vertical track profile which is attributed to the ballast condition. Moreover, the correlation of effects on different geometric parameters are discussed.

4.1 Recovery Effects

Figure 5 shows the recovery values of the standard deviation of track longitudinal level (SD_{LL}) associated with both renewal and tamping. The estimated models for recovery effect of track renewal on SD_{LL} are presented in Equations (9) and (10). To account for the effects of speed, the results are given for two different clusters of turnouts with maximum permissible speeds 120 and 250 km/h. These models suggest that renewal in the turnouts can restore nearly 67% of SD_{LL} before renewal. The coefficient of determination $R^2 = 49\%$ for these models shows that SD_{LL} before and after renewal are not highly correlated.

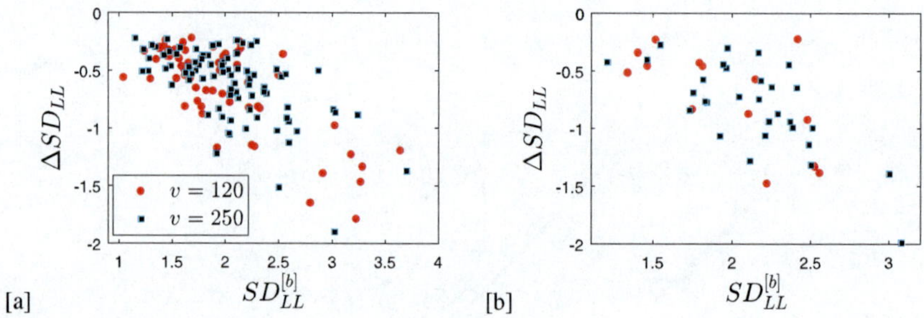

Fig. 5 The recovery values versus track longitudinal quality for turnouts exposed to two different running speeds; **a** tamping effect, **b** renewal effect

$$\Delta SD_{LL} = 0.58 - 0.67 SD_{LL}^{[b]}; \qquad v = 120\ km/h \tag{9}$$

$$\Delta SD_{LL} = 0.62 - 0.67 SD_{LL}^{[b]}; \qquad v = 250\ km/h \tag{10}$$

A tamping recovery model for SD_{LL} is presented in model (11). This regression model with $R^2 = 46\%$ only explains a part of the uncertainty in tamping recovery values. This could be because of high variation in the type and class of tamping. This study did not have access to detailed data on the type and class of tamping. It should be noted that in the current study the tamping events are extracted from the data. Specifically, a tamping intervention is assumed when the standard deviation of track longitudinal level decreases significantly. Model (11) suggests that about 44% of SD_{LL} before tamping can be recovered. In model (11), v^c is a categorical variable for track speed and takes on unity value when $v = 250$ km/h and zero otherwise. Increasing the line speed significantly decreases the amount of possible recovery by tamping.

$$\Delta SD_{LL} = 0.23 - 0.44 SD_{LL}^{[b]} + 0.05 v^c \tag{11}$$

To evaluate the effects of renewal/tamping on ballast quality, the longitudinal level data are analysed with the fractal analysis method. Here the effects of renewal/tamping on Q_{bl} are compared with their effects on SD_{LL}. Please note that the fractal values are always negative quantities, but in order to have a more comprehensive comparison between the fractal values and the standard deviation, the absolute values of the fractal values are assigned to the ballast condition indicator Q_{bl}. Figure 6 shows the recovery values of Q_{bl} associated with both renewal and tamping. A renewal recovery model for ballast quality in turnouts is estimated in model (12). This regression model with $R^2 = 86\%$ can explain a significant part of the uncertainty in renewal recovery values. Model (12) suggests that about 89% of Q_{bl} before renewal can be recovered, which is an expected effect from a renewal intervention. Similar to the effects on ΔSD_{LL}, the line speed does not significantly alter the effect of renewal on ballast quality recovery ΔQ_{bl}.

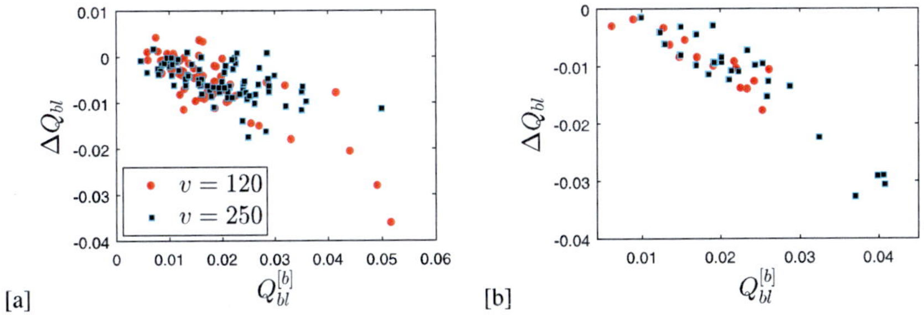

Fig. 6 Ballast recovery versus ballast quality before intervention in turnouts **a** tamping effect, **b** renewal effect

$$\Delta Q_{bl} = 0.0079 - 0.89 Q_{bl}^{[b]} \tag{12}$$

A tamping recovery model for ballast quality in turnouts is estimated in model (13). This regression model with $R^2 = 38\%$ can partly explain the uncertainty involved with the effects of tamping on fractal values. Similar to the effects of tamping on ΔSD_{LL}, increasing the line speed significantly reduces the possible amount of tamping recovery ΔQ_{bl}. Model (13) also suggests that about 36% of Q_{bl} before renewal can be recovered. Compared to the renewal effect, this value is significantly smaller. This result also agrees with the general expectation about tamping effect on ballast quality. The engineering insight about tamping effect implies that if not destructive, tamping could not positively affect ballast quality, ballast fouling condition, or crushed particles. Therefore, the percentage of recovery revealed by model (13) could not be interpreted as the effect of tamping on ballast particles and fouling condition. However, tamping helps restoring some elasticity of the ballast layer, fixing some part of ballast differential settlement, and filling in the voids under the sleepers. These effects may be seen in the fractal values as they may also seen in the standard deviation of longitudinal level.

$$\Delta Q_{bl} = 0.0008 - 0.36 Q_{bl}^{[b]} + 0.0011 v^c \tag{13}$$

Figure 7 shows the recovery values of track alignment and twist associated both with renewal and tamping. These data are used to estimate models for recovery effects on these two parameters in Table 1. As seen, renewal can recover 59% of the degraded quality of track alignment and 98% of the degraded quality of twist. Increasing the line speed also decreases the recovery effect of renewal on track alignment but speed does not significantly affect that of the twist parameter. Tamping, as expected, has a smaller effect than renewal in restoring track alignment and eliminating twist defects. Tamping can only recover 31% of alignment degradation and 56% of the degraded quality of twist. The effect of speed is not the same for the cases of renewal and tamping. Results show that the line speed has significantly less effect on alignment and twist in the case of tamping than in renewal. Speed has different effect on recovery

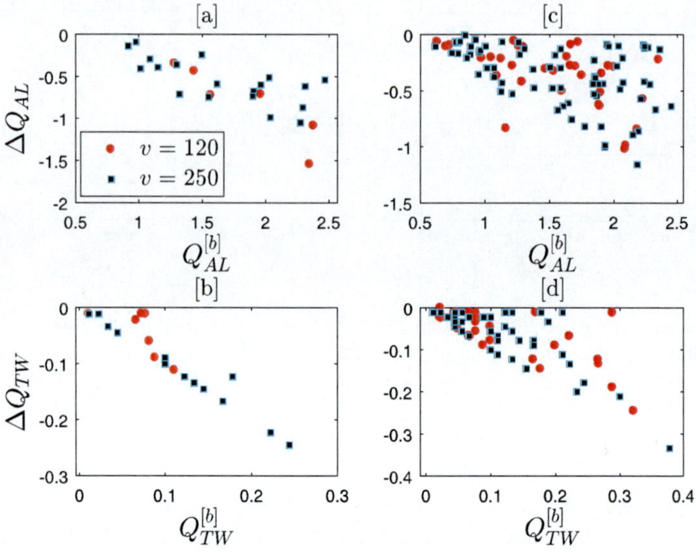

Fig. 7 The recovery effect versus track quality before intervention of **a** renewal on alignment, **b** renewal on twist, **c** tamping on alignment, **d** tamping on twist

Table 1 Recovery effect models of renewal/tamping for track alignment and twist

Intervention	Parameter	R^2	Estimated model
Renewal	Alignment	74%	$\Delta Q_{AL} = 0.17 - 0.59 Q^{[b]}_{AL} + 0.26 v^c$
Renewal	Twist	93%	$\Delta Q_{TW} = 0.03 - 0.98 Q^{[b]}_{TW} - 0.02 v^c$
Tamping	Alignment	38%	$\Delta Q_{AL} = 0.11 - 0.31 Q^{[b]}_{AL} + 0.03 v^c$
Tamping	Twist	58%	$\Delta Q_{TW} = 0.00 - 0.56 Q^{[b]}_{TW} - 0.006 v^c$

of vertical profile and alignment than on recovery of twist, in both cases of tamping and renewal. In high speed lines, less recovery of vertical profile and alignment is possible after renewal whereas more recovery of twist can be expected. For tamping, speed does not show a big impact on recovery of alignment or twist.

The correlation structure of recovery after intervention of these three geometric parameters is drawn in Table 2. As seen, recovery values of track vertical level and horizontal alignment are highly correlated. This correlation is more pronounced after renewal but still significant after tamping. Recoveries of track alignment and twist in turnouts are not significantly correlated. There is a moderate correlation between recovery of longitudinal level and twist.

Table 2 Linear correlation (%) of the recovery after intervention of geometric parameters

Intervention	Longitudinal and alignment	Longitudinal and twist	alignment and twist
Renewal	71	31	20
Tamping	53	18	5

4.2 *Effects on Ballast and Vertical Geometry Degradation Rates*

Figure 8 compares the degradation rate of ballast and vertical track profile before and after both renewal and tamping. The first observation from Fig. 8 is that more than 75% of track renewals has occurred in the turnouts on the high speed lines. The data in Fig. 8a and b do not statistically support the hypothesis that track renewal or tamping decreases geometry degradation rate. Moreover, the rate of change in the fractal values which is an indication of ballast degradation rate is not altered by either tamping or renewal. These hypothesises are statistically tested in Table 3 where a paired t-test is used to compare $\beta_{LL}^{[b]}$ and $\beta_{LL}^{[a]}$ for the case of geometry degradation and to compare $\beta_{bl}^{[b]}$ and $\beta_{bl}^{[a]}$ for the case of ballast degradation. For all the tests in Table 3, the null hypothesis is $\mathcal{H}_0 : \beta_{LL/bl}^{[b]} = \beta_{LL/bl}^{[a]}$ and as seen in this table, none of the null hypothesis can be rejected (p-value>0.05). This observation is a support for the assumption that degradation of geometry and ballast condition is only proportional to the cumulative loading of the track and other operational conditions. Particularly, when the degradation rate of both geometry and ballast are very low (e.g. below their averages), no change in the rates is observed after intervention. It is likely that in these cases, the reason for renewal has been a fast degradation outside the turnout area, but in its close vicinity.

It is however evident from Fig. 8c and d that when degradation rate of ballast before intervention is higher than average (when $\beta_{bl}^{[b]} \geq 0.0005$), the degradation rate after intervention is significantly reduced. This hypothesis is statistically tested in Fig. 9 where the probabilistic approach to modelling degradation rate before and after interventions is applied. The two-parameter lognormal distribution has shown better fit compared to Weibull and Gamma distributions. As seen in Fig. 9a, when the ballast degradation is happening at a relatively high rate, the renewal slows down the degradation process.

The probabilistic approach also confirms the effectiveness of tamping actions in reducing ballast degradation rate (Fig. 9b), when degradation rate is above average. The lognormal distributions fitted to degradation rates before and after tamping are different. The lognormal fit for $\beta_{bl}^{[b]}$ before tamping has $\mu = -6.8$ and $\sigma = 0.65$ and the fit for $\beta_{bl}^{[a]}$ after renewal has $\mu = -7.7$ and $\sigma = 1.2$. The difference of the two distributions suggests that in turnouts when the degradation of ballast layer is happening at very high rate, tamping decreases the degradation rate. This may be

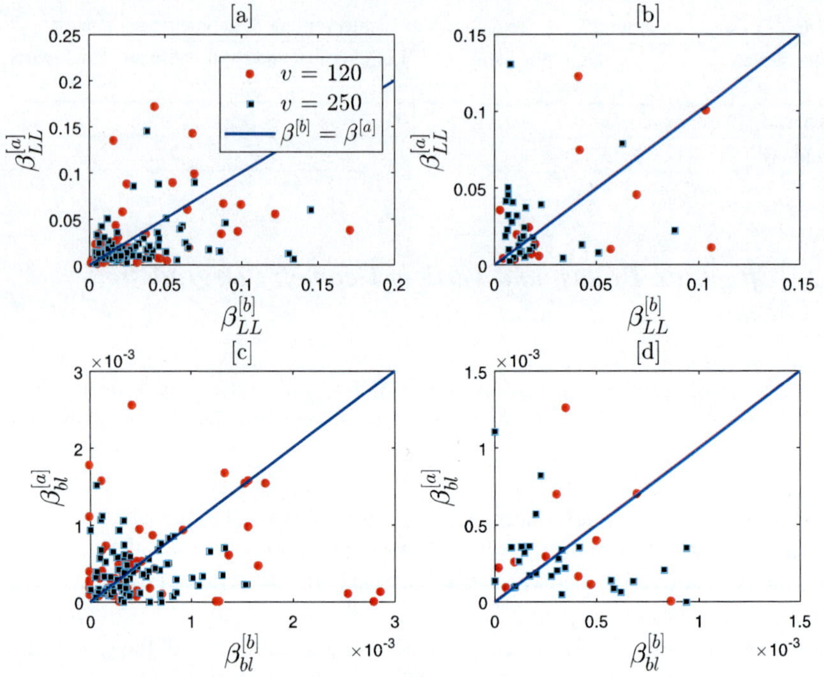

Fig. 8 Comparison of the degradation rate of track longitudinal quality before and after intervention; **a** tamping and SD_{LL}, **b** renewal and SD_{LL}, **c** tamping and fractal values, **d** renewal and fractal values

Table 3 Test of significance for the effects of renewal/tamping on degradation rates

Intervention	Parameter	df	sd	tstat	p-value
Renewal	β_{LL}	45	0.07	−0.24	0.81
Tamping	β_{LL}	143	0.05	1.03	0.31
Renewal	β_{bl}	43	0.001	0.59	0.56
Tamping	β_{bl}	142	0.001	1.38	0.17

because of the leveling effect of tamping in the longitudinal profile which helps reducing dynamic train/track interaction forces [16, 17].

4.3 Comparison with Open Track

In this section, the models for recovery effect of tamping which are available from the related literature are compared with the models generated for turnouts (Table 4). In this table, ΔSD_{LL} shows the change in the standard deviation of track longitudinal

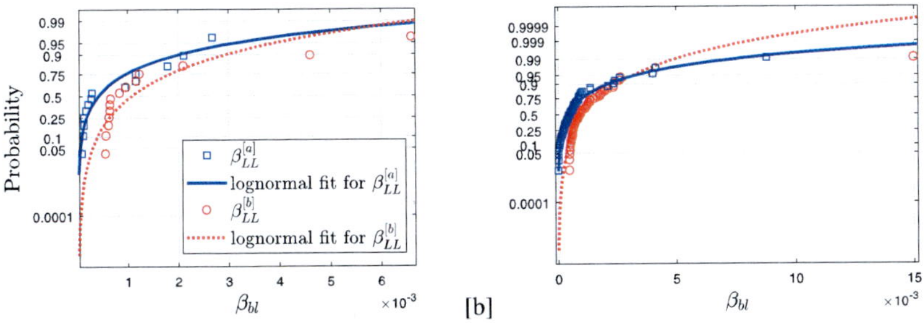

Fig. 9 Probabilistic models for $\beta_{bl}^{[b]}$ and $\beta_{bl}^{[a]}$ **a** renewal, **b** tamping

Table 4 Available models for tamping recovery effect

Source	Model*	Explanation
Caetano and Teixeira [15]	$\Delta SD_{LL} = 0.25 - 0.69SD_{LL}^{[b]}$	First tamping after renewal, the Lisbon - Porto railway line, maximum train running speeds of 220 km/h
Soleimanmeigouni et al. [13]	$\Delta SD_{LL} = 0.13 - 0.57SD_{LL}^{[b]}$	Derived from Weibull parameters, Main Western Line in Sweden, maximum train running speeds of 200 km/h
Wen et al. [18]	$\Delta SD_{LL} = 0.2 - 0.6SD_{LL}^{[b]}$	Od - Fa Corridor in Denmark, maximum train running speeds of 250 km/h
This study	$\Delta SD_{LL} = 0.17 - 0.39SD_{LL}^{[b]}$	Turnouts in Denmark, maximum train running speeds of 250 km/h
This study	$\Delta SD_{LL} = 0.33 - 0.49SD_{LL}^{[b]}$	Turnouts in Denmark, maximum train running speeds of 120 km/h

* SD_{LL} is the standard deviation of longitudinal level and superscript $^{[b]}$ indicates before tamping

level after tamping. The idea is to have a comparison between the responses to tamping of track geometry in turnouts and open track. Although these models are estimated based on data from lines with maximum speed 200–250 km/h, it does not mean that all the sections in the railway line are exposed to the same traffic speed. In this respect, the comparison of recovery models involves some level of uncertainty. The models for the effects of tamping in open track suggest that in open track around 57–69% of the initial degraded level of track vertical profile can be restored. This number for turnouts decreases to 39 and 49%, for speeds 250 and 120 km/h, respectively.

An interesting parameter in the recovery models is the first regression coefficient (y-intercept). This parameter shows the amount of recovery which is independent from the degradation level before intervention. For the case of standard deviation, a positive value of y-intercept shows the level of degradation which cannot be removed by intervention. This parameter may relate to the geographical and weather-related conditions in which the track is located. The sub-ballast or the soil quality are among the important variables to which the y-intercept may relate.

5 Conclusions

This study investigated the effects of renewal and tamping on longitudinal profile and ballast condition in turnouts. Data of loaded geometry recording from 165 turnouts were collected, cleaned and aligned for the analysis. The study put forward an integrated methodology for geometry data analysis which enabled calculation of the recovery values and evaluation of degradation for both longitudinal profile and ballast condition.

Results showed that the impact of renewal/tamping on track geometry and ballast quality should be quantified differently. For example, renewal restores 89% of degradation in the ballast layer (as expected) and 67% of degradation in longitudinal profile. Tamping has minor effect on ballast quality restoration (36%), but it can restore 44% of longitudinal level degradation. Results showed that when the condition of ballast are quantified by the fractal method, the chance of having low quality after renewal is near zero. This was different for the standard deviation of the longitudinal level. This difference in the estimated effect of renewal shows the applicability and usefulness of the fractal method for sub-structure condition evaluation.

The lognormal distribution model was found to have the best fit to degradation rates both before and after intervention. The comparison of ballast degradation rates before and after renewal/tamping showed that renewal reduces the degradation rate of ballast when degradation was fast-paced. Despite the fact that there were turnouts in which tamping had increased the degradation rate, a general hypothesis of increasing effect of tamping on degradation rate could not be accepted.

A significant contribution of traffic speed to recovery of geometric parameters has been observed in turnouts. It has also been shown how the recovery models for track longitudinal level after tamping in open track and turnouts are different. In open track, the recovery models from past studies showed that around 57–69% of the degradation level before tamping can be restored. However, in turnouts exposed to low-speed traffic around 49% of the degradation level before tamping can be restored, but when the turnout is exposed to high-speed traffic, this number drops to 39%.

Acknowledgements The author gratefully acknowledges the Innovation Fund Denmark for supporting the INTELLISWITCH project (Grant no. 4109-00003A) within which this work was performed.

References

1. Quiroga LM, Schnieder E (2012) Monte Carlo simulation of railway track geometry deterioration and restoration. Proc Instit Mech Eng [O] J Risk Reliabil 226(3):274–282
2. Xu P, Liu R, Sun Q, Jiang L (2014) Dynamic-time-warping-based measurement data alignment model for condition-based railroad track maintenance. IEEE Trans Intell Transport Syst 16(2):799–812
3. Hovad E, Andersen J, Thyregod C et al (2018) Spatial alignment of track recording car data in turnouts. In: Fourth international conference on railway technology: research, development and maintenance, vol 1. Sitges, Barcelona, Spain
4. Hovad E, Thyregod C, Andersen JF, Lyndgaard CB, Spooner MP, Rodrigues AFDS, Ersbøll BK (2019) The effect of driving direction on spatially aligned track recording car measurements in turnouts. Int J Rail Transport:1–15
5. Soleimanmeigouni I, Ahmadi A, Arasteh Khouy I, Letot C (2018) Evaluation of the effect of tamping on the track geometry condition: a case study. Proc Instit Mech Eng [F] J Rail Rapid Transit 232(2):408–420
6. EN 13848-5 (2008) Railway applications, track geometry quality part 5: geometric quality levels. European Committee for Standardization (CEN)
7. Andrade A, Teixeira P (2013) Unplanned-maintenance needs related to rail track geometry. Proc ICE Trans 166:1–11. https://doi.org/10.1680/tran.11.00060
8. Mandelbrot BB (1983) The fractal geometry of nature, vol. 173, p. 51. WH freeman, New York
9. Hyslip JP (2002) Fractal analysis of track geometry data. Transp Res Rec 1785(1):50–57
10. Landgraf M, Hansmann F (2018) Fractal analysis as an innovative approach for evaluating the condition of railway tracks. Proc Instit Mech Eng [F] J Rail Rapid Transit, 0954409718795763
11. Landgraf M, Hansmann F, Marschnig S, & Veit P (2014) Track geometry and subtstructure condition-a provable correlation? In: Proceeding of GEORAIL 2014, IFSTTAR, 1785, pp 623-632
12. Audley M, Andrews JD (2013) The effects of tamping on railway track geometry degradation. Proc Instit Mech Eng [F] J Rail Rapid Transit 227(4):376–391
13. Soleimanmeigouni I, Ahmadi A, Kumar U (2018) Track geometry degradation and maintenance modelling: A review. Proc Instit Mech Eng [F] J Rail Rapid Transit 232(1):73–102
14. Martey EN, Attoh-Okine N (2018) Modeling tamping recovery of track geometry using the copula-based approach. Proc Instit Mech Eng [F] J Rail Rapid Transit 232(8):2079–2096
15. Caetano LF, Teixeira PF (2015) Optimisation model to schedule railway track renewal operations: a life-cycle cost approach. Struct Infrastruct Eng 11(11):1524–1536
16. Jenkins HH, Stephenson JE, Clayton GA, Morland GW, Lyon D (1974) The effect of track and vehicle parameters on wheel/rail vertical dynamic forces. Railway Eng J 3(1)
17. Li MXD, Berggren EG, Berg M, Persson I (2008) Assessing track geometry quality based on wavelength spectra and track/vehicle dynamic interaction. Veh Syst Dyn 46(S1):261–276
18. Wen M, Li R, Salling KB (2016) Optimization of preventive condition-based tamping for railway tracks. Eur J Operat Res 252(2):455–465
19. Caetano LF, Teixeira PF (2016) Predictive maintenance model for ballast tamping. J Transport Eng 142(4):04016006

The Organic Growth of an Asset Management System at ProRail

Jan Swier

Abstract The development of asset management at ProRail began in the nineties of the last century, when the railway sector was radically reformed. The rail infrastructure was split from the railway company Nederlandse Spoorwegen and, simultaneously, the engineering office and maintenance company were privatised. Within a few years, the traditional infrastructure organisation with 4000 employees transformed into an independent infrastructure manager with 800 employees. This paper is about the organic way in which an asset management system developed in these circumstances. It describes the essence and principles, the organisation and necessary tools, the results realised and the influence of the circumstances. It closes with seven lessons learned. It demonstrates what is needed to optimally and sustainably manage the performance and cost of assets over the life cycle in accordance to the expectations of the stakeholders.

Executive Summary

Asset management (AM) of physical assets is relatively new. A first guideline was published in 2004 (PAS 55 [2]) and replaced in 2014 by the ISO 55000 [1]. The standard describes well what is needed for an asset management but gives little practical in-formation about how to organize it, where to start, what the critical milestones are, etc. Cost reduction and performance improvements are promised but it is unclear how they are realized. This paper is a summary of a book that does provide that insight. The book is a case study about the development of an asset management system at ProRail, the rail infrastructure manager in the Netherlands.

The development of an AM system affects the entire sector. It is a full culture change and lasted in the Netherlands 10–15 years. The paper briefly explains the most important milestones that led to ProRail now delivering much better performance for lower costs; the holy grail for every infrastructure manager. It explains the essence of asset management and describes the process model, explains how roles are clearly

J. Swier (✉)
Swier Advice Asset Management, 3833 VE Leusden, The Netherlands
e-mail: jsswier@gmail.com

R. Galeazzi et al. (eds.), *Intelligent Quality Assessment of Railway Switches and Crossings*, Springer Series in Reliability Engineering,
https://doi.org/10.1007/978-3-030-62472-9_14

separated and what the importance is, shows why risk management is indispensable and describes the milestones and change in orientation in the organic growth path of an asset management system. The paper closes with seven lessons learned by ProRail.

After reading the paper it is clear that asset management is not a tool but an integral management method that affects the entire sector. The method is not implemented but grows organically provided that the goal is always and will remain to optimally and sustainably manage performance, risk and cost over the entire lifespan. The journey should be the goal, man is the measure and good leadership is a condition.

1 Introduction

The European Union introduced competition in the railway markets by mandating separation of previously monopolistic organizations having business both in train operations and rail infrastructure management. This move was undertaken to improve operating performance and decrease costs for the end-user. In 1995, the Nederlandse Spoorwegen (NS) was the only railway company on the Netherland rail network but 20 years later, the number grew to more than thirty, including passenger and freight operators as well as contractors. From the beginning, the Dutch government opted for a full, legal and administrative separation of Transport and Infrastructure. The business relationship is regulated in an institutional triangle among Government, Operators, and Infrastructure Manager. It is illustrated in Fig. 1.

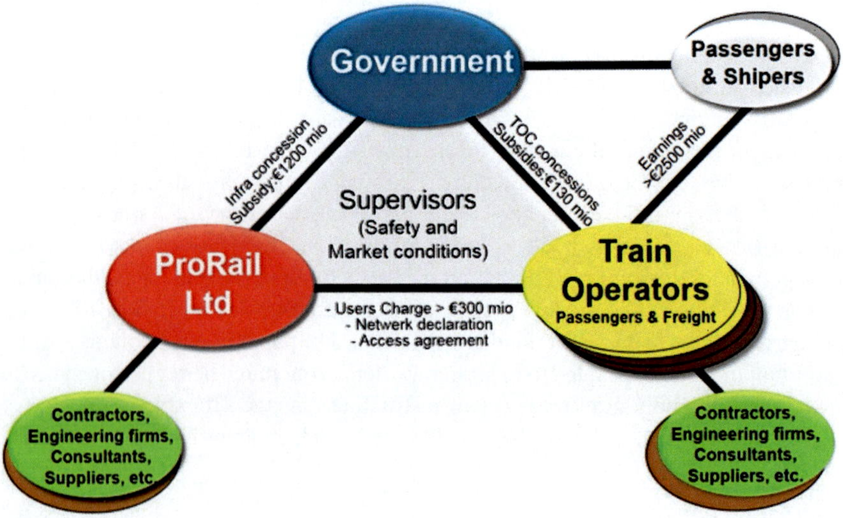

Fig. 1 The institutional triangle for rail transport in the Netherlands

A complicating factor in restructuring was the decision to outsource all execution activities of the infrastructure manager, such as engineering, material supply, personal safety, training and maintenance. As a consequence of this outsourcing, all associated executive knowledge and experiences disappeared and ProRail became a pure asset manager. It was fully dependent on the quality of its service providers and the information received from inspections, reports and systems.

The content of this chapter is an extended summary of the book "*The organic growth of an asset management system. Case ProRail*" by Jan Swier [5].

2 Development of ProRail Vision on Asset Management

The vision of ProRail's rail asset management developed over a period of 20 years, starting in 1995. It is based on the needs and priorities realised over these years as well as the introduction of new management approaches and techniques such as the EN50126 [3] and general management principles such as Plan-Do-Check-Act (PDCA) cycle, quality management (EFQM model), Reliability Centred Maintenance (RCM) and Failure Mode Effect and Criticality Analysis (FMECA).

The development of ProRail vision started years before the appearance of the asset management standards PAS55 and ISO5500. ProRail did not approach asset management as a new technique but rather as a full change of focus and culture of the entire rail infrastructure sector. Asset management was not implemented overnight or as a single programme, but grew organically with the help of external stimuli and partners, actual needs, intermediate milestones and growing maturity. ProRail developed its instruments and skills in an incremental step-by-step manner to build a system for optimal asset management. Outsourcing of maintenance in 1997 forced the organization to rebuild its asset management process and instrumentation from the bottom up without burden of the past. This was not only challenging but also risky. Clarity about the role and position of an asset manager proved to be a fundamental, stable starting point.

2.1 Role of an Asset Manager

The role of an infrastructure asset manager is better defined by distinguishing four key role players in asset management. These include asset user(s), asset owner, asset manager and service providers. In brief an explanation:

- The **asset users** are the train operators and they use the assets as per conditions identified in the access agreement, and pay an access charge for the use of the rail infrastructure network.
- The **asset owner** identifies the needs of all stakeholders, sets the targets and accepted risk levels, and finances capital projects. As is the case in most European

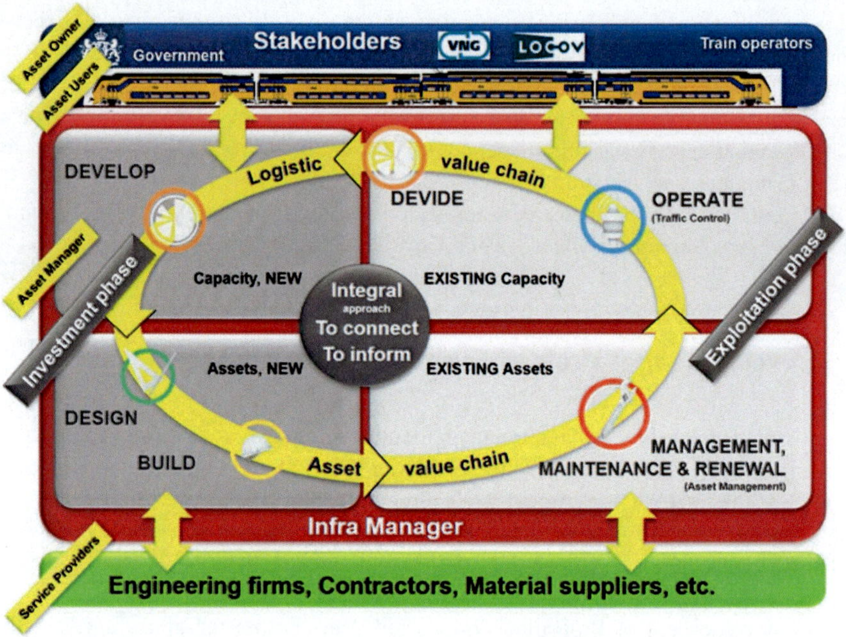

Fig. 2 An asset manager manages two value chains and two life cycle phases

countries, the Government of the Netherlands is the asset owner and financier of the Dutch rail infrastructure network.

- The **asset manager** develops the strategy to achieve targets of the asset owner optimally and sustainably over the life cycle within the financial and risk framework of the asset owner. ProRail is the asset manager.
- The **service providers**, such as contractors, consultants and engineering firms, execute and implement their works as planned by the asset manager.

The four roles can be combined in one organization, but must remain recognizable to keep the associated tasks and responsibilities separate. The Netherlands has chosen for a full separation of the roles. The role of the infrastructure manager ProRail consists of four main tasks/processes (see Fig. 2):

- The development of new network capabilities and the division of existing network capacity.
- The design and building of new assets.
- The maintenance and renewal of the existing assets.
- The operational control of the existing network.

These main tasks/processes cover the whole asset life cycle. It allows the infrastructure manager to connect the requirements and wishes of the asset users/owner about the network, with the service providers who realise them. An asset management system connects, involves and affects all main processes within and outside of the infrastructure manager. The connections are made through specific instruments such as life cycle management, risk management, RAMS (Reliability, Availability, Maintainability and Safety) management, information management and quality management.

2.2 The Essence of Asset Management

Asset management is a professional way of doing maintenance, renewal and more. A good definition for asset management is given by PAS55 [2]:

> Asset management is the systematic and coordinated activities and practices through which an organization optimally and sustainably manages its assets and asset systems, their associated performance, risks and expenditures over their life cycles for the purpose of achieving its organizational strategic plan with long term vision and goals to meet stakeholders' requirements and objectives.

This definition can directly be linked to a conceptual model which ProRail has developed and used for years, and it is illustrated in Fig. 3. This model describes the essence of asset management.

A brief explanation of the model. *Performance* of the assets is the reason of its existence. The main goal of an infrastructure manager is to realize asset functionality with a certain quality according to the expectations of the stakeholders, like

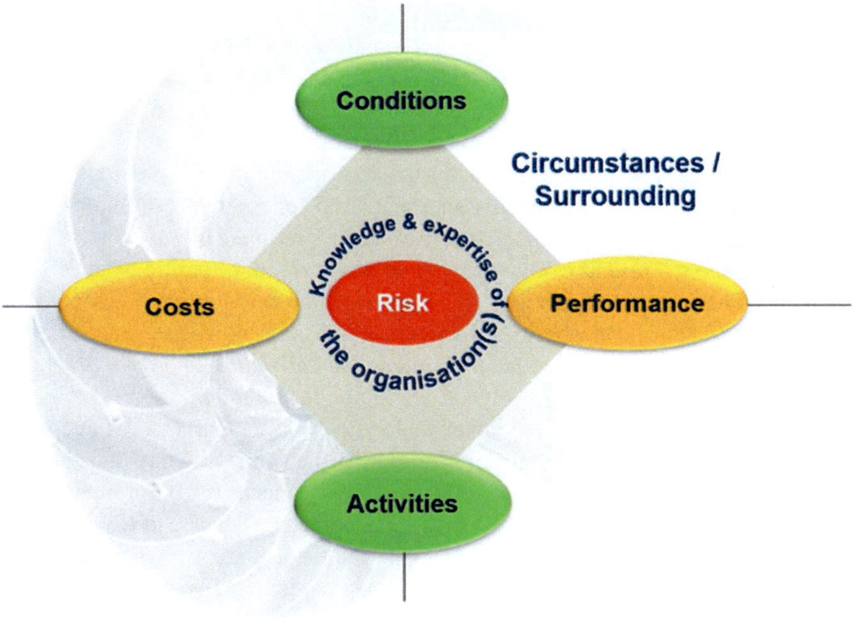

Fig. 3 Model to express the essence of asset management

the government and train operators. A universal way to specify asset quality is with RAMS. *Risk* causes, like wear & tear, vandalism and extreme weather, threaten asset quality. Maintenance & Renewal *activities* are necessary to control the risk causes, and those activities *cost money*. The amount depends on the *conditions* like the quantity of assets, utilisation, working condition, legal requirements, social agreements, etc. *Skills and quality* of the asset manager determine how optimal and sustainable performance, cost and risks can be managed over the life cycle. Dedicated management techniques such as life cycle analysis, risk management and performance management are necessary to establish the relationship between performance and costs, investment and exploitation as well as the logistic and asset value chains. In addition, specific circumstances such as history, culture, finances and politics, influence the capabilities and results of the asset manager. Examples include budget cut, formulation of new rail transport policy or decision to restructure train operations and rail infrastructure as separate companies.

With the evolution of management techniques, maintenance has changed from a craft into a highly professional business, and outsourcing of non-core activities became possible. Despite this, nearly all railway companies in the world perform maintenance on their own because their personnel know how to secure and control safety and reliability. ProRail's unique circumstances forced it to let go of this tradition and become different.

3 Outsourcing of Maintenance

To outsource maintenance is complex because the results to be achieved are difficult to specify and measure. The output is not a tangible product, but the quality one experiences. A complicating factor is that maintenance is a people business and the average lifetime of rail infrastructure is more than fifty years while a maintenance contract runs over a much shorter period. Craftsmanship, experiences and dedication of employees are crucial determinants of the short and long term impact of maintenance. The work they do is neither spectacular nor glamorous, but the impact is considerable: "*If you think maintenance is expensive, try an accident*".

The decision to outsource rail maintenance was made in the period when NS was separated. The aim was to increase efficiency and innovation through competition, to shift from a technical to a business focus and it was in the nineties also fashionable to do it. The slogan was: "*manage on output and at a distance, and the market give you with the best results*". How to do this was not clear and looking back one can say that the decision makers were rather naïve: there was no market and there was a lack of knowledge, experience and competition. The new infrastructure management organization and maintenance contractors were by no means ready for outsourcing and even less ready for public tendering of output contracts. For sure in 1998 the asset manager and contractors did not have the ability to tender and manage maintenance contracts on output and at a distance. However, both parties have worked energetically step-by-step to develop the missing skills, instruments, expertise, processes, etc. Key was the decision how to specify and control performance.

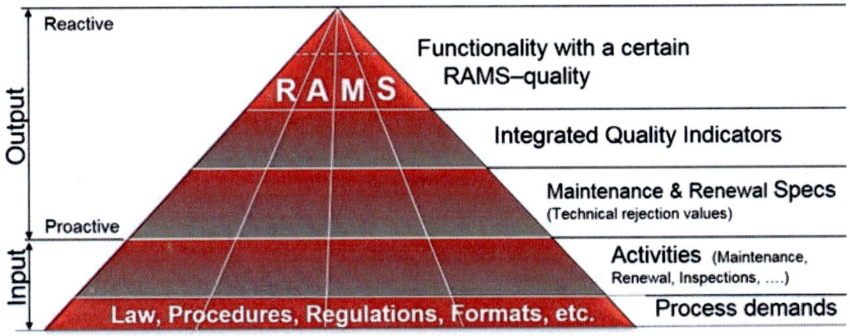

Fig. 4 Performance level specification and control triangle

3.1 Performance Specification and Control

The infrastructure manager of the United Kingdom, Railtrack, was about two years ahead of ProRail in the outsourcing of maintenance but it stopped short in 2002 because of difficulties in controlling costs and performance. The major causes of its difficulty were requirement of a high-level RAMS output and lack of expertise in understanding the relationship of asset condition and performance.

The approach taken in The Netherlands was different. A specification triangle, illustrated in Fig. 4, was formed to distinguish different specification levels. As per this model, the highest asset performance level is RAMS, the output Railtrack used in its contracts. The lowest asset performance level focuses on activities necessary to achieve RAMS. In between these two levels, there are two interconnected output levels: the Integrated Quality Indicator, and Maintenance & Renewal (M&R) specifications. The Integrated Quality Indicator is a combination of several technical specifications into one overall quality indicator. It is comparable to the blood pressure of our body, providing insight into the development of system quality. The M&R specifications define the lowest acceptable technical condition of assets. Systematic measurement of the M&R specifications gives insight in the level of degradation. Asset performance below this level poses high(er) risk of lack of safety and reliability, requiring control activities. It makes a proactive way of steering possible, while steering on RAMS is a reactive way because it steers on what already has happened.

The specification levels are interconnected and lead to different management aims and responsibilities. Based on the maturity of both parties, the infrastructure manager decides if and how the responsibility of a specification level is to be delegated to a contractor. The value of measurement is well expressed by the quote "*Only if you can specify and measure something, you can manage it*". When one knows the degradation level of the asset's quality, the technical limits, and impact of activities on risk control, then the relationship between cost and performance can be determined. It is the principle of risk management, and the holy grail for an asset manager.

Measuring and reporting has become professional business in The Netherlands. The information of specification measurement is not only used for contract control, but also for the planning of maintenance activities and renewal projects. All inspection and measurement information are shared between ProRail and its contractors, and is becoming more easily available electronically.

3.2 The Development of Outsourcing

ProRail started outsourcing of maintenance in 1998 with an input-based maintenance contract known as OPC. The contract was not put out to tender but awarded on the basis of price negotiations about a work plan that had been acquired and M&R specifications. In 2005 ProRail decided to prepare a performance-based maintenance contract, known as PGO. This involved restructuring of the OPC contract specification and general conditions to focus on performance control.

An important question in the development of the PGO was specification and control of performance. A proposed condition of the AM director in that period showed the way: "*I don't want to manage safety on the level of accidents. I want the contractor to demonstrate that it will not happen*". The implication was that he did not want to manage safety on the highly reactive RAMS level but on the proactive technical level of M&R specification. Experience gained in the 1998–2005 period also suggested that there should be a specification for the contractor's contribution in achieving the lowest life cycle expense, or the longest operational life. The requirement was relevant because of the maintenance contract term was five years while the average lifetime of the rail infrastructure was fifty years.

These considerations led to a new PGO specification and a clear risk distribution between client and contractors (known as RAMSHED). It had specified Reliability, Availability and Maintainability ('RAM' of RAMS) health and environment were added because they are legal requirements at a high reactive level, and Safety ('S' of RAMS) and Durability on a proactive level. Health and environment were added because they are legal requirements for an illustration see Fig. 5.

A methodology was devised for determining the proactive level based on risk management instruments previously developed by ProRail. It was the level at which it was possible for the contractor to demonstrate that rail safety and durability risks were under control through preventive activities and wear-and-tear management.

Risk management is vital to control a performance-based contract. It was prescribed as a management technique for the substantiation of the quotation and the development of the work plan, based on failure mode effect and criticality analysis (FMECA). The style of management also had to change to allow contractors the freedom to make their own choices to improve the price-performance ratio. Preconditions for bid include full process control and an accurate specification of the offer which has to be measured and controlled. By means of output specifications and tendering, opportunities and incentives have been created for contractors to distinguish themselves from their competitors.

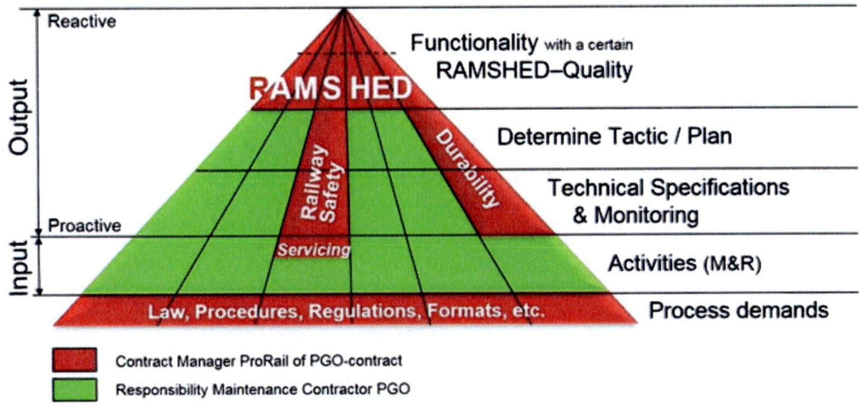

Fig. 5 Demarcation of responsibility areas through the performance contract specifications of the PGO contract

The first PGO-contract was put out to tender in 2007, in accordance with the requirements of the Government of The Netherlands and the Netherlands Competition Authority. ProRail reduced the number of maintenance contracts from 39 to approximately 20. This was expected to reduce overhead costs and tendering frequency as well as double contract volume attracting new market players.

4 Risk Management

Risk implies potential for failure and is defined as (Chance * Impact) of a failure. The kind of risk depends on the business and situation/context. An unacceptable risk is required to be controlled through mitigation activities. Asset risks are RAMSHED performance and costs. ProRail has defined a risk-matrix to identify risk levels and specify which are acceptable or not.

There are various maintenance strategies to control the risks: corrective, preventive, function test and as a last resort modification. Figure 6 illustrates these strategies. Condition based, preventive maintenance has the preference but when it is not possible the alternative is frequency based. Failure based, corrective maintenance is only acceptable when the risk value is about one or less. A new development is maintenance driven by big data, looking at the influence of the circumstances and conditions on the asset performance.

The maintenance strategies are used together and are interchangeable, depending on the risk to be controlled, the technical possibilities and the costs. Their value for the business differs, and the complexity typically increases with effectiveness. Selection of the desired strategy depends on the possibility to specify and monitor the asset's technical condition.

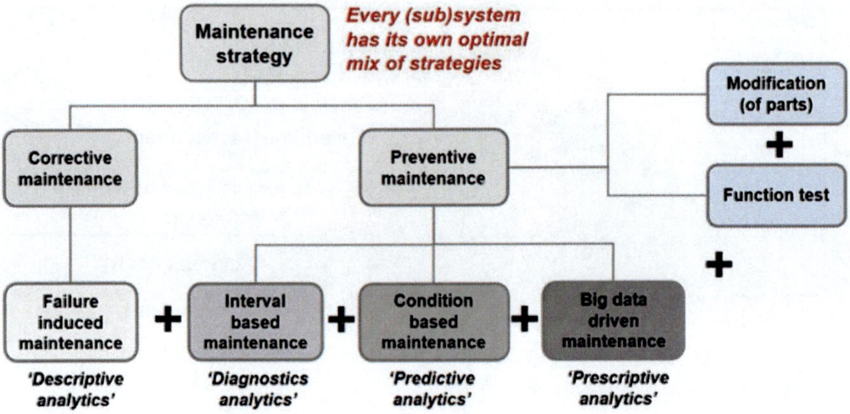

Fig. 6 Maintenance strategy to manage asset performance

4.1 *One Risk Instrument, Multiple Applications*

Risk management is a technique that facilitates the possibility to manage performance and cost explicitly and in relation to each other. Its underlying instrument is risk analysis (FMECA) with a control activity per risk. These are known within ProRail as the M&R Concept. For an illustration see Fig. 7. The M&R Concept is a systematic and structured record of all failure modes and failure causes that might threaten the RAMS quality. In practice, it comprises the entire spectrum of possible failures and possible failure causes and modes.

A risk management strategy is chosen on the basis of the risk value and the control option. Where possible, it contains a specification and a monitor activity to assess the degradation, an M&R activity to control the risk, and a repair activity if the

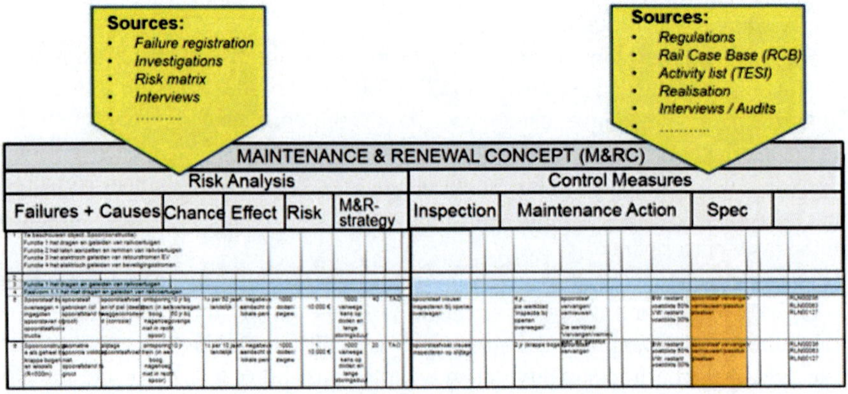

Fig. 7 Structure, content and sources of the M&R Concept

risk nevertheless occurred. ProRail developed M&R Concepts for all systems in the period 2004–2005. In an ideal world, this information would come initially from the designer. The theoretical concept of the designer will change during exploitation in a realistic overview of all possible and exposed risks. ProRail has now collected the necessary risk information through interviews with its own experts and the analysis of the failure registration system and investigation reports of accidents.

In 2010, ProRail decided to improve the quality of the M&R Concepts and developed—together with the maintenance contractors—detailed up-to-date new versions for all systems with additional information about activity-codes, failure causes, realisation information, regulations and contract specifications. The System Asset Management (SAM) programme was launched for this purpose. As of 2015, ProRail had detailed M&R Concepts (which cause more than 80% of the failure risks) for more than 20 asset types. This has created a consistent and detailed M&R knowledge platform for multiple applications.

4.2 Sector Knowledge Platform

The M&R Concept provides knowledge in a systematic and accessible way to manage performance. It is the only instrument by which performance and costs relate to each other, explicitly and coherently. For infrastructure managers such as ProRail, who have outsourced all maintenance activities and manage through performance contracts, M&R Concepts are vital for contract control and continuous performance improvement. The M&R Concept of a system comprises a long list with risks and risk causes, all with a unique ID-number. The length of the list depends on the complexity of the system. For example, an isolated joint has about 50 risks, and a switch construction more than 1000. While the amount of information may be daunting for a layman, it is a gold mine for a maintenance engineer because it provides a systematic, complete and accessible library of all M&R knowledge of a system.

The risks and control activities can be sorted by various criteria such as function, risk control strategy, distribution of risk values, comparable risk causes, comparable control activities, etc. The inspection, maintenance and renewal activities are clustered to form plans, inspection forms and work orders. The plans make it possible to calculate costs, get an overview of needed train-free periods, establish relationship between activities and performance, etc. The challenge for ProRail now is to fully integrate the risk management instrument in the contract management process and its own business process.

5 Organic Growth Model of an Asset Management System

The asset management system development process of ProRail is a tangle of steps: forward, backward, sometimes targeted but often disoriented and forced by circum-

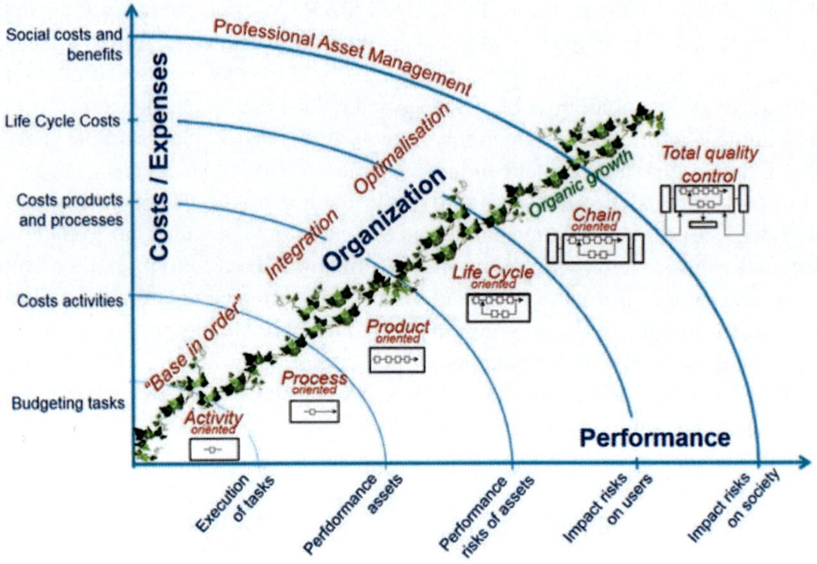

Fig. 8 Organic growth model of an asset management system

stances. It was driven by the available knowledge, expertise and skills of the organisation.

Analysing this process from the European Foundation for Quality Management (EFQM) perspective, leads to the observation that development had the character of an organic growth. The development was in phases and each phase created the conditions for the next one, adding to the maturity of the organisation. For an illustration see Fig. 8. The orientation direction of the asset manager changes at each stage. It not only affects the core skills of all employees in the AM system, but also the culture, structure and systems of the entire organization.

ProRail started with putting the foundation in order ('Base Okay'): asset data-base, financial system, failure registration, dashboard, information standards, specification, contracts, etc. It was an appealing way of prioritising, with focus on M&R activities, cost and performance. Only when the 'Base is okay', the organisation is able to work effectively on the 'Integration' over the life cycle, and only when 'Integration' is achieved, it can reach 'Optimisation' across the entire sector chain.

The organic growth model shows that asset management is not a self-contained management technique for a group of M&R specialists. It involves the whole organisation of the infrastructure manager and all value chain partners. The NEN-ISO 55002 [4] states:

> The asset management system is not self-contained. Part of asset management is the ability to integrate asset management processes, activities and data with those of other processes within the Organization, such as quality, accounting, safety, risk and human resources. Other

business processes should be optimally exploited to avoid unnecessary new work and duplication of existing work and existing data. These interactions with the existing processes are to be communicated clearly to all concerned.

The appendix in [5] provides an insight into the relationship of ProRail's asset management system with three reference frameworks: EFQM, PAS55 and ISO55000. According to the overview provided, coverage is almost 100%.

5.1 Lessons Learned by ProRail

1. *Role separation and role fulfilment have major influence.* Role distribution and involvement of the government determine development of an asset management system. The separation of roles and ability to work together through collaboration requires time, attention and leadership.
2. *The AM system grows organically: the journey is the goal.* Developing a professional asset management system is a journey that is organic in nature. It involves coherent orientation phases leading to overall quality management, optimal performance, as well as minimisation of risks and costs over the entire life cycle.
3. *Begin with making the 'Base Okay' of the M&R organization.* The asset management organisation has to become process-oriented, focused on performance and costs, and equipped with the required tools including ICT and information facilities.
4. *Five AM management techniques need to be secured and managed.* To manage cost and performance optimally and sustainably over the entire life cycle, the connection between the investment and exploitation phases, and that between business goals and realisation needs to be established. The following management techniques and tools must at least be controlled and used: life cycle management, risk management, RAMS management, maintenance management and information management.
5. *Core qualities of an asset manager: integral approach and connectivity.* An infrastructure manager is the connecting link between what stakeholders want and service providers realise, and between the strategic goals of the network and operational activities on objects and systems. The asset manager needs an integral approach and the ability to connect with all stakeholders throughout the life cycle.
6. *Knowledge and information should be secure and shared.* Organize the safeguarding and sharing of asset information and knowledge. The use of an asset information model is essential for exchanging and combining information. To make knowledge accessible and to spread, data and documentation must be easily accessible. All vocational training courses must also be well-secured and organized, and accessible to all employees throughout the whole industry.
7. *Man is the measure.* The employees in the organisation determine the result of the processes it undertakes. They must be empowered, safe, trained and motivated

to perform at their best potential. Managers create the right conditions and know how to use the full potential of their employees by making challenging yet achievable goals at the right moment and give their employees the right attention.

6 Conclusions

Asset management is not new. In the past, it was done by mechanics and called "craftsmanship". The skilled mechanics made implicit risk assessments based on their experience and observations but now the process is professionalised and risk assessment is explicitly done with specific instruments, including ICT systems. Thanks to risk management tools and ICT, the implicit craftsmanship knowledge of the mechanic became explicitly available to the entire sector in a clear and coherent way. This was inevitable because demands for performance improvement and cost reduction increased, and conditions became more complicated. The mechanic could not do it alone any more and needed the necessary assistance from maintenance engineers, planners, logistics providers, inspectors, etc. Today the mechanic is a link in a process chain. He gives and receives accurate asset information through predetermined procedures and systems. ICT and risk management are indispensable for enabling a team of various professionals and organisations to continuously improve performance and reduce costs.

The complete separation of transport and infrastructure, and the procurement of risk-based performance contracts for maintenance, have stimulated the organic growth of an asset management system at ProRail. However, both are not a condition for this.

References

1. British Standard Institution (2014) Asset management: BS ISO 55000:2014. BSI Standards Ltd., London
2. British Standard Institution (2008) PAS 55-1:2008 Asset Management. Part 1: specification of the optimized management of physical assets. BSI Standards Ltd., London
3. European Committee for Electrotechnical Standardization (2017) EN 50126-1 railway applications—the specification and demonstration of reliability, availability, maintainability and safety (RAMS)—part 1: generic RAMS process
4. ISO (2018) ISO 55002 asset management—management systems—guidelines for the application of ISO 55001
5. Swier J (2019) The organic growth of an asset management system: Case ProRai

Ways to Improve the Performance (And Cost) of Switches

Jan Swier

Abstract Switches create most of the failures on the Dutch network and their impact on train service is often large. ProRail started to work together with the maintenance contractors around 2011, to significantly improve the switch performance. In 5 years they have succeeded to reduce the number of train affecting failures by more than 25% while also the maintenance costs dropped. This paper describes the improvements which led to this success. They are in the field of maintenance strategy, risk approach, design, monitoring and control. Decisive was however the renewed attention and focus on the nuts and bolts of the switches, the asset management process, the involvement of operational people and to take into account the constraints. Maintenance is people's work.

Executive Summary

This paper is about ways that have been developed in the Netherlands to improve the performance of switches and crossings (S&Cs). The improvements are very diverse, often very logical and self-evident, on a strategic, tactical and operational level, but what they all have in common is the attention from people to S&Cs to perform better. Decisive was the renewed attention and focus on the nuts and bolts of the switches, the asset management process, the involvement of operational people and to take in account the constraints. The improvement measures can be grouped into six groups:

1. Improvement of the design of poorly performing switch types
2. Maintenance focus on the most vulnerable and poorly performing switches
3. Use risk management as a starting point for the asset management (AM) process and management
4. Have a fundamental, critical look at the layout of yards - which are often designed ages ago—and get rid of switches which are not really necessary
5. Use all technical possibilities to gain insight into the technical condition of individual S&Cs and their degradation
6. Inform the management, planners and operational employees through a condition dashboard for S&Cs: the Maintenance Performance Indicator (MPI).

J. Swier (✉)
Swier Advice Asset Management, 3833 VE Leusden, The Netherlands
e-mail: jsswier@gmail.com

R. Galeazzi et al. (eds.), *Intelligent Quality Assessment of Railway Switches and Crossings*, Springer Series in Reliability Engineering,
https://doi.org/10.1007/978-3-030-62472-9_15

A S&Cs-dashboard/MPI is the tip of the asset management and control iceberg. It provides overview and insight but can only be created if the AM management process and information provision are in order, and the people in the organisation are dedicated to asset performance and the continuous improvement of it.

1 Introduction

Around 2008 ProRail and the maintenance contractors were ready to tender and manage performance-based maintenance contracts, known as the PGO contracts. These imposed much higher demands on the client and contractors than the input-driven OPC contracts which are based on agreed fixed work plans and annual price agreements. By contracting the PGO contracts, contractors were given the opportunity to make their own choices about how the performance can be delivered in the most effective and efficient manner.

The PGO contract has stimulated the development of a more effective organization and attitude than what the OPC contract did. The focus shifted from the settlement of units (e.g. invoice for hours worked) to steering the usefulness and necessity of the maintenance work and reducing the number of train affecting failures (TAOs) and train free periods (TVPs). The shift of focus did take place at the maintenance contractors and at the client ProRail. In a period of 10 years, the total number of TAOs decreased by about 25% and the maintenance costs by more than 30%. The biggest drop in the number of failures was achieved at the switches.

In order to improve the performance of the assets and to reduce the costs, there are few preconditions: a reasonable reliable asset database, a good failure registration and dedicated employees who want to improve the results. Without information about the subject and goals you cannot manage, and without expert and committed employees you cannot improve. Looking back, it is fairly easy to make an inventory of improvement initiatives by ProRail and the maintenance contractors. One should realize however that they are often the result of innovative initiatives where not everyone immediately recognized the benefits. An inventory has been made of eleven improvement initiatives, but we must realize that they often have mutual relationships and dependencies.

2 Differentiation in Maintenance Concept Depending on Usage

The wear behaviour of a switch determines the maintenance that must be carried out. It is wise to make a distinction in maintenance concept depending on the loading of a switch and how often it moves. ProRail has therefore classified switches in four load categories: A, B, C and D. Three for centrally operated switches (A, B and C),

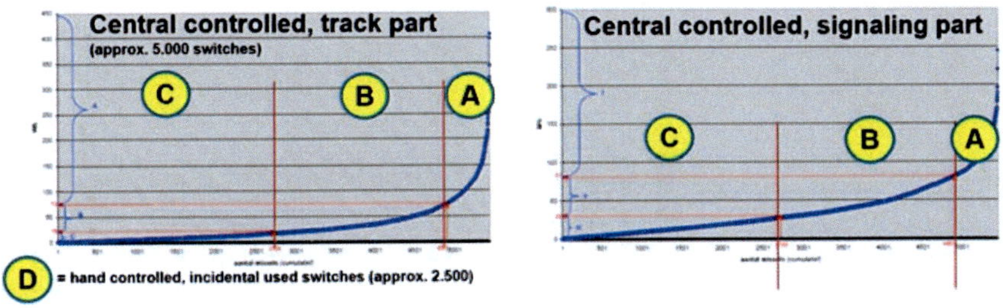

Fig. 1 The load graphs and switch categories for track and signalling part

and one (D) for hand-operated switches. A distinction is made between the load of the switch construction and the signalling part. To determine a category for a switch, four different trend graphs are determined, each with their own assignment of the four load categories. The trend graphs are for:

- Passed axles
- Passed tonnage
- Number of switch tongue movements
- Passed number of axles; straight, curved and against the frog.

The four trend graphs are then assembled, on the basis of selected allocation factors, into a single load graph for the switch construction and one for the signalling part. Both are illustrated in Fig. 1. The load graphs show that less than 10% of the switches are loaded very intensively (category A), about 40% normal (category B) and the remaining 50% hardly (category C). All hand-operated switches, always in side tracks, fall under category D. A specific maintenance and renewal (M&R) concept is developed for each load category, with a different price tag.

Being critical or not for the train process is not a part of the classification. To achieve higher availability for these "golden asset" switches, other measures are possible, such as the continuous monitoring of these switches with the tele-monitoring system POSS, as discussed in Sect. 10.

The classification categories were introduced around 2005 and are part of the asset database. For the risk-based performance contract PGO, however, the category classification is no longer prescribed, but still part of the asset database.

3 Focus on the Most Vulnerable Switch Types

In 2012 ProRail had about 7500 switches. By linking the asset database and failure registration an overview was made of the number of train affecting failures per switch type, see Fig. 2. Approximately 2500 of them are hand-operated switches in side tracks (category D). They cause almost no failures and they have hardly any effect on the train service.

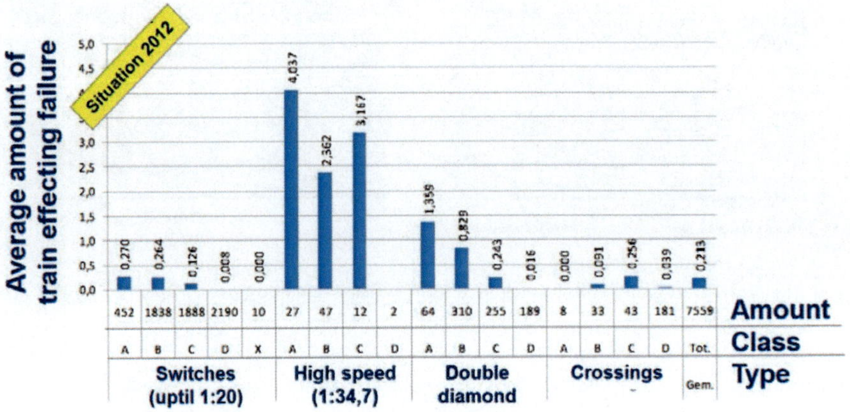

Fig. 2 Average amount of failures per switch type and switch class

What is striking is the high number of train affecting failures for the high speed switches: an average of 2–4 per switch per year. The amount of this switch type is limited to 88, but the failure rate is high and these switches are used intensively in important main lines. It has led to an improved maintenance concept and some modifications, but it became clear that the design had a number of weak points. In particular, the movable frog was vulnerable, as were the four switch machines with associated rods. It led to the development of a new type of high speed switch without movable frog and complex rods, as discussed in Sect. 5.

Also a large part of the 818 double diamond switches have relatively many failures with an annual average of 0.8–1.4. The big advantage of this switch type is that it saves a lot of space, as it combines the functionality of four ordinary points and a crossing, however at the price of an increased complexity and a reduced maximum speed of only 40 km/h. This insight has also led to the decision not to apply the double diamond switch on important yards, and limit the number to a minimum. Section 6 provides an illustration of this changed policy.

4 Focus on Switches with Repetitive Failures

Many switches have no failures at all but there is also a group that has several in a year, often the switches that are used intensively. This at least requires a critical investigation into the effectiveness of the maintenance concept. Today, this is done systematically, and the maintenance contractors adjust their maintenance approach in accordance to the identified causes.

Figure 3 illustrates the situation around 2012. All switch failures in a year are brought together in this graph. On the horizontal axis the switches and on the vertical axis the number of failures per switch. A distinction has been made between failures

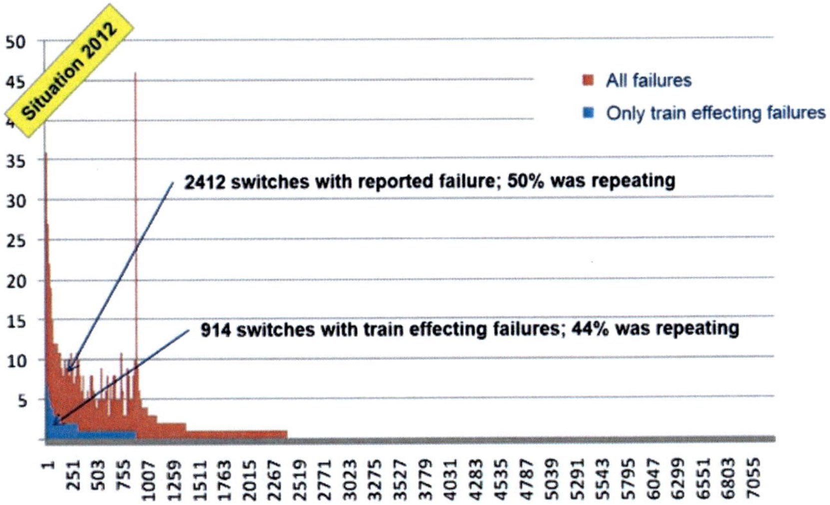

Fig. 3 Overview of all failures per switch in the year 2012

with and without effect on the train service, coloured in blue and red respectively. The switches are grouped by switch category and from zero to "many" failures. The small peak around 900 failures is where category D starts with the switch who has most of the failures in this category. Of the approximately 7500 switches about 2400 (35%) had reported failures and 1400 (20%) had more than one failure in a year, so repetitive failures. A small group even has more than 5 train affecting failures and more than 20 other failures. This group of switches got special attention of ProRail and the maintenance contractors, and contributed to the performance improvement of switches.

5 Improve the Design of (High Speed) Switches

Until recently, ProRail had two types of high speed switches: for 140 and 160 km/h in a curve, with respectively an angle ratio of 1:34.7 and 1:39.1. These switches are complex because they have very long tongues with several drive rods and control points as well as a movable frog, which, according to international regulations, is necessary when the angle ratio is more than 1:30. Due to the large length and several control points, the alternating performance is highly dependent on the stability of the substructure. As a result, the high speed switches have, on average, ten times more failures compared to standard switches, see Fig. 2. Despite modifications and improved maintenance concepts, the high speed switches continued to perform poorly. It was decided to develop a new type, the 1:29 high speed switch, that does not have a movable frog and can still be used at 140 km/h in a curve. At the time, it was decided to fully outsource construction, drive and control, so that the integration is realised by one party. The VAE group, of which Railpro Voestalpine is a part, provided the design according to ProRail's specifications. The 1:29 switch is 135 meters long and is longer than the current switch 1:34.7. This is because the curve in the switch has been optimised such that it begins to faint and gradually becomes sharper. It is also for the first time in the Netherlands that hydraulic drives are used to move the tongue as in Germany and Austria.

The development of the 1:29 high speed switch is a good example of professional asset management. It is an improvement of the asset performance based on a clear specification by an experienced external party who also did the full integration of the different sub-systems. Thanks to an integral tendering procedure, maximum use has been made of existing knowledge and experience in Europe. For the development and manufacturing of the new switch type, companies throughout Europe worked closely together. The result is a high speed switch that performs better, requires less maintenance and, therefore, has (much) lower life cycle costs than the former high speed types.

6 Increase Network Capabilities by Reducing the Amount of Switches

With rail transport, large transport flows can be handled safely in a relatively small space. Government and train operators in the Netherlands expect that rail transport will continue to grow in the coming years. To cope with that and to improve the quality of rail transport, the program "Programma Hoogfrequent Spoor" (PHS) was developed. On the basis of a study conducted by ProRail, the government decided to start 'timetable-free' driving in the conurbation of Western Holland with six intercity trains and six regional trains per hour per direction. The capacity of the existing lines had to be increased for this purpose. The Utrecht junction was an important bottleneck for this. Expansion of the number of tracks was not an option here due to limited space, major potential impact on the environment and financial limitations. In order to solve this bottleneck, the government decided to transform Utrecht into the Flow Station Utrecht (DSSU), see Fig. 4.

Flow-through is achieved by eliminating crossing, conflicting train traffic and handling train service over fixed, conflict-free corridors. As a result around 70% of all switches disappeared, including all double diamond switches that limit the maximum speed at the junction to 40 km/h. By eliminating switches, straightening tracks and reducing signal distances, passenger trains can follow each other with shorter distances (2 instead of 3 between two trains) and the maximum speed on the entire Utrecht junction increased from 40 to 80 km/h. This means that one to two minutes of driving time saving per train is realised and it is possible to schedule 2 * 6 trains per hour per direction. A negative consequence was however that shunting became more difficult but that was accepted. Positive side effects are a reduction of

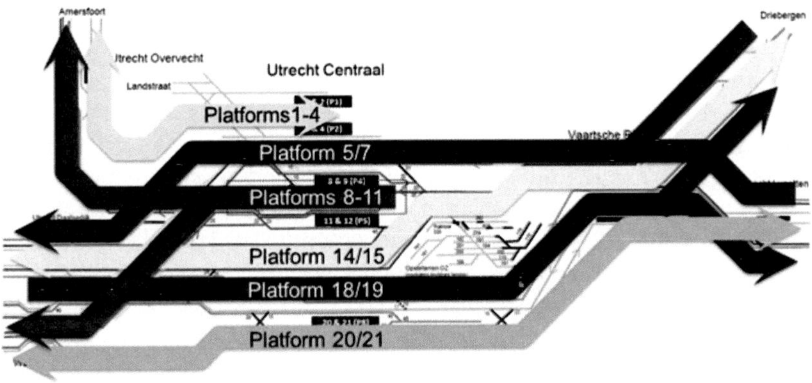

Fig. 4 Flow Station Utrecht (DoorStroomStationUtrecht = DSSU)

the M&R costs and an improvement in the performance of the rail infrastructure. Complexity is reduced not only because of fewer switches but also because of lesser associated catenary, signalling and traffic control systems.

More capacity and quality for lower life cycle costs is the holy grail for every infrastructure manager. It may seem very logical and simple to reduce the number of switches, but it is not. It is a big step to change the design vision of the main junction in the network from maximising flexibility to maximising performance. Traffic control and NS initially did not want to do this, but the budgetary and spatial restrictions forced them to think of this option. To choose the best solution, ProRail worked in design workshops and discussed infrastructure alternatives with all internal and external stakeholders on driving the desired train service, shunting, maintenance and emergencies. Every change was judged on usefulness and necessity and that ultimately led to the final realised layout. The capacity and quality of the current solution is much better than those of the former layout, partly due to the reduction in the number of switches by 70% (from 200 to 60). A result that involved all stakeholders: government, train operators, traffic control, capacity management and asset management. All parties contributed in this textbook example of integral and professional asset management.

7 Use of Risk Management Technique/Instruments

Asset management is about optimal control of costs and performance over the life cycle. Risk management is the only technique to explicitly establish and coherently manage the relationship between costs and performance. A well-known example of risk management is Reliability Centred Maintenance (RCM), for the first time used in the aircraft industry. The same methodology is used by ProRail to develop optimized M&R concepts for all (sub)systems, including all switch types.

Risk is defined as follows: Risk = Chance x Impact. The *Chance* is the annual frequency of the risk event per object when no maintenance is performed multiplied by the number of objects for the entire national network. The *Impact* is the performance lost when a risk occurs. Risk management involves balancing of risks transparently by valuing the chance and the impact of a risk with a weight factor depending the nature of risk. The multiplication results in a risk value per considered risk. It allows risks to be weighed against each other and determines whether the risk is acceptable or not. If it is not acceptable, a maintenance or renewal activity must be planned. The most well-known strategies are: condition based maintenance, frequency based maintenance and failure dependent maintenance. Less known maintenance strategies are: big data based maintenance, a function-test and the modification of components. The choice of a strategy depends on the possibilities, circumstances and size of risk. The strategies are not object but risk-bound; every (sub)system has its own optimal mix of maintenance strategies.

A professional infrastructure manager requires risk management tools, particularly the risk register with information on all possible risks that can and have revealed themselves, quantification of chance and impact, as well as management strategy. The risk analysis (also known as FMECA) is conducted with measures to control the risks. For each risk, a preventive and corrective measure is considered with specification to proactively manage the risk, where possible.

Risk analyses and Control measures are instruments of the risk management process and together they form the maintenance and renewal concept. All maintenance and renewal knowledge are recorded in a structured format to provide reference for continuous improvement of the maintenance process. The M&R concept provide overview and insight. It has three main functions:

1. management tool to manage costs and performance
2. knowledge tool for collecting and sharing all M&R knowledge
3. information source to organise and relate info from disparate sources.

The M&R concepts are used by ProRail to structure system information, test the FMECAs and/or maintenance plans of the contractors, is the source of drafting the information delivery specifications, to check completeness of the M&R specifications in PGO contract, provide overview of M&R specifications with associated monitoring activities, calculate maintenance costs, estimate need for train-free periods and form basis for long-term renewal plan.

A good failure registration system is required to keep risk analyses completed and up-to-date. In addition, (safety) investigations of incidents and interviews with mechanics and technicians are valuable sources. The risks are weighed using the ProRail risk matrix and each risk in the M&R concept is linked to a risk measure. Many of the measures have a unique unit code and cost figure from the rail cost database. The M&R concept contains almost all information for cost calculations, rejection values and references to regulations and work instructions. It is also possible to model the RAMS performance with a selected risk management strategy. In short, the M&R concept has information from very different sources, which is arranged in a logical and systematic way, and brought into relationship with each other. This makes it possible to relate performance (i.e. risks) and costs (i.e. control measures) and manage them together, the holy grail for an asset manager.

8 Introduce Risk-Based Performance Management

An effective and efficient risk based performance process like PGO, is not task-driven from the planning side, but demand-driven from the need and execution side. A prerequisite is a well-organised and transparent process, with the maintenance team in a steering role, supported by a maintenance engineer, work preparator and planner. Figure 5 illustrates the process of specifying performance as well as preparation and support of maintenance execution. In case of ProRail, a service provider performs maintenance engineering and planning; however, in other cases, maintenance can also be the executive organisation of an infrastructure manager. The process remains the same, except that there is no contract.

The PGO contract steering is based on performance specification trees and performance requirements. ProRail provides the underlying generic risk analysis to the branch through the Rail Infrastructure Catalogue (RIC) in the form of Maintenance Risk Analysis (IRA). As of Spring 2017, the IRA and accompanying control measures are placed in a generic performance risk database known as the Central Infrastructure Risk Register (CIRR).

The maintenance contractor provides a contract specific maintenance plan for an offered performance output. The starting point is the generic risk analysis with risk control measures in the IRA/CIRR, possibly supplemented by contract-specific and missing risks by the contractor. For safety and durability, ProRail prescribes technical

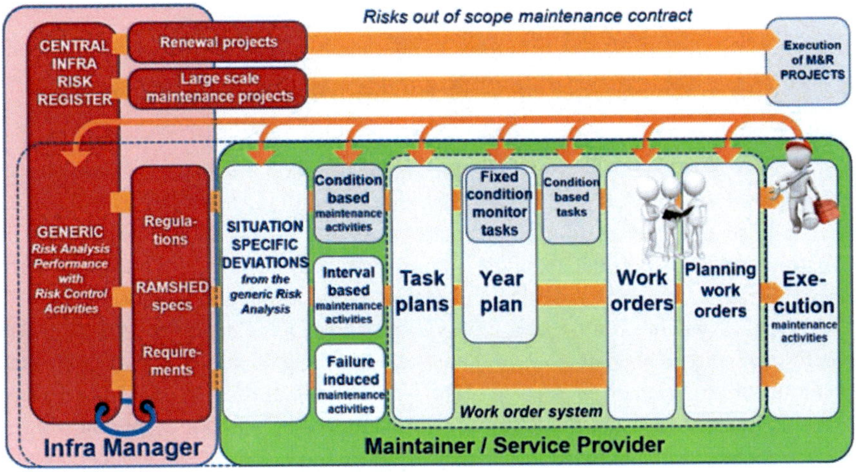

Fig. 5 The process for optimally contracting and managing maintenance

condition based rejection specifications, with associated monitoring activities and, if these are not available, interval based activities.

If the contractor acquires the contract then he puts all activities with a fixed periodical character in an annual year plan. These are all activities with a fixed interval and partly also condition based maintenance activities such as inspections and measurements with a fixed frequency. Failures with a failure induced maintenance strategy are accepted, and therefore no preventive measures are planned. It means that many activities are already planned and fixed a year in advance. Most at signalling and the least at rail and switches. During the year, the activities in the annual year plan are supplemented with condition-based activities planned on the basis of the degradation of the technical condition and the corresponding rejection specifications.

All maintenance activities are elaborated in task plans. These contain a description of the work, the documents needed, time estimate, material and equipment requirement, need for service to hire, etc. The task plans are made location specific in work orders and the work orders are then sent to the planner who plans execution by coordinating a large number of variables such as availability of mechanics, qualifications, shift schedules and day/night rhythms, material deliveries, machine availability, train free periods, etc. This results in preparation of weekly plans per line and team. These contain all the maintenance activities to be carried out and all preconditions arranged. The team leader receives maximum support. He/she knows the current situation and needs, has direct influence on execution because he/she is in direct contact with the work preparator, planner and maintenance engineer, and has the option to create work orders. It means that the ownership for the performance is at the right place and the maintenance process is fully optimised and controlled. Maintenance is only done when it is really needed, and the need is determined by the people who can determine it. This explains why the costs of the PGO contracts are substantially lower than those of OPC (with a cost reduction of more than 30%) and the performance is better.

9 Uniform Measuring of Switch Condition Degradation

A most effective and efficient maintenance strategy is condition based maintenance: maintenance is only done if necessary. The starting point is a technical assessment of the condition of the asset through inspections, measurements and the availability of rejection criteria. If a reject standard is approached or exceeded, maintenance must be performed because the risk is too great that a function may fails.

In case of switches, until recently it was only possible to determine the technical condition by means of an inspection. The switch is systematically measured at critical points according to an inspection standard. Until the rise of ICT, this was always done manually and the measurements were recorded on paper. With the emergence of the ICT, it became possible to do the measurements and recording electronically but it remained manual inspection work of the switch in an unloaded condition.

When ProRail outsourced maintenance, there was a growing need for a national overview of the technical situation of all switches. However, it could not be given. There were several maintenance contractors, many different inspection teams, several types of measuring equipment were used and the measurements were recorded differently. In addition, due to more severe safety rules, the measurements were only allowed to be carried out in planned train free periods, as a result of which the number of train free periods grew rapidly.

This created the need to measure and record the technical condition of switches in a universal way, and that resulted in the development of a measuring train for switches. That solved all objections against the manual method. With a measurement train the measurements are done by a few specialized companies that work on the basis of a certified method according to a uniform measurement process, who report in a national database and measure the switch under loaded condition. Today there are three measuring companies in the Netherlands with a measuring train[1] and ProRail has a national overview of the switch quality.

10 Fixed Tele-Monitoring Systems to Measure Quality Degradation

A new development in the rail infrastructure branch is continuous monitoring of the technical condition of the asset with a tele-monitoring system. It has been used for many years already in the real estate sector, known as a SCADA system. The first application in the Dutch rail infrastructure came from the maintenance contractor Strukton and is known as the POSS-system. POSS is the Dutch abbreviation of *Strukton's Preventive Maintenance and Fault Diagnosis System*. Strukton developed it to optimally perform its maintenance contracts.

POSS continuously and remotely monitors the condition of essential assets in the railway infrastructure and rolling stock. It has proven to be a powerful tool to support asset management, optimise maintenance and increase the availability. POSS allows condition-based maintenance, where repairs are only undertaken when necessary. Maintenance frequency can be reduced as a result, which leads to lower operating costs. The POSS-system has application for switches, level crossings and isolated track sections.

POSS provides trend information from measurements for further analysis on the nature of breakdowns. This helps to assure that recurring breakdowns can be reduced to a minimum. This again leads to lower operating costs. POSS makes it possible to check the condition and quality of objects after commissioning, maintenance and repair work. It is therefore a useful tool for quality assurance. Finally, the event data is converted into information for different user roles, from management to engineer.

[1] Eurailscout (Strukton/SNCF), Inspectation (VolkerRail) and Raildata/RILA (Fugro).

11 Explore the Possibilities of Flexible, Cost Efficient LoRa Sensors

LoRa stands for Long Range, Low Power. This Internet of Things connection has been specially developed to exchange small amounts of data between objects and systems. A single LoRa module can send more than 10 years worth of data using just two penlight batteries. This provides an economical way to gain insights into the status of 'things'. The possibilities of LoRa are endless. Efficiency can be increased or maintenance issues detected occurring in 'things' ahead of time, thereby preventing any incidents.

KPN is the first telecom provider in the Netherlands who cover the whole country with its network allowing as many things to be connected as possible. A single sensor costs only around 100 Euro and the connection to the network is very easy. It is a practical, energy-efficient and cost-efficient solution to exchange small bits of data. ProRail explores the possibilities of the sensors together with the maintenance contractors. ProRail has provided a number of sensors to each of the four contractors to gain knowledge and experience. They may be used anywhere: switches, level crossings, insulated welds, etc. The expectation is that this will quickly become a standard for the continuous measurement of condition degradation of assets.

12 A Condition Dashboard for Switches: Maintenance Performance Indicator

According to the new management concession 2015–2025, ProRail must report to the national government on the state of infrastructure once every three years. This is what the central technicians group at ProRail do. They compose a report that describes the system for each system, the conservation policy, the current performance and developments, followed by a final analysis with conclusions. Such a report is 60–70 pages thick. The writers of the report, and the management of ProRail and national government are not really satisfied with the result. The scope and target group are unclear, as a result of which the content does not adequately match expectations. The report contains too many technical substantive details and gives too little insight into the state of the infrastructure, in particular the infrastructure with a poorer technical condition that influences the performance. The search was for a better assessment method and reporting form. The Swiss colleagues have inspired ProRail with their "Network Condition Report". The method is developed by Optima Nexus (Zug, Switzerland). The method and report are good and is therefore briefly described here.

The rail infrastructure consists of nine systems that differ greatly in terms of technology, performance, lifespan, wear patterns, etc. In order to be able to determine their condition in an unambiguous way, a universal condition measuring device is used that runs from 6–1, see Fig. 6.

Fig. 6 Universal quality standardization; a uniform yardstick

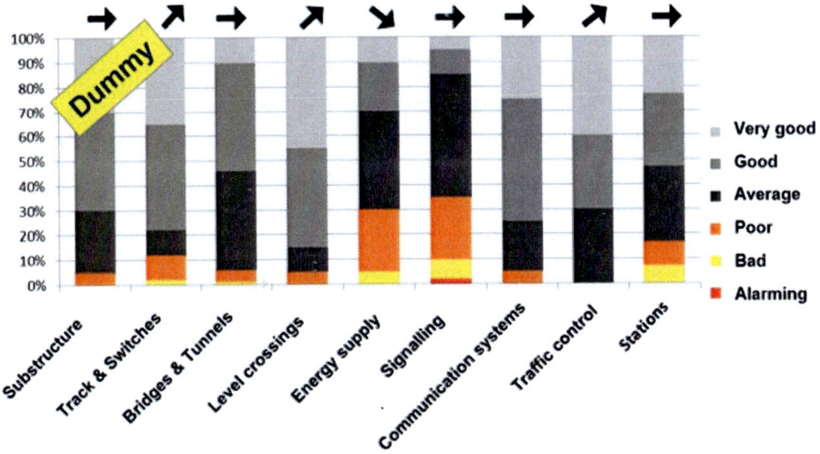

Fig. 7 Example of a graph in the Network Condition Report of SBB

With this universal measuring rod, the condition of sub-systems is determined on the basis of four criteria: the technical quality, availability, safety and remaining technical life. The assessment of the criteria is based on underlying characteristic technical parameters that are added together with a weighting percentage. For each parameter, criteria and (sub)system, not only the current condition is determined, but also the intended goal. This allows a relative comparison between goal and realization. For the drafting of a report, an estimated 350 parameters of approximately 50 subsystems are analysed, divided over the four assessment criteria. For each subsystem, it is estimated that the number of parameters varies between five and twelve. The availability of information is decisive. For the technical quality, these are measurements and inspections, for availability the failure registration, for safety there are various sources and for the life cycle use is made of the asset database and financial depreciation database. For an example of the used reporting graphs in the "Network condition report", see Fig. 7. In the report, a brief explanation with conclusions of no more than an A4 is provided with each graph.

13 Conclusions

For the sustainable optimization of the performance and costs over the life cycle, an infrastructure manager must have a professional asset management system. It all starts by putting "the house" of the management M&R organisation in order

(Pro-Rail mentions that the Asset Management business unit). The traditional, often internally-oriented M&R organisation had to evolve with specific attention to five development area's:

1. From management of tasks to processes
2. From management of technique to the performance of railway lines
3. From a financial focus on budgets to costs
4. Implementation of powerful management techniques and tools
5. Organize and secure ICT and information management.

The INTELLISWITCH project focuses on the conditions and maintenance of S&Cs in the Danish railway system. The primary goal is to formulate a Maintenance Performance Indicator (MPI) that informs about the situation of the individual S&Cs. The five work packages from INTELLISWITCH go beyond that. These are about developing the basic conditions to optimally manage the performance of S&Cs. The INTELLISWITCH project therefore makes an important contribution to the further professionalisation of the infrastructure manager Banedanmark.